D0297016

# GOOD HOUSEKEEPING

# FREEZER RECIPES

# GOOD HOUSEKEEPING

# FREEZER RECIPES

COMPILED BY
GOOD HOUSEKEEPING INSTITUTE

**EBURY PRESS**
LONDON

First published 1974
by Ebury Press
Chestergate House, Vauxhall Bridge Road
London SW1V 1HF

ISBN 0 85223 051 6

Edited by Gill Edden

Photographs by Stephen Baker

Colour photograph facing page 177 by Bryce Attwell

Filmset and printed in Great Britain by
BAS Printers Limited, Wallop, Hampshire
and bound by Webb Son & Co Ltd
London and Glamorgan

# Contents

# Colour Plates

# Foreword

The experts at Good Housekeeping Institute have been at the freezer again and we are proud to present you with our second book of freezer recipes. There are over 600 recipes in this book, all freezer tested.

If you already own a freezer and perhaps our first highly successful *Home Freezer Cook Book* too, you may have enough confidence to carry out your own experiments. But there's still that nagging anxiety that Moules à la bruxelloise may not freeze well. Should you always thaw frozen meat before you cook it? And what's the best and safest way to wrap a Gâteau St. Honoré? (Yes, that does freeze perfectly.) Why risk a failure when GHI has done the experimenting for you? There are answers to all these questions in this book as well as full packaging, thawing and reheating directions with every recipe —absolutely no guess work.

Margaret Coombes, Good Housekeeping Institute's cookery editor, assisted by Beverley Jones, Frances Pratt and Diana Wilkins are responsible for the cookery expertise.

If you have queries about any of the recipes in this book write to us at Good Housekeeping Institute, Chestergate House, Vauxhall Bridge Road, London SW1V 1HF.

Carol Macartney
Director

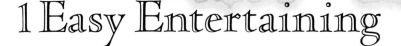

# 1 Easy Entertaining

With a freezer to help you, you can entertain your friends at the drop of a hat. Coffee mornings, buffet parties, cocktail evenings—for these you need a variety of foods that can all be prepared in advance and frozen. Choose from our selection and add your own ideas.

## COFFEE MORNINGS

### Chocolate pecan fingers

3 oz. plain chocolate
8 fl. oz. milk
3 oz. butter
6 oz. plain flour
2 level tsps. baking powder
1 level tsp. bicarbonate of soda
3 oz. caster sugar
1 egg
few drops vanilla essence

*For frosting*
4 tbsps. milk
2 oz. butter
2 oz. light soft brown sugar
1 oz. chocolate dots
2 egg yolks
2 oz. shelled pecan nuts, roughly chopped

*Makes 16*

**TO MAKE:** Grease a cake tin $11\frac{1}{8}$ in. by $7\frac{1}{8}$ in. by $1\frac{1}{4}$ in. deep. Put the broken chocolate, milk and butter in a small pan and heat without boiling, stirring until the chocolate is thoroughly blended. Cool for 5 min. Sift the flour, baking powder and bicarbonate of soda together in a bowl and add the sugar. Pour on the chocolate mixture, add the egg and vanilla essence and beat with a rotary or wire whisk until smooth. Pour into the prepared tin. Bake in the oven at 350°F (mark 4) for about 30 min., until just firm. Turn out and cool on a wire rack.

To make the frosting, place the milk, butter and sugar in a saucepan, heat gently to melt the butter and dissolve the sugar. Add the chocolate and allow to melt. Off the heat, beat in the egg yolks and the nuts. Return to the heat and cook well until the mixture thickens. Cool quickly and spread over the cake.

**TO PACK AND FREEZE:** See notes on page 218.

**TO USE:** Unwrap, place on a wire rack and thaw at room temperature for 3–4 hr. Cut into sixteen fingers as soon as possible to hasten thawing.

### Caraway teabread

8 oz. plain flour
1 level tbsp. baking powder
$\frac{1}{2}$ level tsp. salt
2 oz. butter
4 oz. caster sugar
1 level tbsp. caraway seeds
1 egg, beaten
6 tbsps. milk
$\frac{1}{2}$ tsp. vanilla essence

**TO MAKE:** Sift flour, baking powder and salt together into a basin. Rub in the butter with the fingertips until it resembles fine crumbs. Stir in the sugar and caraway seeds.

Mix to a soft dough with the egg, milk and vanilla essence. Turn the mixture into a $7\frac{3}{4}$-in. by 4-in. (top measurement) loaf tin lined with greased greaseproof paper. Bake in the oven at 350°F (mark 4) for about 40 min. until firm and golden. Turn on to a wire rack and cool.

**TO PACK AND FREEZE:** See notes on page 218.

**TO USE:** Unwrap, place on a wire rack and allow to thaw at room temperature for 2–3 hr. Slice as soon as possible for quicker thawing. Serve buttered.

# Coconut honey rolls

8 oz. plain flour
2 level tsps. baking powder
pinch salt
3 oz. butter
2 oz. caster sugar
1 egg, beaten
1 tbsp. milk
1 oz. butter
1 oz. Demerara sugar
2 oz. clear honey
1 oz. desiccated coconut

*To complete from freezer*
3 tbsps. clear honey
1 tbsp. lemon juice

*Makes 12*

**TO MAKE:** Sift the flour, baking powder and salt into a basin. Rub in the butter until the mixture resembles fine crumbs. Add the sugar and mix to a soft dough with the egg and milk. Lightly knead the mixture on a floured board and roll out to an oblong about 12 in. by 6 in. Melt the butter in a saucepan, stir in the sugar, honey and coconut and spread this mixture over the dough. Roll the dough up like a Swiss roll from the long side and cut into 12. Place the slices on a greased baking tray, cut side down and bake in the oven at 400°F (mark 6) for about 20 min., until golden. Cool on a wire rack.

**TO PACK AND FREEZE:** See notes on page 218.

**TO USE:** Heat the honey and lemon juice in a saucepan. Unwrap rolls, place them on a wire rack, glaze with the honey and leave for about 2 hr. at room temperature before serving.

# Cherry marshmallow tops

8 oz. self-raising flour
pinch salt
4 oz. icing sugar, sifted
4 oz. butter
1 egg, beaten
2 oz. glacé cherries, chopped
16 marshmallows, cut into quarters

*Makes about 16*

**TO MAKE:** Sift the flour and salt together into a bowl. Rub in the butter with the fingertips until the mixture resembles fine crumbs and add the sugar, then mix to a firm dough with the beaten egg. Form into 16 balls, place them on greased baking sheets and flatten slightly. Bake in the oven at 375°F (mark 5) for about 10 min. Remove from the oven and press chopped marshmallows and cherries on top of the biscuits. Return them to the oven for further 5–10 min., or until the marshmallows have melted, and the biscuits are a golden brown.

**TO PACK AND FREEZE:** See notes on page 218.

**TO USE:** Unwrap, place on a wire rack and thaw at room temperature for 1 hr. before serving.

# Almond crunch cake

4 oz. butter
4 oz. caster sugar
2 eggs, beaten
3 oz. self-raising flour
1 oz. ground almonds
few drops almond essence
a little milk

*For topping*
4 oz. flaked almonds
2 oz. butter
3 level tbsps. caster sugar
1 tbsp. cream
1 level tbsp. flour

**TO MAKE:** Cream the butter and sugar together in a basin until light and fluffy.

Beat in the eggs a little at a time. Sift the flour and almonds together and fold into the creamed mixture. Add the essence and a little milk if necessary, to make a soft dropping consistency. Turn the mixture into an 8–9 in. greased sandwich tin and bake in the oven at 375°F (mark 5) for about 30 min., until the cake is firm to the touch in the centre. Meanwhile heat the topping ingredients together in a saucepan. Spread over the cake and return to the oven for a further 10–15 min. Leave to cool a little before turning out on to a wire rack.

**TO PACK AND FREEZE:** See notes on page 218.

**TO USE:** Unwrap, place on a wire rack and thaw at room temperature for about 4 hr.

# Caramel shortbread

6 oz. plain flour
2 oz. caster sugar
6 oz. butter
2 oz. light soft brown sugar
1 large can condensed milk
4 oz. plain chocolate

*Makes about 16*

**TO MAKE:** Sift the flour into a basin. Add the caster sugar and rub in 4 oz. butter until the mixture resembles fine crumbs. Press the mixture into a Swiss roll tin measuring $11\frac{1}{8}$ in. by $7\frac{1}{8}$ in. Bake in the oven at 325°F (mark 3) for about 30 min. until just coloured. Allow to cool. Heat the remaining butter and brown sugar together in a saucepan. Add the condensed milk and heat gently until the sugar dissolves. Bring to the boil and stir continually until the mixture is a creamy fudge colour. Pour over the shortbread and level the surface. Cool. Melt the chocolate in a basin over hot water and pour over the caramel. Tap the base of the tin on a working surface to level the chocolate and leave to set.

**TO PACK AND FREEZE:** Freeze unwrapped to harden the chocolate. Remove from the tin and cut into fingers. Wrap in a polythene bag in a single layer. Seal and label.

**TO USE:** Thaw unwrapped at room temperature for about 1 hr. before serving.

# Fresh cherry muffins

10 oz. plain flour
$2\frac{1}{2}$ level tsps. baking powder
pinch salt
2 oz. caster sugar
1 egg, separated
$\frac{1}{4}$ pt. plus 2 tbsps. milk
$\frac{1}{2}$ lb. fresh red cherries, stoned and chopped

*Makes 24*

**TO MAKE:** Sift the flour, baking powder and salt into a basin. Add the sugar. Mix together the egg yolk and milk and stir into the dry ingredients. Whisk the egg white stiffly and fold into the dough with the cherries. Spoon the mixture into 24 greased bun tins and bake in the oven at 400°F (mark 6) for about 20 min., until risen and golden brown. Cool on a wire rack.

**TO PACK AND FREEZE:** See notes on page 218.

**TO USE:** Unwrap and place on a baking sheet. Cover the muffins with foil and reheat from frozen in the oven at 400°F (mark 6) for 20 min. Split and serve warm with butter.

# COCKTAIL IDEAS

## Cheese cubes

1 in. cubes of white bread
beaten egg
a little milk
grated cheese

**TO MAKE:** Dip the bread cubes into the egg mixed with a little milk, then into the grated cheese. Place on a baking sheet.

**TO PACK AND FREEZE:** Freeze unwrapped until firm. Wrap in a polythene bag, seal and label. Return to the freezer.

**TO USE:** Unwrap the cubes and place on a baking sheet in the oven at 400°F (mark 6) for 15–20 min. Serve in a bowl, with chutney handed separately.

# Savoury choux buns

1½ oz. butter or margarine
¼ pt. water
2½ oz. plain flour, sifted
2 eggs, beaten

*To complete buns from freezer*
4 oz. cream cheese
2 oz. softened butter
1 tsp. lemon juice
Marmite
chopped parsley

*Makes 24*

**TO MAKE:** Place the butter and water in a saucepan, heat until the butter melts and then bring to the boil. Remove from the heat, shoot in all the flour, beat well and cook gently until the mixture forms a ball and comes away from the sides of the pan. Cool slightly, then beat in the eggs a little at a time.

Using a ½-in. plain vegetable nozzle, pipe out about 24 small walnut-sized balls of paste on to greased baking sheets. Bake in the oven at 400°F (mark 6) for 15–20 min. until golden and crisp. Cool on a wire rack.

**TO PACK AND FREEZE:** See notes on page 218.

**TO USE:** Unwrap the buns and place on baking sheets. Refresh from frozen in the oven at 350°F (mark 4) for about 10 min. Cool on wire racks. For the filling, cream together the cheese, butter and lemon juice. Fill a forcing bag fitted with a small plain vegetable nozzle, make a hole in the base of each choux bun with the nozzle and pipe the filling into the buns. Glaze with a little Marmite and sprinkle with chopped parsley.

# Ham mustard pin wheels

(*see picture opposite*)

4 oz. butter, softened
2 level tsps. made mustard
4 oz. cooked shoulder ham, sliced

*Makes about 30*

**TO MAKE:** Mix the butter and mustard together in a basin until well blended. Spread the mixture evenly on the ham slices and roll each up like a Swiss roll.

**TO PACK AND FREEZE:** Pack the rolls in a rigid container, secure the lid on top, or cover with foil, seal, label and freeze.

**TO USE:** Unwrap but keep loosely covered and thaw at room temperature for about 2 hr.; slice thinly and serve.

# Cointreau dates

(*see picture opposite*)

8 oz. whole dates
6 oz. soft cream cheese
grated rind 1 orange
2 tbsps. Cointreau

*To complete dish from freezer*
grated rind of 1 orange

*Makes about 20*

**TO MAKE:** Split the dates, but do not cut them completely in half; remove the stones. Soften the cheese and blend it with the rind and Cointreau in a basin. Spoon or pipe a little of the mixture into each date.

**TO PACK AND FREEZE:** Place the stuffed dates on a baking sheet, open freeze until firm. Wrap in a polythene bag, seal and label. Return to the freezer.

**TO USE:** Unwrap and allow to thaw for 2–3 hr. Sprinkle with freshly grated orange rind. Serve in paper cases or on cocktail sticks.

# Anchovy twists

(*see picture opposite*)

scraps of puff pastry (about 3 oz.)
2-oz. can anchovy fillets
lemon juice
beaten egg

*Makes about 25*

**TO MAKE:** Roll out the pastry ⅛ in. thick and cut into strips the length of each anchovy. Lay a strip of anchovy on each pastry strip and sprinkle with lemon juice. Twist the two together, brush with beaten

left, *Cointreau dates;* top, *Maraschino roll-ups;* centre, *Sablés;* right, *Ham mustard pin wheels;* bottom, *Anchovy twists*

egg, place on baking sheets and bake in the oven at 450°F (mark 8) for 5–8 min. Cool on a wire rack.

**TO PACK AND FREEZE:** See notes on page 218.

**TO USE:** Unwrap the twists, allow to thaw for 1 hr. or place in the oven still frozen and refresh at 350°F (mark 4) for about 5 min.

## Maraschino roll-ups

(*see picture above*)

**8-oz. bottle Maraschino cherries**
**12–15 rashers streaky bacon, rinded**

*Makes about 30*

**TO MAKE:** Drain the cherries. Cut each rasher of bacon in 2–3 pieces. Roll a piece of bacon round each cherry and secure with a wooden cocktail stick.

**TO PACK AND FREEZE:** Wrap in foil in a single layer, overwrap in polythene, seal and label.

**TO USE:** Unwrap the roll-ups, place frozen under a preheated grill until crisp. Serve on cocktail sticks.

## Pâté fleurons

**7½-oz. pkt. frozen puff pastry, thawed**
**beaten egg**
**4½-oz. tube liver pâté**
**2 oz. butter**

*Makes 50*

Roll out the pastry to about ⅛ in. thick.

Using a 1½-in. fluted cutter, stamp out as many rounds as possible, re-rolling as necessary. Brush each round with beaten egg and fold over into a semicircle. Place on baking sheets. Allow the fleurons to stand in a cool place for at least 30 min. Brush again with beaten egg before baking in the oven at 400°F (mark 6) for about 15 min. With a sharp knife, almost cut through the pastry to allow steam to escape. Leave to cool on a wire rack. Combine the pâté with the butter and beat well. Fill a piping bag fitted with a no. 8 star icing nozzle. Pipe in a 'shell' shape down the centre of each fleuron.

Alternatively, freeze the cooked fleuron cases; when required, unwrap them and refresh from frozen in the oven at 400°F (mark 6) for 5–8 min. before filling with pâté.

## Sablés

(*see picture page 13*)

4 oz. plain flour
4 oz. butter
4 oz. mature Cheddar cheese, grated
pinch salt and dry mustard
freshly ground black pepper
beaten egg
a few chopped walnuts and almond halves

*Makes about 60*

TO MAKE: Sift the flour into a basin. Rub the butter into the flour with the fingertips until the mixture resembles fine crumbs, add the cheese and seasoning and work together to form a dough. Roll out the dough and cut into bite-size pieces. Brush with beaten egg and press the chopped nuts over the surface. Place the sablés on baking sheets and bake in the oven at 400°F (mark 6) for about 10 min. Cool on a wire rack.

TO PACK AND FREEZE: See notes on page 218.

TO USE: Unwrap and allow to 'come to' at room temperature for about 30 min.

## Talmouse

7½-oz. pkt. 'boil in the bag' smoked haddock fillets
¼ pt. frozen béchamel sauce (see page 62)
salt and pepper
13-oz. pkt. frozen puff pastry, thawed
beaten egg

*Makes about 12*

Cook the fish as directed on the packet. When cold, discard the skin and bones and flake the flesh. Heat the sauce gently in a saucepan until thawed and mix in the fish. Adjust the seasoning. Thinly roll out the puff pastry and stamp out as many rounds as possible, using a 3-in. plain cutter, re-rolling as necessary. Brush the rim of each with beaten egg and place a little fish into the centre of each. Shape up the pastry into a tricorn. Seal the edges firmly, brush with more beaten egg, place on greased baking sheets and bake in the oven at 400°F (mark 6) for about 20 min. until golden. Serve hot.

# SLIMMERS' RECIPES

## Orange pots

3 large oranges
2 oz. stoned dates, chopped
2 oz. dried prunes, stoned and chopped
2 oz. dried apricots, chopped
2 oz. dried figs, chopped

*Serves 3*

TO MAKE: Remove 'lids' from the oranges. Scoop out the flesh with a sharp knife or spoon and extract any remaining juice. Place the orange flesh and juice in a saucepan with the dried fruits and simmer for 15–20 min., until the juice is absorbed and the fruit is tender. Cool, put mixture in a blender to chop roughly, pour the mixture into the orange peel pots. Replace the lids.

**TO PACK AND FREEZE:** Place the orange pots on a plate and open freeze until solid. Wrap in foil, overwrap in polythene. Seal and label. Return to the freezer.

**TO USE:** Unwrap, thaw at room temperature for 3–4 hr.

# Grapefruit granita

2 large grapefruit
well flavoured chicken stock
1½ level tsps. powdered gelatine
4 oz. cucumber, peeled and chopped finely
½ tsp. fresh chopped mint
freshly ground black pepper
1 egg white, beaten stiffly

*Serves 8*

**TO MAKE:** Squeeze the grapefruit, extract all the juice and make it up to 1 pt. with stock. Place a little of the grapefruit stock in a saucepan, sprinkle the gelatine over and heat very gently to dissolve. Stir the remaining juice into the gelatine mixture. Mix well, add the cucumber, mint and pepper. Turn the mixture into a freezer container and freeze until mushy. Remove from the freezer and fold in the egg white.

**TO PACK AND FREEZE:** Pour the mixture into a 1½-pt. capacity rigid foil container. Freeze until firm, cover with a lid, seal and label. Return to the freezer.

**TO USE:** Allow to 'come to' at room temperature for ¾–1 hr. Crush well with a fork and serve in stemmed glasses. Garnish with fresh mint leaves or cucumber slices. Serve as a starter.

# Frozen pineapple cocktail

16-oz. can crushed pineapple
½ pt. unsweetened orange juice
½ pt. unsweetened grapefruit juice
½ pt. low-calorie ginger ale
liquid sweetener
mint to garnish

*Serves 4–6*

**TO MAKE:** Mix together the crushed pineapple, fruit juices, and ginger ale. Add 2–3 drops of liquid sweetener.

**TO PACK AND FREEZE:** Pour into a rigid foil freezer tray or plastic container. Cover with a lid, and seal. Label and if necessary overwrap in a polythene bag. Freeze until solid.

**TO USE:** Place the cocktail in the refrigerator ½–1 hr. before serving, so that the ice softens slightly. Spoon into glass dishes and garnish with mint.

# Apple stuffed kippers

2 pairs frozen kippers
1 oz. butter
¼ lb. frozen button mushrooms
¼ lb. tomatoes, skinned
1 apple peeled and cored
juice of ½ lemon
¼ level tsp. salt

*Serves 4*

Leave the kippers covered to thaw for 3–4 hr. in the refrigerator. Carefully remove the bones, fold in half and place, tails uppermost, down the centre of an oblong ovenproof dish. Pull the tails back, so that the kippers are opened ready for filling. To make the stuffing, melt the butter in a saucepan, toss the frozen mushrooms in the butter for a few seconds and chop them roughly. Cut the tomatoes and apple into ½-in. pieces, add to the mushrooms and mix well. Add the lemon juice and salt. Fill the kippers with this mixture and fold the tails over again. Cover with foil and bake in the oven at 425°F (mark 7) for 20–25 min. Serve hot.

# Strawberry delight

1½ lb. frozen strawberries, half-thawed
¼ lb. cottage cheese
¼ pt. fat-free yoghurt
liquid sweetener

*Serves 6*

Divide the strawberries between 6 tall glasses. Beat together the cottage cheese and yoghurt, adding sweetener to taste. Pour over the strawberries and serve.

## Slimmers' fish bake

2 frozen haddock fillets
1 oz. butter
6 oz. cucumber, peeled and seeded
2 large tomatoes, skinned and seeded
1 small green pepper, seeded
$\frac{1}{4}$ pt. soured cream
salt and pepper

*Serves 2*

Place fish in an ovenproof dish and dot with flakes of butter. Chop the cucumber, tomatoes and pepper. Sprinkle these over the fish, cover the dish with foil and bake in the oven at 425°F (mark 7) for about 30 min. Remove the fish from the oven.

Place each fillet on an individual serving dish; keep warm. Add the soured cream to the vegetables, mix well and season to taste; turn the mixture into a saucepan and heat gently through. Divide the mixture between the fish and serve immediately.

# DISHES FOR A SUMMER BUFFET

## Summer orange soup

6-fl. oz. can frozen orange juice
2 level tbsps. tapioca
2 oranges
1 banana

*Serves 4–6*

Make up the orange juice as directed on the can. Add the tapioca and heat gently in a saucepan for 10–15 min., until the tapioca is transparent. Pour into a serving dish to cool, then chill until required. Just before serving, peel the oranges and banana, slice the fruit thinly and float in the soup.

## Prawn and cucumber mayonnaise

5 tbsps. mayonnaise
1 tbsp. dry white wine
1 tsp. lemon juice
6 oz. frozen prawns, thawed
3 oz. cucumber
salt and pepper
lettuce
lemon twist
paprika pepper

*Serves 2–3*

Mix the mayonnaise, wine and lemon juice

together in a bowl. Fold through the prawns. Cut the cucumber in half lengthwise, remove the seeds and dice the flesh. Add to the prawns and adjust the seasoning. Shred the lettuce and arrange on an oblong serving dish or small platter. Spoon the prawn mixture down the centre. Garnish with a lemon twist and paprika. Serve chilled.

## Cheesey Danish croissants

1 oz. onion, skinned and chopped
$\frac{1}{2}$ oz. butter
1 oz. Parmesan cheese, grated
2 oz. mature Cheddar cheese, grated
1 oz. fresh white breadcrumbs
salt and pepper
1 egg, beaten
10-oz. pkt. frozen Danish pastry
  (5 sheets), thawed
poppy seeds

*Makes 10*

**TO MAKE:** Sauté the onion in the butter. Place the onion, cheeses, breadcrumbs and seasoning in a basin and bind with the egg. (Reserve a little egg for glazing.) Spread the cheese mixture on to the sheets of pastry and cut each in half diagonally. Roll each triangle of pastry from the wide side and shape into a crescent. Brush with the remaining beaten egg and sprinkle with poppy seeds. Bake in the oven at 400°F (mark 6) for about 30 min. Allow to cool on a wire rack.

**TO PACK AND FREEZE:** See notes on page 218.

**TO USE:** Leave wrapped for 1–2 hr. at room temperature, place on a baking sheet, cover with foil and refresh in the oven at 425°F (mark 7) for 5 min. Alternatively place while still frozen in a single layer wrapped in foil in the oven at 350°F (mark 4) for 15–20 min.

# Devilled ham cornets

7½-oz. pkt. frozen puff pastry, thawed
1 egg, beaten
poppy seeds

*For filling*
4 level tbsps. thick mayonnaise
1½ level tbsps. whipped double cream
1½ level tbsps. chutney
¾ level tsp. made mustard
3 oz. lean cooked ham, chopped
1 level tbsp. powdered aspic
2 tbsps. hot water

*Makes 12*

**TO MAKE:** Roll out the pastry to an oblong 15 in. by 8 in. Brush with beaten egg and cut 12 long ribbons from the pastry with a sharp knife. Wind each strip round a cream horn tin, glazed side uppermost, starting at the tip and overlapping slightly; place on a damp baking sheet. Sprinkle with poppy seeds. Bake in the oven at 425°F (mark 7) for about 8 min. until golden. Cool a little and remove from the tins by twisting the pastry carefully. Leave to cool on a wire rack.

To make the filling mix together all the ingredients except the aspic and water in a bowl. Dissolve the aspic powder in the water and stir into the ham mixture. Leave in the refrigerator to cool.

**TO PACK AND FREEZE:** Open freeze the horns on a baking sheet until firm. Wrap in a polythene bag, seal and label. Place the filling in a rigid container, cover with the lid, overwrap, seal and label. Freeze.

**TO USE:** Allow filling to thaw covered at room temperature for about 3 hr. Unwrap the horns and refresh from frozen on a baking sheet in the oven at 350°F (mark 4) for 5–10 min. Cool. Place filling in a forcing bag fitted with a plain vegetable nozzle and pipe into horns. Garnish with parsley.

# Chicken and lemon mille feuilles

2½–3 lb. frozen chicken, thawed
2 onions, skinned and quartered
thinly pared rind and juice of 1 lemon
2 bayleaves
7½-oz. pkt. frozen puff pastry, thawed
1 oz. butter
1 oz. flour
2 egg yolks
salt and pepper
¼ pt. double cream, whipped

*To complete dish from freezer*
lemon slices
parsley

*Serves 6*

**TO MAKE:** Place the chicken, onion, lemon rind and bayleaves in a large saucepan; add enough water to come half way up the legs, cover the saucepan, bring to the boil and simmer gently for 1 hr. Drain the chicken, reserve ¾ pt. stock and cool. Divide the pastry into 3 equal parts and roll each out to an oblong about 8 in. by 5 in.; prick and bake on damp baking sheets in the oven at 450°F (mark 8) for about 10 min., until crisp and golden brown. Cool on a wire rack. Remove the flesh from the chicken and chop it. Melt the butter in a saucepan, add the flour and cook for 1–2 min. Add ¾ pt. stock, stirring all the time, bring to the boil and simmer for a few min. Remove from the heat, add the lemon juice, egg yolks, and chopped chicken; beat well and adjust seasoning. When cold, stir in the whipped cream.

**TO PACK AND FREEZE:** Wrap the pastry layers in foil, overwrap in polythene; seal, label and freeze. Place the sauce in a rigid container, leave ½ in. headspace. Cover, seal, label and freeze.

17

**TO USE:** Thaw filling for 8 hr. at room temperature, then beat well. Unwrap the pastry, place it on a baking sheet and refresh in the oven at 350°F (mark 4) for about 5–10 min. Cool. Trim each layer to the same size. Sandwich the layers with filling and garnish with lemon slices and parsley.

# Chicken mayonnaise flan

8 oz. shortcrust pastry (i.e. made with
   8 oz. flour etc.)
3-lb. oven-ready chicken, cooked
4 oz. streaky bacon rashers, rinded and
   chopped
½ pt. thick mayonnaise
¼ pt. soured cream
1 level tsp. salt
freshly ground black pepper
1 tsp. lemon juice

*To complete dish from freezer*
8 small tomatoes, skinned
black olives
10 spring onions, trimmed
2 tsps. salad oil
1 tsp. white vinegar

*Serves 8*

**TO MAKE:** Roll out the pastry and use it to line an 11 in. diameter flan ring. Bake blind in the oven at 400°F (mark 6) for about 25 min. Remove the baking beans and cook for a further 10 min. to dry the base. Remove from the flan ring and cool. Quickly fry the bacon. Drain off the fat and cool. Discard the chicken skin and carve the flesh into small pieces. In a bowl, mix together the mayonnaise, soured cream, seasoning and lemon juice. Combine with the chicken and bacon and spoon into the flan case.

**TO PACK AND FREEZE:** Open freeze then cover with foil, overwrap with polythene, seal and label.

**TO USE:** Remove overwrap. Thaw the flan overnight in the refrigerator and spoon a little extra mayonnaise through the filling if necessary. Cut the tomatoes in half, remove the seeds and drain the tomato halves on absorbent paper. Arrange them round the flan, cut side down with an olive between each. Scissor-snip the onions into thin slices and toss them in the oil and vinegar. Leave in the marinade until just before serving. Drain and spoon into the centre of the flan.

# Cheese pâté roulade

4 eggs, separated
1 oz. cornflour
pinch each salt, mustard and paprika pepper
1 oz. Parmesan cheese, freshly grated
8 oz. soft liver pâté
1 tbsp. chopped parsley

*Serves 8*

**TO MAKE:** Line a large Swiss roll tin (11¾ in. long) with greaseproof paper and grease well. Beat the egg yolks, cornflour and seasonings in a basin until creamy; stir in the cheese. Stiffly whisk the egg whites and fold them into the yolks. Pour the mixture into the prepared tin and bake in the oven at 400°F (mark 6) for about 10 min., until golden. Turn out on to dampened greaseproof paper and roll up with paper between like a Swiss roll. Leave to cool on a wire rack. Mix the pâté and parsley together. Unroll the roulade, remove the paper, spread with pâté and re-roll.

**TO PACK AND FREEZE:** Wrap in foil, overwrap, seal, label and freeze.

**TO USE:** To serve cold, unwrap and thaw at room temperature for about 3 hr. Slice as soon as possible. To serve hot, remove the overwrap and reheat in the oven at 400°F (mark 6) for 30–40 min.

# Salami cornucopias

3-oz. pkt. cream cheese
1 level tbsp. creamed horseradish
½ level tsp. made mustard
pinch salt
10 slices salami

*To complete dish from freezer*
capers

*Makes 10*

**TO MAKE:** Blend the cheese with the horseradish, mustard and a little salt, in a basin. Shape the salami slices into cornets.

Place the cheese mixture in a piping bag fitted with a small star vegetable nozzle and pipe a swirl into each cornet.

**TO PACK AND FREEZE:** Open freeze on a baking sheet until firm, wrap in foil, over-wrap in polythene, seal and label. Return to the freezer.

**TO USE:** Unwrap, allow to thaw at room temperature for 2–3 hr. Garnish each cornucopia with a caper.

# Crumbed pork roll

4 lb. belly of pork
3 level tsps. dried fines herbes
freshly ground black pepper
1 lb. pork sausage meat
2 tbsps. dry sherry
2 level tsps. dried sage
1 clove garlic, skinned and crushed
1 level tsp. salt
4 oz. tongue, sliced
4 oz. large whole gherkins
2 bayleaves
3 pt. light stock
4 oz. buttered breadcrumbs (4 oz. fresh
  white breadcrumbs fried in 2 oz. butter
  until golden)

*Makes 30 slices*

**TO MAKE:** Bone the meat and reserve the bones. Lay the belly of pork skin side up on a flat surface; using a sharp knife, remove the skin, turn meat over and trim off any excess fat. Sprinkle with fines herbes and pepper.

In a bowl combine the sausage meat, sherry, sage, garlic and salt. Spread this filling evenly over the pork. Place the tongue in a single layer over the filling and lay the gherkins in a single or double line down the centre. Roll up tightly from the long side. Place skewers or cocktail sticks in to hold the shape while oversewing seam with a trussing needle and thread. Place the joint on a double thickness of muslin or an old clean tea towel. Wrap around tightly and secure the ends like a cracker. Tie the roll in three places with string and weigh it. Place the bones in a large pan, lay the roll on top with the bayleaves. Add stock to come about two-thirds the way up the pork. Bring to the boil, reduce to a slow bubble, cover and allow 30 min. per lb.; turn the meat halfway through the cooking time and replenish the liquid if necessary. Remove the meat, discard the wrapping and lightly scrape away any surface jellied fat. Cut away the string when cold and coat the pork in buttered crumbs.

**TO PACK AND FREEZE:** Wrap in foil in suitably sized pieces. (This facilitates thawing.) Overwrap in polythene, seal, label and freeze.

**TO USE:** Thaw for up to 24 hr. in the refrigerator depending on the size of the pieces. Slice while still firm.

# Chocolate mandarin gâteau

(*see picture overleaf*)

4 oz. butter or margarine
4 oz. caster sugar
2 large eggs, beaten
3½ oz. self-raising flour
½ oz. cocoa
11-oz. can mandarin oranges
4 tbsps. sieved apricot jam
6 fl. oz. double cream
4 oz. plain chocolate or chocolate dots
2 eggs, separated
2 tbsps. Grand Marnier

*Serves 8–10*

**TO MAKE:** Grease and line the base of an 8½-in. straight-sided sandwich tin. In a bowl, cream together the butter and sugar until light and fluffy. Add the eggs gradually, beating well between additions. Sift the flour and cocoa and fold in. Turn into prepared tin; level the surface and bake in the oven at 350°F (mark 4) for about 25 min. Turn out the cake and cool on a wire rack, base side uppermost. Drain the mandarins, add 2 tbsps. juice to the sieved apricot jam and heat until of a thin glaze consistency. Reserve 10 mandarins, chop the rest and scatter over the sponge base, together with 2 tbsps. juice. Brush the glaze over the fruit and around the sides of the sponge.

Reserve 2 tbsps. cream for the filling and whip the remainder until it holds its shape. Fit a nylon forcing bag with a star vege-

*Chocolate mandarin gâteau*

table nozzle and pipe a thick cream border around the edge of the sponge.

Melt the chocolate in a bowl over a pan of hot water. Remove from the heat and beat in the egg yolks, Grand Marnier and 2 tbsps. reserved cream. Beat well together. Whisk the egg whites until stiff, fold them into the chocolate mixture until evenly incorporated. Allow to stand for a while until the mixture begins to thicken. Pour the filling into the centre of the cream border, and swirl the top with a palette knife when it begins to set. When set, decorate with mandarins, grouped in bunches around the top.

**TO PACK AND FREEZE:** See notes on page 218.

**TO USE:** Unwrap, place the gâteau on a serving plate and allow to thaw at room temperature for 4–6 hr.

# Timbale aux fruits

(*see colour picture facing page 32*)

2 eggs
4 oz. caster sugar
few drops vanilla essence
3 oz. plain flour, sifted

*To complete dish from freezer*
6 tbsps. apricot jam
2 lb. frozen fruit, thawed, (e.g. raspberries, strawberries, etc.)
1 tbsp. lemon juice
whipped cream

*Serves 8*

**TO MAKE:** Grease a 1½-pt. capacity fluted ring mould. Whisk the eggs, sugar and essence in a basin until thick and creamy. Re-sift the flour over the egg and fold in with a metal spoon. Turn the mixture into the mould and bake in the oven at 350°F (mark 4) for 35–40 min. until firm. Remove from the mould and cool on a wire rack.

**TO PACK AND FREEZE:** See notes on page 218.

**TO USE:** Unwrap the timbale and place on a serving plate, thaw at room temperature for 2 hr. Put the jam, 3 tbsps. juice from the thawed fruit and the lemon juice into a saucepan. Heat gently, sieve and then brush over the sponge. Fill the centre of the timbale with the fruit and serve with whipped cream.

# EASY STARTERS

## Terrine of rabbit

1½ lb. rabbit, jointed
1 lb. chicken livers
4 oz. pork fat
10 oz. pie veal
4 oz. streaky bacon
4 oz. onion, skinned
1 clove garlic, skinned
5 tbsps. port
½ level tsp. ground mace
grated rind of ½ lemon
1 tbsp. lemon juice
salt and freshly ground black pepper
lemon slices
2 bayleaves

*Serves 12*

**TO MAKE:** Cut away as much of the flesh of the rabbit from the carcass as possible. Roughly chop the livers, removing any gristle. Cut the pork fat and the veal into manageable pieces. Remove the rind and any bones from the bacon. Combine all the meats and put through a mincer, along with the onion and garlic. In a bowl combine the meat mixture with the port, mace, lemon rind and juice. Adjust seasoning.

Press the mixture into a 2½-pt. capacity terrine or ovenproof casserole. Cover and cook in a water bath in the oven at 325°F (mark 3) for about 3 hr., or until fully cooked and the juices run clear. Strain off the juices and reserve. Place a flat plate or double layer of foil over the surface and weight down. Chill the terrine until set. Reduce the meat juices by boiling to concentrate and strengthen the jelly, leave to cool and when half set pour over the terrine. Garnish with lemon slices and bayleaves.

**TO PACK AND FREEZE:** See notes on page 93.

**TO USE:** Allow to thaw for 24 hr. at room temperature or up to 48 hr. in the refrigerator. Unwrap, slice and serve with toast.

## Pepper anchovy rolls

3 oz. cream cheese
2-oz. can anchovy fillets, drained and
    chopped
1 cap canned pimiento, chopped
2 level tbsps. soured cream
freshly ground black pepper
16 slices white bread from a large loaf,
    crusts removed

*To complete dish from freezer*
melted butter

*Makes 16*

**TO MAKE:** Combine all the ingredients except the bread in a basin. Spread the filling on the bread and roll each slice up from the long end. Cut each roll in two.

**TO PACK AND FREEZE:** Pack in a rigid container with non-stick paper between each roll. Seal, label and freeze.

**TO USE:** Thaw wrapped at room temperature for 1–1½ hr. Unwrap, brush with butter, and brown under a hot grill until golden. Serve hot.

# Danish cheese mousse

4 oz. Danish blue cheese
$\frac{1}{4}$ pt. double cream
$\frac{1}{2}$ oz. shelled walnuts, chopped
$\frac{1}{4}$ oz. powdered gelatine
2 tbsps. water
salt, pepper and dry mustard
1 egg white
8 walnut halves

*To complete dish from freezer*
1 lettuce
fresh grapes or mandarin oranges

*Serves 8*

**TO MAKE:** Grate the cheese. Half whip the cream, add the nuts and fold in the cheese. Sprinkle the gelatine over water in a saucepan, heat gently to dissolve it, then stir it into the cheese mixture; season to taste with salt, pepper and mustard. Stiffly whisk the egg white and fold into the mixture. Press it into a small container and garnish with walnut halves so that each portion will have a nut.

**TO PACK AND FREEZE:** Wrap in polythene, overwrap with foil, seal, label and freeze.

**TO USE:** Unwrap, remove from the container whilst still frozen. Cover lightly and allow the mousse to thaw at room temperature for 3 hr. Cut into 8 pieces. Serve on a bed of lettuce, garnished with fresh grapes or mandarin oranges.

# Cauliflower ramekins

1 medium sized cauliflower
1 pt. water or chicken stock
1 bunch watercress
$2\frac{1}{2}$ oz. butter
4 level tbsps. flour
$\frac{1}{2}$ pt. milk
salt and freshly ground black pepper

*To complete dish from freezer*
1 oz. fresh white breadcrumbs
2 oz. grated cheese
butter

*Serves 6*

**TO MAKE:** Divide the cauliflower into florets and place them in a pan with the water or stock. Cover and boil gently for 10 min. Wash and trim the watercress, chop it and add to the cauliflower, cook for a further 5 min. Drain and reserve the vegetable water. Melt the butter, stir in the flour and cook for 1–2 min. Make $\frac{1}{2}$ pt. vegetable water up to 1 pt. with the milk and beat the liquid into the roux, away from the heat. Purée the cauliflower mixture with a little of the unthickened sauce, in the blender. Return it to the pan, bring to the boil and cook for 2 min., stirring. Adjust the seasoning. Cool and pour into a foil container.

**TO PACK AND FREEZE:** Open freeze before packing.

**TO USE:** Leave wrapped and thaw overnight in the refrigerator. Unwrap, stir and divide between 6 ramekin dishes. Scatter the breadcrumbs, mixed with the cheese, over the top. Place the dishes on a baking sheet, dot with butter and bake in the oven at 400°F (mark 6) for 30 min.

# Citrus tomato ice

(*see picture opposite*)

$1\frac{1}{2}$ lb. ripe tomatoes
1 clove garlic, skinned and crushed
grated rind and juice of 1 lemon
1 level tsp. dried basil
1 level tbsp. caster sugar
1 level tbsp. tomato paste
$\frac{1}{2}$ level tsp. salt
freshly ground black pepper

*To complete dish from freezer*
4 avocados (optional)
cucumber slices (optional)

*Serves 8*

**TO MAKE:** Quarter the tomatoes and place them in a large saucepan with all the other ingredients. Bring to the boil, lower the heat, cover and simmer for 10–15 min. Pass through a sieve; cool quickly.

**TO PACK AND FREEZE:** Pour the mixture into a shallow container, allowing $\frac{1}{2}$ in. headspace. Cover with foil, seal, label and freeze.

**TO USE:** Leave the ice at room temperature for about 45 min. to 'come to'. Crush it and

*Citrus tomato ice in avocado*

serve in halves of avocado, or in stemmed glasses and garnished with cucumber slices.

# Greek aubergine appetiser

1½ lb. aubergines, trimmed and wiped
½ pt. olive oil
2 oz. onion, skinned and chopped
3 tbsps. lemon juice
1 clove garlic, skinned and crushed
½ level tsp. dried oregano
2 tomatoes, skinned and chopped
salt and pepper

*Serves 8*

TO MAKE: Slice the aubergines thickly, heat half the oil in a large frying pan and sauté the aubergines until barely coloured, but soft. Pound the flesh in a bowl with the onion, lemon juice, garlic, oregano, tomatoes, salt and pepper. Gradually beat in the remaining oil until the mixture is smooth and thick. (Alternatively purée all ingredients together in a blender.)

TO PACK AND FREEZE: Cool the purée quickly. Pour into preformed polythene bags or waxed containers in ½ pt. quantities; seal, label and freeze.

TO USE: Turn the purée out into a dish and allow to thaw at room temperature, or heat in saucepan to thaw, then cool. Serve in individual pots with buttered toast cut into fingers.

# Marinated kippers

3 pairs frozen kippers
1 onion, skinned and sliced
2 bayleaves
4 tbsps. wine vinegar
4 tbsps. olive oil
1 level tsp. caster sugar
lemon wedges and parsley for garnish

*Serves 6*

Place the frozen kippers, flesh side down, in a large shallow container. Scatter the onions and bayleaves on the top. Mix the remaining ingredients and pour over the kippers. Cover and refrigerate for 24 hr. Remove the skin and bones from the fish and cut the flesh into strips.

Pile the fish on to a serving dish and garnish with lemon and parsley. Serve with freshly made toast.

# MAIN COURSE DISHES

## Veal balls with fresh tomato sauce

*For veal balls*
2 lb. veal, minced
2 oz. onion, skinned and finely chopped
2 oz. fresh white breadcrumbs
2 eggs, beaten
2 tbsps. olive oil
1 tbsp. wine vinegar
2 tbsps. finely chopped parsley
$\frac{1}{2}$–1 tsp. chopped mint
$\frac{1}{2}$ level tsp. dried oregano
2 cloves garlic, skinned and crushed
2 tbsps. boiling water
salt and pepper
flour for coating
oil or lard for frying

*For tomato sauce*
6 oz. onion, skinned and sliced
2 tbsps. olive oil
1 lb. ripe tomatoes, skinned and chopped
1 level tsp. sugar
$\frac{1}{4}$ level tsp. ground cinnamon
$\frac{1}{2}$ pt. stock or water
salt and pepper

*Serves 6*

**TO MAKE:** Combine all the ingredients for the veal balls. Divide into walnut-size pieces and shape into small balls. Flour a baking sheet, place the balls in a single layer and coat them in flour by shaking the sheet backwards and forwards a few times. Fry them in hot fat for about 5 min. until evenly browned. Drain and cool.

For the sauce, sauté the onion in oil until soft. Add the rest of the ingredients and simmer until the tomatoes are broken down. Put through a sieve or purée in a blender. Cool the sauce quickly.

**TO PACK AND FREEZE:** Place meat balls and tomato sauce in a large rigid foil container or preformed polythene bag. Seal, label and freeze.

**TO USE:** Unwrap and place in a large flame-proof casserole. Cover and reheat from frozen on top of the cooker for 30–40 min., until gently bubbling.

## Paupiettes of pork

2 6-oz. pork escalopes
2 oz. butter
2 oz. lean streaky bacon, rinded and diced
4 oz. onion, skinned and chopped
2 oz. fresh white breadcrumbs
4 oz. pork sausage meat
dried thyme
salt and freshly ground black pepper
1 level tbsp. flour
1 tbsp. cooking oil
3 tbsps. dry white wine
$\frac{1}{4}$ level tsp. paprika pepper
$\frac{1}{2}$ pt. chicken stock
1 tbsp. cream

*Serves 2*

**TO MAKE:** Beat the escalopes between sheets of damp greaseproof or non-stick paper until fairly thin.

Melt 1 oz. butter in a small pan, add the bacon and 1 oz. onion and sauté until soft. Cool. Combine with the breadcrumbs, sausage meat and $\frac{1}{4}$ level tsp. thyme. Season with salt and pepper and spread over the meat. Roll up and tie with string. Toss the rolls in flour. Heat the oil in a small pan and add 1 oz. butter, when it begins to colour fry the pork until evenly browned. Transfer to a small casserole.

Reheat the pan juices. Add the remaining 3 oz. onion and sauté until transparent. Stir in any excess flour with the wine, a pinch of thyme, paprika, stock and cream. Bring to the boil and pour over the meat. Cover and cook in the oven at 400°F (mark 6) for about 1¼ hr. Skim off the excess fat using absorbent paper. Cool quickly.

TO PACK AND FREEZE: See notes on page 92.

TO USE: Unwrap the paupiettes and replace in the casserole or reheat in the foil container from frozen in the oven at 375°F (mark 5) for 1½–1¾ hr. until bubbling. Transfer to serving dish. Garnish with chopped parsley.

# Filet de porc chasseur

2 lb. pork fillet
2 tbsps. cooking oil
2½ oz. butter
8 oz. onions, skinned and chopped
8 oz. button mushrooms
3 level tbsps. flour
¼ pt. light stock
¼ pt. dry white wine
salt and freshly ground black pepper

*To complete dish from freezer*
chopped parsley
croûtons

*Serves 6*

TO MAKE: Cut the pork into 1–1½ in. pieces. Heat the oil in a frying pan, add the pork and cook quickly to brown and seal surface. Remove from the pan and transfer to a casserole. Heat 2 oz. butter in the frying pan, add the onions and cook slowly until soft. Add the mushrooms and quickly sauté; remove them while still crisp and place over the meat. Blend the flour into the remaining pan juices, adding the remaining ½ oz. butter, and gradually add the stock and wine. Blend

to a smooth consistency. Bring to the boil and simmer for 2–3 min. Adjust the seasoning and pour the sauce into the casserole. Cover and cook in the oven at 350°F (mark 4) for about 1¾ hr., until pork is tender.

TO PACK AND FREEZE: See notes on page 92.

TO USE: Unwrap and reheat in the oven at 375°F (mark 5) for 1½–1¾ hr. Transfer to a serving dish. Sprinkle with parsley and surround with croûtons.

# Steak and wine pie

2¼ lb. lean chuck steak
½ lb. kidney
1½ oz. plain flour
salt and pepper
3 tbsps. cooking oil
3½ oz. butter
1 large onion, skinned and chopped
6 oz. mushrooms, sliced
2 cloves of garlic, skinned and crushed (optional)
1 pt. beef stock
½ pt. red wine

*To complete dish from freezer*
13-oz. pkt. frozen puff pastry, thawed
beaten egg

*Serves 6*

TO MAKE: Trim the meat and cut it into 1-in. pieces. Prepare the kidney and toss both in seasoned flour. Heat the oil in a large frying-pan, add 2 oz. butter and when it is on the point of turning brown add the meat and fry briskly until brown on all sides. Stir in any excess flour. Turn into a large casserole. Melt the remaining butter and fry the onion for a few min. Add the mushrooms and garlic and fry for a further few min. Pour the stock and wine into the pan, bring to the boil and pour over the meat. Cover and cook in the oven at 325°F (mark 3) for 1½–2 hr. until tender. Strain off the juices and boil to reduce by one-third. Return to the casserole and cool quickly.

TO PACK AND FREEZE: See notes on page 92. Use a 4-pt. pie dish as a preformer.

**TO USE:** Unwrap and turn the meat into a 4-pt. pie dish. Cover and reheat from frozen in the oven at 375°F (mark 5) for about 1½ hr. Meanwhile roll out the pastry. Remove the meat from the oven, add a pie funnel and cover the pie dish with a pastry lid. Brush with beaten egg and place in the oven at 450°F (mark 8) for about 20 min., until the pastry is golden and the filling is bubbling.

# Casserole Stroganoff

3½ lb. thick flank beef
14-oz. can tomatoes
2 beef stock cubes
½ lb. onions, skinned and quartered
½ lb. carrots, peeled and halved
1 lb. button mushrooms
2 oz. butter

*To complete dish from freezer*
½ pt. soured cream
chopped parsley

*Serves 6*

**TO MAKE:** Trim the fat from the beef and cut the meat into thin strips. Place the fat in a covered dish to render down whilst the beef is cooking. Pour the can of tomatoes into a wide flameproof 4-pt. casserole; the contents should be in a shallow layer. Crumble in the stock cubes. Arrange the meat in the centre, with the carrots and onions at the side. Cover tightly and cook in the oven at 325°F (mark 3) for 2 hr. Remove the lid and gently turn the meat in the juice. Cover, reduce the oven temperature to 300°F (mark 2) and cook for a further 1 hr., until tender. Slice the mushrooms thickly and sauté in melted butter. Discard the onion and carrots using a draining spoon. Add the mushrooms to the meat. Cool quickly.

*Note:* Discarded vegetables may be used to make soup.

**TO PACK AND FREEZE:** See notes on page 92.

**TO USE:** Unwrap and reheat from frozen in the oven at 375°F (mark 5) for about 1¾ hr. Remove from the oven and stir in the soured cream. Sprinkle with chopped parsley and serve with noodles or rice.

# Sausage spirals

8 oz. plain flour
1 level tsp. dry mustard
½ level tsp. chilli powder
pinch salt
2 oz. margarine
2 oz. lard
water
4 level tbsps. chutney
8 thick pork sausages
a little beaten egg

*Makes 8*

**TO MAKE:** Sift the flour with the seasonings and rub in the fats with the fingertips until the mixture resembles fine breadcrumbs. Add about 8 tsps. water to mix to an elastic dough. Roll out the pastry to an oblong about 8 in. by 15 in. Spread with the chutney. Cut into 8 long strips. Spiral the pastry, chutney side inside, around the sausages. Brush with beaten egg to glaze and place on a baking sheet. Bake in the oven at 375°F (mark 5) for about 45 min. Cool on a wire rack.

**TO PACK AND FREEZE:** See notes on page 92.

**TO USE:** Unwrap, place in a single layer on a baking sheet, cover with foil and refresh from frozen in the oven at 400°F (mark 6) for 40–45 min.

# Duckling with pineapple
(*see picture opposite*)

2 3½-lb. ducklings
2 onions, skinned
4 cloves
water
flour
2 tbsps. cooking oil
1½ oz. butter
1 level tbsp. ground ginger
2 level tbsps. clear honey
15-oz. can crushed pineapple, drained
1 chicken stock cube
2 level tbsps. cornflour
juice ½ lemon
4 tbsps. brandy

*To complete dish from freezer*
8 cocktail cherries

*Serves 8*

*Duckling with pineapple*

**TO MAKE:** Joint each duckling into 8 and skin. Make stock with the giblets, 1 onion stuck with cloves and water to cover. Simmer, covered, for about 45 min.

Dredge the joints with flour. In a large shallow pan, heat the oil and butter; add the joints, flesh side down, and fry for 10 min. until golden. Place them in a large casserole. If necessary, fry the duck a little at a time. Finely chop the remaining onion and sauté in the reheated pan juices. Stir in the ginger, honey and pineapple.

Crumble the stock cube into 1 pt. strained giblet stock. Blend the cornflour with a little pineapple juice, add with the stock and lemon juice to the pan juices. Bring to the boil, stirring. Pour over the duck, cover, and cook in the oven at 325°F (mark 3) for about 2 hr., until tender. Remove the excess fat from the casserole. Warm the brandy, ignite it and pour into the casserole. Cool quickly.

**TO PACK AND FREEZE:** See notes on page 128.

**TO USE:** Unwrap and reheat from frozen in the oven at 375°F (mark 5) for about 1½ hr. Add the cherries just before serving.

# Veal birds

(*see picture overleaf*)

4 veal escalopes
4 slices lean cooked ham
4 slices Danbo cheese
salt and pepper
3 oz. butter
4 oz. button mushrooms, sliced
1 level tsp. cornflour
water
4 tbsps. single cream

*Serves 4*

*Veal birds*

**TO MAKE:** Beat the veal between sheets of non-stick paper. Place a slice of ham and then a slice of cheese on top. Roll up and secure with wooden cocktail sticks. Season with salt and pepper. Melt 2 oz. butter in a wide shallow flameproof casserole and brown the escalopes evenly. Cover and cook in the oven at 375°F (mark 5) for about 25 min. Sauté the mushrooms in the remaining butter. Blend the cornflour with a little water in a pan, stir in the cooking juices drained from the casserole, bring to the boil then reduce heat and stir in the cream. Adjust seasoning. Place the 'birds' in a rigid foil container and top with the drained mushrooms. Spoon the sauce over and cool quickly.

**TO PACK AND FREEZE:** Cover with a lid or foil, seal, label and freeze.

**TO USE:** Unwrap, cover and reheat from frozen at 350°F (mark 4) for about $1\frac{1}{4}$ hr. Transfer to a serving dish. Garnish with chopped parsley.

# Ham breaded cutlets

2 oz. fresh white breadcrumbs
2 oz. lean cooked ham, finely chopped
6 frozen lamb cutlets
seasoned flour
1 egg, beaten
2 oz. lard

*Makes 6*

Mix together the breadcrumbs and ham. Dip the cutlets in seasoned flour, then in beaten egg. Press the breadcrumb mixture on to the sides of the cutlets. Melt the lard in a roasting tin, place the cutlets in the tin and bake in the oven at 375°F (mark 5) for about 1 hr. until golden and cooked through.

# Beef patties

$\frac{1}{2}$ oz. butter
4 oz. onion, skinned and chopped
1 lb. fresh minced beef
4 oz. carrot, peeled and grated
1 beef stock cube
1 level tbsp. flour
$\frac{1}{4}$ pt. water
pinch salt
freshly ground black pepper
1 lb. plain flour
4 oz. margarine
4 oz. lard

*Makes 24*

**TO MAKE:** Melt the butter in a saucepan, add the onion and sauté for a few min. until transparent; add the beef and continue to cook for 10 min. Add the carrot, crumbled stock cube, flour and water. Bring to the boil, adjust seasoning and reduce the heat. Cover and cook gently for 20 min. Cool quickly. Meanwhile sift the flour and salt into a large basin, rub in the margarine and lard with the fingertips until the mixture resembles breadcrumbs. Mix with about 5 tbsps. water to give a firm but manageable dough. Roll out about half the pastry and use to line 24 $2\frac{1}{2}$-in. diameter (top measurement) bun tins, using a 3-in. plain cutter. Divide the meat filling between the pastry-lined tins and either cover with plain pastry lids, cut with a smaller cutter, or prepare fancy tops. To do this, roll out the pastry into strips the length of the sheet of tins. With a pastry wheel cut $\frac{1}{2}$-in. ribbons. Arrange these across the filled cases, overlapping slightly, and brush each with beaten egg. With a plain cutter the size of the pans, stamp out each lid, remove trimmings and use for re-rolling. Bake in the oven at 400°F (mark 6) for about 25 min. Cool on a wire rack.

**TO PACK AND FREEZE:** Wrap in a single layer in polythene bags (6 per bag) seal, label and freeze.

**TO USE:** Unwrap and allow to thaw at room temperature for 2–3 hr. Serve cold or warm.

# Fried salami chicken

20 frozen chicken drumsticks, thawed
6–8 oz. salami, skinned and sliced
  (40 slices)
4–5 large eggs, beaten
1 large white loaf made into breadcrumbs
oil for frying

*Makes 20*

Make an incision to the bone along one side of each drumstick and loosen the flesh around the bone. Place 2 slices salami around the bone, pull the chicken flesh together and secure with cocktail sticks. Dip the chicken first into beaten egg then breadcrumbs and pat the crumbs in well. Repeat egg and breadcrumbing to give a good coating. Deep fry in hot oil (360°F) for 7–10 min. until golden brown and flesh is cooked through. Drain on absorbent paper. Serve cold.

# Whisky prawns

2 oz. butter
4 oz. onion, skinned and chopped
1 clove garlic, skinned and crushed
8 oz. frozen prawns
3 tbsps. whisky
salt and pepper
1 tbsp. lemon juice
chopped parsley

*Serves 2–3*

Melt the butter in a large saucepan, sauté the onion and garlic until transparent but not browned. Add the frozen prawns, cover and cook gently for about 5 min. Add the remaining ingredients and heat through. Serve with boiled rice and green salad.

# Pork chops in gingered cider

4 frozen pork chops
1 oz. butter
6 oz. onions, skinned and chopped
2 level tbsps. Demerara sugar
1 level tbsp. tomato paste
1 level tbsp. flour
2 level tsps. ground ginger
$\frac{1}{2}$ pt. cider
salt and pepper

*Serves 4*

Place the frozen pork chops in a casserole in one layer. Melt the butter in a pan and sauté the onions until transparent, stir in the sugar, tomato paste, flour, ginger, cider, salt and pepper. Bring to the boil and pour over the chops. Cover and cook in the oven at 400°F (mark 6) for about $1\frac{1}{2}$ hr. turning after 45 min. Remove from the oven, place on a serving plate and spoon over the sauce. Serve with creamy mashed potatoes and a green vegetable.

# Herb stuffed kidneys

1 oz. butter
1 oz. onion finely chopped
8 level tbsps. fresh white breadcrumbs
1 level tsp. dried mixed herbs
1 small egg, beaten
salt and freshly ground black pepper
4 frozen lamb kidneys, thawed
4 rashers streaky bacon, rinded and stretched

*Serves 2*

Melt $\frac{1}{2}$ oz. butter and sauté the onions until transparent. Add crumbs, herbs, egg and seasoning. Skin the kidneys, cut through but do not completely separate the halves and remove the core. Place $\frac{1}{4}$ of the stuffing in each kidney. Roll each in a bacon rasher, secure with a cocktail stick and place in a baking dish. Dot with the remaining butter

and cook in the oven at 375°F (mark 5) for about 1 hr.

# Salmon with horseradish ice

(*see picture opposite*)

*For horseradish ice*
$\frac{1}{4}$ pt. double cream
freshly grated horseradish

4 frozen salmon steaks (1–1$\frac{1}{2}$ in. thick)
2 oz. butter
4 tbsps. dry white wine

*Serves 4*

Whip the cream until it leaves a trail and stir in horseradish to taste. Turn into a small rigid container and freeze until firm. Place each salmon steak on a large piece of foil, dot $\frac{1}{2}$ oz. butter on each and sprinkle with 1 tbsp. wine. Wrap the foil into loose parcels and place them in a steamer over hot water; steam for about 45 min. Skin the salmon and serve hot or cold with the horseradish ice. New potatoes and a tossed green salad make a good accompaniment to this dish.

# Lamb cutlets korintsas

8 frozen lamb cutlets
1 recipe quantity apricot and currant sauce (see page 65)

*Serves 4*

Place the cutlets and block of sauce in a large casserole. Cover and cook in the oven at 350°F (mark 4) for about $1\frac{1}{2}$ hr. until the chops are tender and sauce bubbling. Stir occasionally during cooking.

# Savoury chicken vol-au-vent

2 13-oz. pkts. frozen puff pastry, thawed
1 egg, beaten

*For filling*
1 lb. cooked white chicken or turkey meat
$\frac{3}{4}$ pt. white sauce
4 oz. streaky bacon rashers, grilled and chopped
salt and pepper

*Serves 8*

*Salmon with horseradish ice*

**TO MAKE:** Place 1 piece of puff pastry on the top of the other, sealing them together with a little of the beaten egg. Roll the pastry out to about 1 in. thick and cut a circle 10 in. in diameter (a large dinner plate would be ideal to cut around). With a small, sharp knife mark a circle $\frac{1}{2}$–$\frac{3}{4}$ in. inside the larger one, to form a lid, cutting about half way through the pastry. Prick the pastry all over with a fork, then brush with beaten egg. Place the pastry on a damp baking sheet and bake in the oven at 450°F (mark 8) for 30–35 min., covering it with damp greaseproof paper when it is sufficiently

brown. When the pastry is cooked, remove the lid and scoop out any soft pastry from inside. Dry out the case in the oven for a further 5–10 min. To make the filling, cut the chicken into finger-sized pieces and add to the sauce with the bacon. Heat through, adding a little more seasoning if necessary.

**TO PACK AND FREEZE:** Freeze the case in a rigid container, seal and label. Freeze the filling separately in a pre-formed polythene bag.

**TO USE:** Unwrap the pastry case, put it on a baking sheet and heat through from frozen in the oven at 350°F (mark 4) for about 45 min. Meanwhile turn the filling into a saucepan and heat, stirring continuously, until bubbling. Spoon the filling into the hot pastry case, replace the lid and serve hot with a crisp green salad or vegetables cooked and tossed in butter.

# Trout with lemon sauce

4 frozen trout
2 onions, skinned and sliced
salt and pepper
$\frac{1}{2}$ pt. water
juice of 2 lemons
grated rind of 1 lemon
2 level tsps. arrowroot
2 egg yolks, beaten

*Serves 4*

Place the frozen trout in an ovenproof serving dish, cover with onions, sprinkle with salt and pepper and pour $\frac{1}{2}$ pt. water around the fish. Poach uncovered in the oven at 400°F (mark 6) for 1 hr. Drain the fish and keep it warm. Reserve the stock. Blend the lemon juice, rind and arrowroot together in a saucepan, and strain the stock into this mixture. Bring to the boil and stir until thickened. Remove from the heat, stir in the egg yolks and then cook for a little longer without boiling. Turn into a warm sauce boat and serve with the trout.

# Noisettes au gratin

6 frozen noisettes of lamb
lard, butter or oil
6 florets of frozen cauliflower
2 oz. mature Cheddar cheese, grated

*For sauce*
4 tbsps. white wine vinegar
1 shallot, skinned and finely chopped
2 egg yolks, beaten
3 oz. butter, softened
salt and pepper

*Serves 6*

Place the noisettes in a frying pan with a little fat or oil and cook gently for 20–30 min. turning once until cooked through. Meanwhile cook the cauliflower according to packet instructions, so that cauliflower and noisettes are ready at the same time. Place the vinegar and shallot in a saucepan and boil until reduced by half. Strain on to the egg yolks in a basin and cook over hot water until thick. Immediately the sauce starts to thicken remove from the pan of hot water, stir in the softened butter and season to taste. Place the noisettes on an ovenproof serving dish, top with a floret of cauliflower and sprinkle with a little grated cheese. Brown quickly under the grill. Serve the sauce separately.

# Sausage bake

$\frac{1}{2}$ lb. apples, peeled and sliced
$\frac{1}{2}$ lb. onions, skinned and sliced
1 lb. frozen sausages
$\frac{1}{4}$ pt. light stock, hot
salt and pepper

*Serves 3–4*

Place the apples and onions in the base of an ovenproof casserole. Lay the frozen sausages on top, pour over the hot stock and season. Cook uncovered in the oven at 400°F (mark 6) for $1–1\frac{1}{4}$ hr. until the sausages are well browned.

# Chicken suprême julienne

2 1-lb. pkts. frozen chicken breasts
2 oz. seasoned flour
2 tbsps. cooking oil
2 oz. butter
4 oz. onion, skinned and finely chopped
2 level tsps. tomato paste
1 level tsp. redcurrant jelly
$\frac{1}{4}$ level tsp. dried oregano
4 tbsps. dry Vermouth
1 pt. chicken stock
salt and freshly ground black pepper
2 oz. sliced ham
2 oz. sliced tongue
chopped parsley

*Serves 6*

Thoroughly thaw the chicken breasts and coat them in flour. Heat the oil and butter in a large frying pan or paella-type dish. When

*Dinner party: Timbale aux fruits* (page 20); *Veal Marsala with duchesse potatoes* (pages 51 and 163); *Trout en papillote* (page 90); *Haricots verts and carrots*

foaming, add the chicken and fry gently until golden on both sides. Remove, drain on absorbent paper and place in a large shallow casserole. Add the onion to the reheated pan juices and sauté until soft. Stir in any excess flour, the tomato paste, jelly, oregano and Vermouth. Pour in the stock and bring it to the boil, stirring. Adjust the seasoning and pour it over chicken; cover and cook in the oven at 325°F (mark 3) for about 1¼ hr. Meanwhile slice the ham and tongue into strips. Remove the chicken from the casserole, keep warm. Reduce the juices by half by rapid boiling, add the ham and tongue and spoon over the chicken. Sprinkle with parsley.

# Smoked haddock with mushrooms

1 lb. frozen smoked haddock
10½-oz. can condensed mushroom soup
4 oz. frozen mushrooms
1 tbsp. lemon juice
2 tbsps. chopped parsley
salt and pepper

*Serves 3–4*

Lay the frozen fish in a shallow ovenproof casserole. Mix the soup with ½ can of water. Add the frozen mushrooms, lemon juice and parsley. Adjust seasoning and pour over the fish. Cover and cook in the oven at 375°F (mark 5) for 40–45 min., until the fish is

cooked and flakes easily when tested with a sharp knife.

# Poussins Marsala

3 frozen poussins (approx. 1¾ lb. each), thawed and halved
2 oz. butter
¼ pt. Marsala or sherry
¼ pt. double cream
¼ level tsp. paprika pepper
1 clove garlic, skinned and crushed
4 oz. Cheddar cheese, grated
salt and pepper
chopped parsley
sprigs of rosemary, optional

*Serves 6*

Wipe the poussin halves and fry in butter for about 5 min. until golden brown. Drain off the excess fat from the pan. Replace the chicken halves flesh side down and pour over the Marsala. Top with a tightly fitting lid or kitchen foil and bubble gently for about 40 min., until the juices run clear when the chicken flesh is punctured with a fork.

Meanwhile whip the cream and fold in the paprika, garlic and cheese. Season with salt and pepper. Transfer the poussin halves and juices to a flameproof dish. Spoon over the cheese topping and flash under a fierce grill until the cheese melts and browns. Serve sprinkled with parsley. Garnish with rosemary sprigs.

# DESSERTS

# Dundee dessert

½ pt. double cream
½ lb. cottage cheese
grated rind ½ lemon
3 level tbsps. Curaçao marmalade

*Serves 4*

**TO MAKE:** Whip the cream in a basin until thick enough to hold its shape. Fold in the remaining ingredients with a metal spoon. Spoon the mixture into 4 individual serving dishes suitable for the freezer.

**TO PACK AND FREEZE:** Open freeze until

firm then wrap individually in foil. Overwrap in polythene, seal, label and return to the freezer.

**TO USE:** Unwrap and allow to thaw for 2 hr. at room temperature or 4 hr. in the refrigerator.

# Tarte noël

(*see picture below*)

4 oz. plain flour
pinch salt
2 oz. butter, softened
2 oz. caster sugar
2 egg yolks
6 level tbsps. mincemeat
4 oz. ground almonds
4 oz. caster sugar
2 egg whites
almond essence
2 oz. flaked almonds

*To complete dish from freezer*
4 tbsps. apricot jam

*Serves 6–8*

*Tarte noël*

**TO MAKE:** Sift together the flour and salt on a pastry board. Make a well in the centre and into it put the butter, sugar and yolks; using the fingertips, pinch and work the pastry together until well blended. Put the pastry in a cool place to relax for 1 hr. Roll it out and use it to line a $7\frac{1}{2}$-in. loose bottomed French fluted flan ring and spread the mincemeat on the base.

Blend together the ground almonds, sugar, egg whites and a few drops of essence. Pour this over the mincemeat. Cover with the flaked almonds. Bake the flan in the oven at 350°F (mark 4) for about 1 hr. until firm in the centre. Allow to cool in the tin.

**TO PACK AND FREEZE:** Remove the flan carefully from the tin, wrap in foil, overwrap in polythene, seal, label and freeze.

**TO USE:** Unwrap and allow to thaw for 3–4 hr. at room temperature. Heat the jam in a saucepan and brush over the almonds to glaze.

# Gâteau St. Honoré

*For pastry base*
4 oz. plain flour
pinch salt
2 oz. butter, at room temperature
1 oz. caster sugar
1 egg yolk
$\frac{1}{4}$ tsp. vanilla essence

*For choux buns*
1 oz. butter or margarine
$\frac{1}{4}$ pt. water
$2\frac{1}{2}$ oz. plain flour
pinch salt
2 eggs, beaten
$\frac{1}{4}$ pt. double cream

*To complete dish from freezer*
1 pt. milk
4 oz. caster sugar
$1\frac{1}{2}$ oz. plain flour
1 oz. cornflour
2 eggs
2 egg yolks
vanilla essence
8 oz. sugar
8 tbsps. water

*Serves 8*

**TO MAKE:** For the pastry base, sift the flour and salt into a mound, on a working surface. Make a well in the centre, add the butter and sugar and cream together with the fingertips. Add the egg yolk and essence and mix to a soft dough, bringing the flour into the centre gradually. Wrap and chill until a manageable consistency. Roll out the pastry to $8\frac{1}{2}$ in. diameter on a floured surface. Place it on a baking sheet, prick with a fork, crimp the edge with the fingers and bake in the oven at 350°F (mark 4) for about 20 min. Cool on the baking sheet until beginning to firm, then lift carefully on to a wire rack. With a little melted lard, grease two baking sheets. Press the rim of an $8\frac{1}{2}$ in. cake tin on to a floured surface and lift on to one baking sheet to make a guide circle.

For the choux pastry, put the butter and water in a small saucepan and heat until the butter has melted, bring to the boil. Sift the flour and salt on to a piece of greaseproof paper. Draw the pan aside and shoot in all the flour at once. Beat thoroughly with a wooden spoon until the mixture is smooth and comes away from the sides of the pan.

Cool slightly, then beat in the eggs a little at a time. Spoon the choux mixture into a forcing bag fitted with a $\frac{1}{2}$-in. plain nozzle. With the floured ring as a guide, pipe two thirds of the paste in a circle. Pipe 16 small choux buns on the other sheet. Bake both buns and ring in the oven at 425°F (mark 7) for about 25 min. When they are cooked pierce the base of each to allow steam to escape and cool on a wire rack. When cool fill the buns with whipped cream.

**TO PACK AND FREEZE:** Wrap the pastry base in foil, overwrap in polythene, seal and label. Place choux buns and ring in a rigid container, cover, seal and label. Freeze.

**TO USE:** Unwrap and allow pastry and buns to thaw at room temperature for 2–3 hr. Heat the milk in a pan. Mix together the caster sugar, flour, cornflour and eggs and stir in a little of the hot milk. Return the mixture to the saucepan, stir and heat until the mixture thickens and just comes to the boil. Add essence to taste. Cool slightly. Dissolve the sugar in the water in a small pan and boil rapidly until the thermometer reads 260°F (or until a drop of syrup hardens when placed in cold water). Using tongs, dip the top and sides of the buns into the glaze, and position the buns on the ring. Put the pastry base on a flat serving plate. Pour the remaining glaze around the edge of it and carefully sit the choux ring on it. Fill the centre with warm pastry cream. Serve cold.

# Marron parfait

2 oz. each currants, sultanas, candied peel and glacé cherries
6 tbsps. Maraschino or rum
4 egg yolks
8 oz. icing sugar
1 pt. double cream
4 oz. unsweetened chestnut purée
cooking oil

*To complete dish from freezer*
whipped cream
marrons glacés

*Serves 8–10*

**TO MAKE:** Soak the currants and sultanas in warm water until they swell and drain well. Chop the peel and cherries. Marinate the fruits in spirit for 1 hr. turning twice. Whisk the egg yolks with the sugar until pale and fluffy. Heat three quarters of the cream in a double boiler or a bowl over a pan of hot water; do not let cream come to the boil. Whisk the cream gradually into the egg, then stir the mixture continuously over gentle heat until of a coating consistency, but do not boil. Remove from the heat. In a small bowl whisk together the chestnut purée and a little egg custard; return to the bulk of the custard and blend together; allow to cool. Turn into a shallow foil container and freeze until thick and slushy. Brush a 2–2½ pt. bombe mould or pudding basin lightly with a little oil. Whip the remaining cream until stiff and combine thoroughly with the chilled custard. Finally add the marinated fruit and place in the bombe mould.

**TO PACK AND FREEZE:** Freeze covered until firm. Wrap in foil, overwrap, seal, label, and return to the freezer.

**TO USE:** Unmould and allow to 'come to' at room temperature for 1 hr. Decorate with whipped cream and marrons glacés.

# Chestnut roll

3 oz. caster sugar
3 eggs
3 oz. plain flour
2 oz. ground almonds
1 oz. plain chocolate, grated

*For filling*
2 oz. butter, softened
4 oz. icing sugar
2 egg yolks
1 tbsp. brandy
1 tsp. vanilla essence
8-oz. can sweetened chestnut purée

*Serves 8*

**TO MAKE:** Place the sugar and eggs in a large basin and whisk until thick and creamy. Sift the flour and ground almonds and fold into the egg mixture with the chocolate. Turn into a Swiss roll tin 13¾ in. by 9¼ in. lined with greased greaseproof paper. Bake in the oven at 375°F (mark 5) for about 20 min. until firm and golden. Turn out on

to a piece of greaseproof paper and roll up with the paper inside. Wrap in a clean damp teatowel and cool on a wire rack. Meanwhile cream together the butter and icing sugar until soft. Add the yolks, brandy, essence and purée and beat well. When the roll is cool, unwrap, spread with filling, and re-roll without the paper.

**TO PACK AND FREEZE:** See notes on page 218.

**TO USE:** Unwrap, place on a wire rack and thaw for 3–4 hr. at room temperature. Slice as soon as possible to hasten thawing.

# Coffee cream flan

8 oz. digestive biscuits
3 oz. butter, melted
2 eggs, separated
3 oz. caster sugar
1½–2 level tbsps. instant coffee granules
4 tbsps. water
2 tbsps. Tia Maria
2 level tsps. powdered gelatine

*To complete dish from freezer*
¼ pt. double cream
1 tbsp. milk
36 hazelnuts, toasted and skinned

*Serves 6–8*

**TO MAKE:** Crush the biscuits, blend the crumbs with the butter using a fork and line a deep 8-in. fluted flan ring placed on a foil lined baking tray. Press the crumbs well on to the base and up the sides. Chill. Whisk the egg yolks and sugar in a deep bowl until thick and creamy. Blend together the instant coffee, 2 tbsps. water and the liqueur and gradually whisk into the egg mixture. Sprinkle the gelatine over 2 tbsps. water in a small saucepan, heat very gently to dissolve. Cool and add to the coffee mixture. When the coffee mixture is beginning to set, fold in the stiffly whisked egg whites. Turn into the crumb case.

**TO PACK AND FREEZE:** Open freeze until firm, remove flan ring, wrap in foil, overwrap in polythene, seal and label. Return to the freezer.

**TO USE:** Unwrap, place on a serving plate and thaw at room temperature for 6 hr. Whip the cream with the milk until it just holds its shape. Pipe whirls of cream round the edge of the flan and decorate with hazelnuts.

# Pots au mocha

6 oz. plain chocolate
2 oz. butter
4 eggs, separated
1 tbsp. coffee essence
1 tbsp. warm water

*Serves 8*

**TO MAKE:** Melt the chocolate and butter in a basin over hot water. When melted, move from the heat and allow to cool slightly before stirring in the egg yolks, coffee essence and water. Whisk the egg whites stiffly and fold into the chocolate mixture. Pour into 8 individual serving dishes suitable for the freezer.

**TO PACK AND FREEZE:** Freeze uncovered on a tray until firm. When firm, put into a rigid container in a single layer. Seal, label and return to the freezer.

**TO USE:** Unwrap and allow to thaw at room temperature for about 1 hr. Decorate with a whirl of whipped cream if desired.

# Apricot topped cheesecake

6 oz. digestive biscuits, crushed
2 oz. margarine, melted
8 oz. cottage cheese, sieved
2 oz. caster sugar
finely grated rind and juice 2 lemons
1 small can evaporated milk, chilled

*To complete dish from freezer*
15-oz. can apricots
4 tbsps. apricot jam
whipped cream

*Serves 6*

**TO MAKE:** Mix together the crumbs and margarine, press half into the base of a 7-in., round, $1\frac{1}{2}$-pt. capacity rigid container. Chill. Combine the cottage cheese, sugar, lemon rind and juice in a bowl. Beat well. Whisk the evaporated milk in a deep bowl until really thick and fold into the cheese mixture,

turn into the container, top with the remaining crumb mixture and press down.

**TO PACK AND FREEZE:** Open freeze until firm then turn out, wrap in foil, overwrap with polythene, seal and label.

**TO USE:** Unwrap and place on a serving plate. Thaw for about 6 hr. at room temperature. Decorate with apricots, heat the jam in a saucepan with 2 tbsps. apricot syrup, sieve and use to glaze the apricots. Decorate with whipped cream.

# Apfel strudel

8 oz. plain flour
$\frac{1}{2}$ level tsp. salt
1 egg, slightly beaten
2 tbsps. cooking oil
4 tbsps. lukewarm water
$1\frac{1}{2}$ oz. seedless raisins
$1\frac{1}{2}$ oz. currants
3 oz. caster sugar
$\frac{1}{2}$ level tsp. ground cinnamon
$2\frac{1}{2}$ lb. cooking apples, peeled, cored, and thinly sliced
$1\frac{1}{2}$ oz. butter, melted
4 oz. ground almonds

*To complete dish from freezer*
icing sugar
double cream

*Serves 4–6*

**TO MAKE:** Sift the flour and salt into a large bowl, make a well in the centre and pour in the egg and oil. Add the water gradually, stirring with a fork to make a soft, sticky dough. Work the dough in the bowl until it leaves the sides, turn it out on to a lightly floured surface and knead for 15 min. Form into a ball, place in a warmed bowl and cover with a cloth. Leave to 'rest' in a warm place for an hour. Add the raisins, currants, sugar, and cinnamon to the apples and mix thoroughly.

Warm the rolling pin. Spread a clean old cotton tablecloth on the table and sprinkle lightly with 1–2 tbsps. flour. Place the dough on the cloth and roll out into a rectangle about $\frac{1}{8}$ in. thick, lifting and turning it to prevent it sticking to the cloth. Gently stretch the dough, working from the centre to the outside and using the backs of the hands, until it is paper-thin. Trim the edges to give an oblong about 27 in. by 24 in. Arrange the dough with one long side towards you, brush it with melted butter and sprinkle with ground almonds. Spread the apple mixture over the dough, leaving a 2-in. border uncovered all round the edge. Fold these pastry edges over the apple mixture, towards the centre. Lift the corners of the cloth nearest to you up and over the pastry, causing the strudel to roll up, but stop after each turn, to pat it into shape and to keep the roll even. Brush the roll with the remaining melted butter and slide it on to a lightly buttered baking sheet. Bake in the oven at 375°F (mark 5) for about 40 min., or until golden brown. Cool on a wire rack.

**TO PACK AND FREEZE:** See notes on page 218.

**TO USE:** Unwrap and allow the strudel to thaw for 3–4 hr. at room temperature. Then place it on a baking sheet in the oven at 350°F (mark 4) for 20 min. Dust with icing sugar and serve with cream.

# Lemon soufflé

3 egg yolks
2 oz. caster sugar
grated rind and juice of 2 lemons
$\frac{1}{2}$ pkt. lemon jelly
2 tbsps. water
$\frac{1}{4}$ pt. double cream, whipped
3 egg whites

*To complete dish from freezer*
whole sprigs of redcurrants, dipped in
beaten egg white and caster sugar

*Serves 4*

**TO MAKE:** Prepare a 5-in. soufflé dish with a band of double greaseproof paper or foil forming a collar 2 in. above the rim. Whisk together the egg yolks, sugar, lemon rind and juice in a deep bowl over hot water, until really thick. Dissolve the jelly in the water.

Either place both in a small pan over a low heat or put them in a small basin over hot water. Stir well and leave to cool. When lukewarm, whisk into the egg yolk mixture. Leave in a cool place; when half set, fold in the cream. Lastly fold into the stiffly whisked egg whites. Turn the mixture into the prepared soufflé dish and place it on a baking sheet.

**TO PACK AND FREEZE:** Freeze until firm. Leave the collar in place, wrap in foil and overwrap with polythene. Seal and label. Return to the freezer.

**TO USE:** Remove the wrappings but cover loosely with a polythene bag. Thaw at room temperature for about 4 hr. or in the refrigerator overnight. Remove the collar and decorate the soufflé with sprigs of redcurrants.

# Lemon wine syllabub

grated rind and juice of 1 large lemon
$\frac{1}{4}$ pt. dry white wine
3 oz. caster sugar
$\frac{1}{2}$ pt. double cream

*Serves 6*

**TO MAKE:** Put the lemon rind and juice, wine and sugar in a basin. Place in the refrigerator for 3–4 hr. Add the cream to the wine mixture and whip well until the cream leaves a trail.

**TO PACK AND FREEZE:** Spoon the syllabub into individual dishes, open freeze then wrap in foil, overwrap in polythene, seal, label and return to the freezer.

**TO USE:** Unwrap the dishes and allow to thaw for 1–2 hr. at room temperature or about 4 hr. in the refrigerator. (Serve syllabub on the day of thawing or it will start to separate.) Serve with boudoir biscuits or sponge drops.

# 2 Quick Meals

Most of the recipes in this chapter can be served in about $\frac{1}{2}$ hr. Some use home or commercially frozen raw foods from the freezer; others are cooked for the freezer and need very little reheating. The plate meals in the 'Main meal' section are particularly useful in families where one person comes home at irregular hours and may need a meal unexpectedly. To wrap these meals for the freezer, cover the whole plate or tray with foil or the cardboard lid supplied; overwrap with polythene, seal, label and freeze. To use, remove the polythene bag and cardboard lid if used and cover loosely with foil, or just loosen the foil covering if that was used originally. Reheat the meals from frozen according to the recipe.

## BASIC COOKING TIMES FROM FROZEN FOR MEAT AND FISH PORTIONS

**Minced beef:** Sauté 4 oz. onions in a little fat. Add 1 lb. mince with $\frac{1}{2}$ pt. hot stock. Bring to boil, cover and simmer for 30 min. Season to taste.

**Sausages:** (a) *Thick variety with skins:* Fry for 15 min. Brown quickly on all sides; reduce heat, cover and allow to cook through. Grill for 15–20 min.

Before cooking, separate sausages with a sharp knife being careful not to split the skins.

(b) *Thinner skinless variety:* Fry 10–15 min. Grill 10–15 min.

*Note:* Or follow manufacturer's instructions.

**Pork chops:** Bake at 350°F (mark 4) for $1\frac{1}{4}$–$1\frac{1}{2}$ hr. (Grilling and frying not recommended)

**Plaice fillets:** Coat with egg and breadcrumbs when slightly thawed. Deep fry for 3 min. or shallow fry for 6–7 min., turning once. Bake with a little liquid for 25–30 min.

in the oven at 425°F (mark 7).

**Breaded scampi:** Deep fry for 3–4 min. depending on size, or shallow fry 6–8 min., turning once.

**Lamb chops:** Depending on thickness: Grill 15–25 min., beginning with a high heat and then reducing to cook through. Fry 15–25 min., beginning with high heat and then reducing to cook through.

**Lamb cutlets:** Depending on thickness: Grill 10–15 min. Fry 15 min.

**Noisettes of lamb:** Grill 20 min. Fry 30 min.

**Veal escalopes:** Fry coated with egg and breadcrumbs 4–5 min. Fry plain 3–4 min.

**Liver:** Fry 20 min. Bake in the oven with sauce at 375°F (mark 5) for 1 hr.

# STARTERS, SNACKS AND SUPPER DISHES

## Prawn and cheese ramekins

½ pt. frozen béchamel sauce (page 62)
4 oz. frozen prawns
2 eggs, hard-boiled, peeled and chopped
2 oz. Cheddar cheese, grated
1 tbsp. lemon juice
salt and pepper
pinch of dry mustard
pinch of paprika pepper
parsley to garnish

*Serves 6*

Heat the sauce and prawns together in a saucepan. When hot remove from the heat and add the eggs, 1 oz. cheese, lemon juice and seasonings. Divide the mixture between 6 ramekins. Top each with a little of the remaining grated cheese and pop under a preheated grill to brown. Garnish with sprigs of parsley and serve at once.

## Fish fingers with spinach mayonnaise

6 oz. frozen leaf spinach
16 frozen fish fingers
6 level tbsps. thick mayonnaise
2 tbsps. milk
pinch grated nutmeg
1 tbsp. lemon juice
salt and pepper

*Serves 4*

Gently heat the spinach in a saucepan until thawed. Cool it and purée in a blender. Cook the fish fingers according to the directions on the packet. Mix the purée with remaining ingredients and serve as a separate sauce with the hot fish fingers.

## Shrimp scramble

3-oz carton frozen potted shrimps
2 slices bread (¼–½ in. thick)
1 oz. butter
2 tomatoes, halved
4 eggs
2 tbsps. milk
salt and pepper
parsley to garnish

*Serves 2*

Empty the carton of shrimps into a pan and heat through gently. Toast and butter the bread and grill the tomato halves. Beat the eggs well, add the milk, season and pour into the pan with the shrimps until they 'scramble'; take care not to overcook. Pile the mixture on to the toast and garnish with the tomatoes and parsley.

*Note:* Prawns may be substituted for the potted shrimps, but ½ oz. butter or margarine should be added when heating them through.

## Farmhouse supper

(*see picture opposite*)

½ pt. frozen cheese sauce (page 63), thawed
4 frozen large broccoli spears
4 slices cooked ham
1 oz. fresh white breadcrumbs
1 oz. Cheddar cheese, grated
1 oz. butter

*Serves 4*

Re-heat the sauce. Cook the broccoli according to the directions on the packet. Roll the broccoli in the ham slices and place them in an ovenproof casserole in a single layer. Cover with the hot sauce. Sprinkle with the crumbs and then the cheese. Dot with flakes of butter and brown under a preheated grill until bubbling.

## Devilled mushrooms

(*see colour picture facing page 33*)

12 oz. frozen button mushrooms
2 oz. butter
2 tbsps. lemon juice
salt and pepper
¼ pt. double cream, whipped
2 tbsps. Worcestershire sauce
1 level tsp. French mustard
2 oz. mature Cheddar cheese, grated
grated nutmeg
chopped parsley to garnish

*Serves 6*

Sauté the mushrooms in butter until thawed. Add the lemon juice and season. Divide the mushrooms between 6 ramekins. Keep warm. Stir all the remaining ingredients into the

*Farmhouse supper*

whipped cream, spoon it over the mushrooms and place the dishes under a preheated grill for 2–3 min. until golden brown and bubbling. Garnish with parsley.

# Chicken liver croustades

2 large slices bread
butter
4 oz. frozen chicken livers
2 oz. frozen button mushrooms
1 tbsp. sherry
salt and pepper
parsley to garnish

*Serves 2*

Cut the crusts from the bread and butter the slices. Press the bread, butter side down, into individual Yorkshire pudding tins. Bake in the oven at 375°F (mark 5) for 20 min. Meanwhile melt 1 oz. butter in a frying pan. Add the frozen livers and mushrooms. Cover and simmer for 15–20 min. Add the sherry and season. Place the hot bread cases on individual plates, fill with liver and garnish with parsley.

# Oeufs florentines

½ pt. frozen cheese sauce (page 63)
12 oz. frozen leaf spinach
4 eggs
1 tomato, sliced
1½ oz. grated Cheddar cheese

*Serves 4*

Reheat the cheese sauce from frozen in a saucepan. Place the spinach in a very little boiling salted water and cook until tender. Poach the eggs. Drain the spinach well and press out all the water. Chop it roughly and place in the base of a buttered shallow ovenproof dish. Lay the poached eggs on top and pour the hot cheese sauce over. Arrange the tomato slices on the sauce, sprinkle with grated cheese, and brown under a hot grill.

# Sausage pilaff

1 lb. frozen sausages
8 oz. long grain rice
2 tbsps. cooking oil
3 oz. onion, skinned and chopped
½ red pepper, seeded and sliced
½ green pepper, seeded and sliced
2 oz. mushrooms, sliced
salt and pepper

*Serves 3–4*

Separate the sausages carefully with a sharp knife and cook under a hot grill for about 20 min. Cook the rice in boiling salted water for about 12 min., and drain well. Heat the oil, sauté the onion and peppers for a few min. Add the mushrooms and season. When cooked remove from the pan. Drain the rice and combine with the onion and peppers. Slice each sausage into about 6 and add to the rice mixture. Pile into a serving dish.

# Hot German salad

1 lb. frozen sausages
1½ lb. potatoes, peeled and cut into
 ½-in. dice
4 onions, skinned and chopped
salt and pepper
6 tbsps. mayonnaise
2 tbsps. milk
chopped parsley

*Serves 3–4*

Cook the sausages according to notes at beginning of chapter and slice each into 6 or 8. Meanwhile, cook the diced potatoes in boiling water for 10–15 min. Drain well. In a mixing bowl, combine the hot sausage, hot potato and raw onion. Season the mayonnaise well and mix with the milk; pour it over the sausage mixture. Toss lightly and serve sprinkled with chopped parsley.

White wine may be used instead of milk.

# Gnocchi

½ pt. milk
2 oz. semolina
salt and pepper
pinch grated nutmeg
paprika pepper
1 egg, beaten
1 oz. Parmesan cheese, grated

*To complete dish from freezer*
1½–2 oz. Cheddar cheese, grated

*Serves 4*

**TO MAKE:** Bring the milk to the boil, sprinkle in the semolina and seasonings and stir over a gentle heat until the mixture is really thick. Remove from the heat, beat well until smooth, then stir in the egg and cheese. Spread the mixture about ¼–½ in. thick in a shallow buttered dish and allow to cool. Cut into 8 pieces.

**TO PACK AND FREEZE:** Freeze the gnocchi unwrapped until firm on a foil plate. Wrap in foil, overwrap in polythene. Seal and label. Return to the freezer.

**TO USE:** Remove polythene wrappings and place, still covered with foil, in the oven. Reheat at 375°F (mark 5) for 45 min. Just before serving, sprinkle with the extra cheese and brown quickly under the grill or in the oven, turned up to its hottest.

# Sweetcorn special

12 oz. frozen sweetcorn kernels
3 eggs, hard-boiled
2 oz. butter
salt and pepper
3 oz. Cheddar cheese, grated
3 tomatoes, halved
melted butter

*Serves 3*

Cook sweetcorn in boiling salted water for about 5 min. Drain and shell the eggs whilst hot. Drain the corn and toss it in the butter. Pour the corn into a shallow ovenproof dish, halve the eggs and place them on top. Season well and cover with the grated cheese. Garnish with the tomatoes. Brush with a little melted butter and place under a hot grill for 4–5 min. to brown.

# Spinach pancakes

4 oz. plain flour
large pinch salt
pinch grated nutmeg
freshly ground black pepper
1 egg
½ pt. milk
2 tsps. cooking oil
2 heaped tbsps. spinach purée, well drained
oil or lard for frying

*Serves 4*

**TO MAKE:** Sift the flour with the salt, nutmeg and pepper. Make a well in the centre and drop in the egg with a little of the milk. Stir the flour into the egg. Add half the milk and beat to a batter, gradually adding the remaining milk. Beat the batter thoroughly and add 2 tsps. oil. Stir in the spinach purée. Heat the oil or lard in a small frying pan or omelette pan and spoon in a little of the mixture to make each pancake. Cool.

**TO PACK AND FREEZE:** Interleave the cold pancakes with waxed paper and then wrap them all in foil. Place in a polythene bag, seal, label and freeze.

**TO USE:** Unwrap the pancakes and reheat each one separately for about $\frac{1}{2}$ min. in a little melted butter in the frying pan.

**Suggestion for serving:** Heat 4 oz. chopped prawns in $\frac{3}{4}$ pt. cheese sauce. Place the reheated, rolled up pancakes in a shallow serving dish and pour over the sauce.

# Crisp-fried mushroom balls

$1\frac{1}{2}$ level tsps. powdered gelatine
$\frac{1}{2}$ pt. chicken stock
butter
1 oz. flour
2 oz. Gruyère or Edam cheese, grated
4 oz. mushrooms, chopped
3 oz. cooked pork or ham, finely chopped
1 tsp. chopped parsley
salt and freshly ground black pepper
1 egg, beaten
3–4 oz. fresh white breadcrumbs

*Makes 8*

**TO MAKE:** Sprinkle the gelatine over the stock. In a saucepan, melt 1 oz. butter, stir in the flour and cook for 1 min. Gradually add the stock, beating between each addition. Bring almost to the boil, stirring continuously. Add the grated cheese. Sauté the mushrooms in $\frac{1}{2}$ oz. butter. Drain off any excess liquor and add the mushrooms to the sauce mixture. Fold in the chopped pork or ham and parsley. Adjust the seasoning. Turn the mixture out on to a plate and chill well until really cold, then divide into 8 and roll each portion into a ball. Dip each ball in beaten egg and then coat with crumbs, patting them well in. Repeat to give two coatings.

**TO PACK AND FREEZE:** Open freeze until firm. Wrap in a polythene bag, seal and label. Return to the freezer.

**TO USE:** Unwrap and cook gently from frozen in deep fat for about 10–15 min. until golden on the outside and warm in the centre.

# Scampi fiesta

$1\frac{1}{2}$ oz. margarine
6-oz. pkt. frozen scampi (not breaded)
4 oz. long grain rice
2 oz. onion, skinned and chopped
1 green pepper, seeded and chopped
6 oz. mushrooms, sliced
2 tsps. lemon juice
4 tbsps. white wine
2 tbsps. brandy
$\frac{1}{4}$ pt. single cream
1 egg yolk
salt and pepper
lemon wedges
parsley

*Serves 4*

Melt 1 oz. margarine in a frying pan, add the frozen scampi and cook gently for 6–7 min. Cook the rice in boiling salted water and drain. Melt $\frac{1}{2}$ oz. margarine in a separate pan, sauté the onion, pepper and half the mushrooms. Add the other half of the mushrooms to the scampi with the lemon juice and wine. Continue to cook until the mushrooms are tender. Warm the brandy, ignite it and pour it over the scampi. Blend the cream with the egg yolk, stir into the scampi, season and reheat gently without boiling. Mix the rice with the vegetables and arrange around the sides of a serving dish. Turn the scampi into the centre of the rice and add the lemon and parsley as a garnish.

# Scampi amande

(*see colour picture facing page 33*)

fat for deep fat frying
8 oz. frozen breaded scampi or prawns
2 oz. butter
2 oz. almonds
1 banana
juice 1 lemon
salt and pepper

*Serves 2*

Heat the fat to 360°F. Deep fry the scampi from frozen for 3 min.; drain and keep warm. Melt the butter in a frying pan and sauté the almonds until golden. Peel and cut the banana in thick diagonal slices; add to the frying pan and warm through. Add the scampi to the pan with the lemon juice and seasonings. Serve immediately.

# Prawn lumacine

10-oz. can condensed chicken soup
¼ pt. water
6 oz. frozen prawns
4 oz. frozen peas
4 oz. frozen sweetcorn kernels
2 oz. lumacine, cooked
2 level tsps. curry powder
2 tbsps. sherry
salt and pepper

*Serves 2–3*

Mix the soup and water together in a saucepan. Add all the remaining ingredients, cover and heat through gently for about 20 min. until the mixture is bubbling and the vegetables tender. Turn into a serving dish.

*Note:* Lumacine are small pasta snail-like shapes. As an alternative use shell or similar pasta.

# Prawns au vin

4 oz. long grain rice
2 tbsps. cooking oil
4 oz. frozen prawns
1 small onion, skinned and thinly sliced
3 tomatoes, skinned and chopped
2 tbsps. dry white wine
salt and pepper
cayenne pepper

*Serves 2*

Cook the rice in boiling salted water; drain and keep it warm. Meanwhile heat the oil in a frying pan and fry the prawns and onion. Add the tomatoes, wine and seasonings to the prawn mixture. Cook for 7 min. Put the rice on a hot serving dish and pour the prawns over.

# Sweet and sour prawns

6 oz. long grain rice
sweet and sour sauce (see page 51)
6 oz. frozen prawns, thawed
4 oz. frozen peas

Cook the rice in boiling salted water until tender. Prepare the sauce and add the prawns and peas. Cover and simmer for about 6 min. Drain the rice and arrange round the edge of a shallow serving dish. Pour the prawns into the centre and serve at once.

# Curried prawn mayonnaise

(*see picture opposite*)

8 oz. frozen prawns
¼ pt. thick mayonnaise
4 level tbsps. fruit purée (e.g. apple, apricot)
1 level tbsp. curry powder
1 tbsp. lemon juice
salt and pepper
4 oz. long grain rice, cooked
lemon wedges and chopped parsley

*Serves 2–3*

Thaw the prawns for about 10 hr. in the refrigerator; dry them on absorbent paper. Mix all the other ingredients together to make the curry sauce. Reserve one third of the sauce and fold the prawns into the remainder. Serve the curry on a bed of cold cooked rice, mixed with a little chopped parsley. Pour the reserved sauce over to mask the prawns. Garnish with lemon wedges and parsley.

# Quick paella

2 oz. butter or margarine
4 oz. onion, skinned and chopped
8 oz. long grain rice
¾–1 pt. chicken stock
1 green pepper, seeded and sliced
3½-oz. can pimiento, drained and sliced
4 oz. frozen peas
7½-oz. can tuna steak, drained and flaked
8 oz. frozen prawns
1 tomato
1 lemon

*Serves 4*

Heat the butter or margarine in a large frying pan or paella dish. Sauté the onion until golden, then add the rice. Cook gently for 1–2 min., slowly pour in the stock, add the pepper and pimiento, bring to the boil, stirring, and simmer uncovered for about 15 min. Add the peas, tuna and prawns and

*Curried prawn mayonnaise*

cook uncovered for a further 15–20 min., or until all the stock has been absorbed. Fork through occasionally to prevent the rice sticking. Cut the tomato and lemon into wedges and use to garnish.

# Fish finger bake

butter
16 frozen fish fingers
6 oz. mushrooms, sliced
8-oz. can tomatoes, drained
salt and pepper
1 tbsp. lemon juice
2 oz. cheese, grated

*Serves 3–4*

Butter a shallow ovenproof dish large enough to hold the fish fingers in a single layer. Place the fingers in the base. Put the mushrooms, tomatoes, seasoning and lemon juice in a saucepan. Heat through and stir

well. Spread the tomato mixture over the fish fingers and bake uncovered in the oven at 350°F (mark 4) for 25–30 min. Remove from the oven, sprinkle with the cheese and brown under a hot grill.

# Smoked fish supper

½ pt. frozen cheese sauce (page 63)
6 oz. shell pasta
4 oz. frozen peas
3 eggs
12-oz. pkt. boil-in-the-bag frozen
    smoked haddock fillets
salt and pepper
1½ oz. cheese, grated
1 tomato, sliced

*Serves 3–4*

Place the frozen sauce in a saucepan to heat through. Cook the pasta in boiling salted water for 10–15 min., adding the peas after

*Snow-capped fish cakes*

5 min. Hard-boil the eggs, drain, shell and slice whilst hot. Cook the smoked haddock according to the instructions on the packet, then remove the bag, drain, skin and flake the fish. Add the fish to the cheese sauce. Drain the pasta and peas. Place the pasta and peas in a shallow ovenproof dish, season; cover with the sliced egg and pour over the cheese sauce. Top with the grated cheese and pop under a hot grill to brown. Garnish with freshly sliced tomato.

# Snow-capped fish cakes

(*see picture above*)

6 frozen fish cakes
2 egg whites
4 level tbsps. prawn cocktail sauce or
   tartare sauce

*Serves 2–3*

Grill the fish cakes according to the directions on the packet. Whisk the egg whites stiffly and fold into the sauce with a metal spoon. Top each cake with the egg mixture and return it to the grill to brown. Serve immediately.

# Individual haddock soufflés

7½-oz. pkt. boil-in-the-bag frozen
   smoked haddock fillets
½ oz. butter
2 oz. onion, skinned and finely chopped
½ pt. frozen béchamel sauce (page 62)
2 eggs, separated

*Makes 6*

Cook the haddock as directed on the packet. Melt ½ oz. butter and sauté the onion until soft. Keep to one side. Reheat the sauce in the saucepan, add the onion. Remove from the heat and beat in the egg yolks. Drain, skin and flake the fish. Divide between 6 individual 3½-in. soufflé dishes. Stiffly whisk the egg whites and then fold into the sauce mixture. Top the fish with the egg mixture. Bake in the oven at 425°F (mark 7) for 15 min. until well risen and golden. Serve immediately.

# Fisherman's spaghetti

6 oz. spaghetti
7½-oz. pkt. boil-in-the-bag frozen
  smoked haddock fillets
4 oz. Cheddar cheese, grated
¼ pt. soured cream, warmed
2 oz. butter
3 tomatoes, skinned and chopped
2 tbsps. chopped parsley
freshly ground black pepper

*Serves 3*

Cook the spaghetti in boiling salted water for 12 min. Meanwhile cook the smoked haddock as directed on the packet. Drain, skin and flake the fish. Drain the spaghetti. Combine all the ingredients together in a large dish, mixing well. Adjust seasoning and serve immediately with a green salad.

# Haddock and spaghetti au gratin

8-oz. pkt. frozen onions in white sauce
4 oz. frozen mixed vegetables
12-oz. pkt. boil-in-the-bag frozen
  smoked haddock
4 oz. spaghetti
2 oz. Cheddar cheese, grated
1 tomato, sliced

*Serves 3*

Cook the onions in white sauce according to the instructions on the packet, adding the frozen mixed vegetables to cook in the same saucepan. Cook the smoked haddock according to the instructions on the packet. Cook the spaghetti in boiling salted water until tender. Drain and flake the fish and add it to the vegetables and sauce. Pour the mixture into the base of an ovenproof dish. Drain the spaghetti well and pile it on top of the fish mixture. Top with the grated cheese and pop under a hot grill to brown. Serve garnished with the sliced tomato.

# Plaice fillets in wine sauce

4 frozen plaice fillets
¼ pt. dry white wine
½ oz. butter
1 level tbsp. flour
6 tbsps. milk
salt and pepper

*Serves 2–3*

Put the fish fillets in a shallow ovenproof casserole just large enough to take them. Add the wine, cover and cook in the oven at 425°F (mark 7) for 20–30 min. until the fish flakes easily. Drain the fish and reserve the stock. Keep the fish warm. Melt the butter in a saucepan, add the flour and cook gently for 1–2 min. Add the strained stock, milk, salt and pepper. Cook gently for a few min. Pour the sauce over the fish. Serve with asparagus tips or broccoli.

# Spanish plaice

½ oz. margarine or butter
½ red pepper, seeded and chopped
2 oz. onion, skinned and chopped
3 rashers streaky bacon, rinded and
  chopped
8 oz. frozen haricots verts (or frozen
  whole green beans)
12 oz. frozen plaice fillets
1 egg, beaten
breadcrumbs for coating
salt and pepper
lemon wedges

*Serves 3*

Heat the margarine in a pan, sauté the pepper, onion and bacon for 4–5 min. Cook the haricots verts in boiling salted water until just tender. Separate the plaice fillets and coat them with beaten egg and breadcrumbs whilst still frozen (or buy them ready breaded). Fry the fish in shallow fat for about 7 min., turning once, or in deep fat for about 3 min. Drain the beans, place in the base of a serving dish and add the peppers, onion and bacon. Season well. Drain the fish and lay it over the vegetables. Garnish with lemon wedges.

# Plaice and cheese bake

butter or margarine
13-oz. pkt. frozen plaice fillets
½ pt. milk
1 small can evaporated milk
2 eggs, beaten
3 oz. Cheddar cheese, grated
salt and pepper

*Serves 4*

Butter a 1½-pt. casserole. Skin the frozen plaice (see photograph on page 75), cut into strips with scissors. Lay the fish in the casserole. Warm the milk and evaporated milk. Remove from the heat and add to the eggs, with the remaining ingredients. Pour this custard over the fish. Bake in the oven at 325°F (mark 3) for about 1 hr. until firm and golden. Serve immediately.

# Plaice fillets in cream

4 frozen plaice fillets
¼ pt. double cream
¼ pt. milk
1 level tsp. cornflour
1 egg yolk
salt and freshly ground black pepper
2–3 oz. white grapes, skinned and pipped

*Serves 2*

Put the fish in a shallow ovenproof dish. Mix the cream and milk together and pour over the fish. Bake in the oven at 425°F (mark 7) for 25–30 min. Drain the fish and keep it warm. Reserve the cream juices. Mix the cornflour with the egg yolk and stir into the cream; reheat gently to thicken. Season to taste. Pour the sauce over the fish, garnish with grapes and serve with creamed potatoes and a green vegetable.

# Cod provençale

1 recipe quantity provençale sauce
  (page 64)
12 oz. frozen cod steaks
melted margarine or butter
salt and pepper
lemon twist
chopped parsley

*Serves 4*

Place the block of sauce in a saucepan to heat through. Brush the cod steaks with melted fat, season and grill for 15 min., turning once. Lay the cod in a shallow serving dish and pour the sauce over. Garnish with the lemon and parsley. Serve with creamed potatoes and a tomato salad.

# Country cod steaks

½ pt. frozen béchamel sauce (page 62)
12 oz. frozen cod steaks
12 oz. frozen sliced green beans
½ oz. margarine or butter
2 oz. onion, skinned and chopped
2 tomatoes, skinned
2 oz. mushrooms, sliced
salt and pepper
lemon and parsley to garnish

*Serves 2–3*

Place the frozen block of sauce to heat through in a saucepan until bubbling. Grill the cod steaks for 15–20 min., turning once. Cook the beans in boiling salted water until just tender. Melt the margarine or butter, sauté the onion, thickly sliced tomatoes and mushrooms. Drain the beans and add the onion, tomato and mushrooms. Mix well and season. Place the beans in the base of a shallow serving dish, arrange the cod steaks on the top and pour over the white sauce. Garnish with lemon and parsley.

# Peppered sausage pancakes

½ lb. frozen sausages or chipolatas
1½ oz. margarine or butter
3 oz. onion, skinned and chopped
1 tomato, skinned and chopped
2 green peppers, seeded and finely sliced
4 7-in. frozen pancakes
salt and pepper

*Serves 2*

Separate the sausages and cook from frozen under a hot grill for 15–20 min. Melt 1 oz. fat and sauté the onion, tomato and peppers. Keep these hot. Melt ½ oz. fat in a pan and reheat the pancakes one by one, keeping them hot on a warm plate. When the sausages are cooked, cut them into ½-in. slices. Adjust seasoning. Set aside a little of the pepper mixture and add the sausages to the rest in the pan. Divide this mixture between the pancakes, fold over and serve with the reserved pepper mixture for garnish.

# Sausages Italian style

½ recipe quantity frozen tomato sauce
  (page 67)
1 lb. frozen sausages
1 lb. frozen cauliflower florets
salt and pepper
2 tbsps. grated Parmesan cheese

*Serves 4*

Place the block of sauce in a saucepan to heat through. Separate the sausages with a sharp knife, being careful not to split the skins. (Skinless variety could also be used). Place them under a hot grill and cook for 15–20 min. Cook the cauliflower gently in boiling salted water for about 7 min. Drain well and place the cauliflower in the base of a shallow ovenproof dish. Season well and cover with the cooked sausages. Pour over the tomato sauce and top with Parmesan cheese. Pop under a hot grill for a few min. until the sauce is bubbling. Serve at once.

*Note:* Spaghetti or noodles could be used instead of cauliflower.

# Peppered beefburgers with rice

6 oz. long grain rice
1 tbsp. cooking oil
½ green pepper, seeded and finely sliced
½ red pepper, seeded and finely sliced
½ onion, skinned and chopped
2 oz. mushrooms, sliced
6-oz. can condensed tomato soup
6 fl. oz. water
salt and pepper
4 frozen beefburgers

*Serves 3*

Cook the rice in boiling salted water until tender. Heat the oil in a pan and sauté the

peppers, onion and mushrooms. Mix the soup and water, add to the pan and bring to the boil. Adjust the seasoning, cover and simmer for 5 min. Cook the beefburgers according to manufacturer's instructions and cut them into quarters or strips. Add to the sauce. Drain the rice and arrange around the edge of a warmed shallow serving dish. Pour the beefburger and pepper sauce into the centre. Serve with a tossed green salad or buttered green beans.

# Cottage beefburger

4 frozen beefburgers
4 oz. frozen mixed vegetables
8-oz. can tomatoes, drained
2½ oz. instant mashed potato
½ oz. margarine
1 egg, beaten
salt and pepper
1½ oz. Cheddar cheese, grated

*Serves 3–4*

Grill the beefburgers according to manufacturer's instructions. Cook the vegetables in boiling salted water, drain and add the tomatoes to heat through. Make up the potato and beat in the margarine and egg. Season the tomato mixture and put it in the base of an oven-proof dish. Cut each cooked beefburger into 4 and place on top of the vegetables. Pile the potato on top and fork over. Sprinkle with grated cheese and place under a hot grill to brown.

# Poor man's Stroganoff

1 oz. butter
4 frozen beefburgers
4 oz. onion, skinned and finely sliced
4 oz. frozen sliced mushrooms
salt and pepper
¼ pt. soured cream

*Serves 2*

Melt the butter in a large frying pan and fry the beefburgers according to the manufacturer's instructions, with the onions and mushrooms, for about 7 min. Season to taste. Quarter the beefburgers, stir in the cream and serve immediately with boiled rice or noodles.

# Chilli burgers

1 oz. butter
½ green pepper, seeded and sliced
4 oz. frozen mixed vegetables
8-oz. can tomatoes, drained
½ level tsp. chilli powder
1 tbsp. lemon juice
salt and pepper
4 frozen beefburgers

*Serves 2*

Melt butter and sauté the pepper in a saucepan, add the vegetables, tomatoes and seasonings, cover and simmer for 10–15 min. Cook the beefburgers according to the manufacturer's instructions. Serve the beefburgers on the sauce.

# Minced beef with vegetables

1 oz. butter
4 oz. onion, skinned and chopped
1 lb. frozen minced beef
½ pt. beef stock
salt and pepper
4 oz. frozen mixed vegetables

*Serves 3*

Melt the butter in a saucepan and sauté the onion until transparent. Add the meat, cover with the stock and season well. Bring to the boil, reduce the heat, cover and simmer for 30 min., stirring occasionally. About 10 min. before the end of cooking time, add the vegetables. Serve with mashed potatoes.

# Freezer beef risotto

1 oz. butter
4 oz. onion, skinned and chopped
4 oz. long grain rice
1 lb. frozen minced beef
4 oz. frozen mixed vegetables
2 oz. sultanas
1 pt. beef stock
salt and pepper
gravy browning (optional)

*Serves 4*

Melt the butter in a saucepan and sauté the onion and rice until beginning to brown. Add all the other ingredients, bring to the boil,

reduce the heat, cover and simmer for 30–40 min., stirring occasionally until the meat is cooked and the liquid is absorbed.

# Cutlets portuguese

1½ oz. butter or margarine
2 frozen lamb cutlets
2 tomatoes, halved
¼ pt. water
2 level tbsps. tomato paste
1 clove garlic, skinned and crushed
1 level tsp. dried mixed herbs
salt and pepper
2 level tbsps. packet sage and onion stuffing
1 level tsp. cornflour

*Serves 2*

Melt 1 oz. butter or margarine and fry the frozen cutlets for about 15 min. Meanwhile scoop out the pulp from the tomatoes and put it into a pan with the water, tomato paste, garlic, herbs, salt and pepper. Bring to the boil, cover and simmer for 5 min. Mix the packet stuffing with 3–4 tbsps. boiling water and pile it into the tomato halves. Dot with the rest of the butter or margarine and grill for 4–5 min. Remove the cutlets from the frying pan and keep them warm on a serving dish. Pour off the excess fat from the pan. Sieve the tomato sauce mixture and blend a little of it with the cornflour in the frying pan, add the remainder and bring to the boil, stirring to loosen any sediment. Cook gently for 2 min. and adjust the seasoning. Pour the sauce over the cutlets and garnish with the stuffed tomatoes. Serve with boiled rice.

# Barbecue chops

4 frozen lamb chops

*For sauce*
2 tbsps. cooking oil
6 oz. onion, skinned and chopped
2 level tbsps. tomato paste
4 tbsps. vinegar
¼ level tsp. dried thyme
dash Worcestershire sauce
3 level tbsps. clear honey
¼ pt. stock

*Serves 4*

Place the frozen lamb chops under a hot grill,

brown on both sides, reduce the heat and cook for 20–25 min. Heat the oil in a saucepan, and sauté the onion without browning. Stir in the tomato paste and add all the remaining ingredients. Bring to the boil and simmer for 10 min. Brush the chops with a little sauce to glaze. Place on a serving dish and serve the remaining sauce separately. Accompany by chipped potatoes and sweetcorn kernels.

## Sweet and sour chops

4 frozen lamb chops
cooking oil

*For sweet and sour sauce*
½ lb. tomatoes, skinned
11-oz. can pineapple rings or cubes, drained
1½ level tbsps. cornflour
½ pt. water
3 oz. sugar
4 tbsps. cider vinegar
3 tbsps. soy sauce
1 green pepper, seeded and thinly sliced
salt and freshly ground black pepper

*Serves 4*

Place the frozen chops in a frying pan with a little oil and fry each side over a high heat to seal and brown. Reduce the heat and cook for 20–25 min. until almost tender. Meanwhile make the sauce. Purée the tomatoes and pineapple in a blender. Blend the cornflour with a little of the water in a saucepan. Add the sugar, vinegar, soy sauce and the rest of the water. Bring to the boil, stirring, reduce the heat and simmer for 5 min. Add the tomato and pineapple purée and pepper, stir well and simmer for 5 min. Adjust seasoning. Drain the fat from the chops in the frying pan and pour in the sweet and sour sauce. Let the chops complete cooking in the sauce. Serve with rice.

## Pan-fried liver with onions

½ oz. butter or margarine
6 oz. onions, skinned and chopped
½ lb. frozen sliced lambs' liver
salt and pepper

*Serves 2*

Melt the butter or margarine in a frying pan. Add the onions, frozen liver and seasonings. Brown the meat on both sides over a high heat, cover and simmer for 20 min., turning the liver occasionally as it separates. Remove from the pan, drain and serve with boiled rice.

## Rognoni tripolata

¾ lb. frozen sheeps' kidneys, thawed
1 oz. butter or margarine
grated rind of ½ lemon
¼ level tsp. salt
freshly ground black pepper
2 tsps. lemon juice
1 level tbsp. cornflour
¼ pt. beef stock
2 tbsps. Marsala
chopped parsley

*Serves 2*

Skin the kidneys and cut them in half. Remove the cores and slice thinly. Melt the butter or margarine in a pan, add lemon rind, salt and pepper. Sauté the kidneys gently for 2–3 min. then add the lemon juice. Cream the cornflour with 3 tbsps. stock until smooth. Add remaining stock and blend together before adding to the kidneys. Cook gently for about 10 min., stirring occasionally. Increase the heat to thicken the sauce and cook for a further 2–3 min. Remove from the heat, and stir in the Marsala. Sprinkle with chopped parsley and serve with rice or noodles.

## Veal Marsala

(*see colour picture facing page 32*)

2 frozen veal escalopes
salt and pepper
1 tbsp. lemon juice
1 oz. butter
3 oz. frozen mushrooms
2 tbsps. Marsala
4 tbsps. double cream

*Serves 2*

Season the frozen escalopes with salt, pepper and lemon juice. Melt the butter in a frying pan, sauté the frozen mushrooms, add the veal and cook for 4–5 min. each side.

Add the Marsala and heat through. Remove from the heat, stir in the cream and serve immediately.

## Wiener Schnitzel

2 frozen veal escalopes
seasoned flour
beaten egg
fresh white breadcrumbs
2 oz. butter

*Serves 2*

Rub the frozen escalopes with flour, dip them in egg and then breadcrumbs, pressing them on well. Melt the butter in a frying pan, and fry the escalopes for 4–5 min. on each side until golden brown.

## Aubergine cheese soufflé

(*see picture below*)

½ pt. frozen aubergine purée
¼ pt. frozen cheese sauce (page 63)
3 eggs, separated

*Serves 4–5*

Thaw the purée and sauce together in a saucepan. Remove from the heat and beat in the egg yolks. Stiffly beat the egg whites, fold into the hot sauce mixture and pour into a 3-pt. buttered soufflé dish. Place it on a baking sheet and bake in the oven at 375°F (mark 5) for 30–40 min., until well risen and golden. Serve immediately.

*Aubergine cheese soufflé*

## Onion cartwheel flan

(*see picture below*)

8-oz. pkt. frozen onions in white sauce
2 eggs, beaten
milk
salt and pepper
9½-in. frozen baked shortcrust flan case
½ lb. frozen thin pork sausages

*Serves 4–6*

Cook the onions in sauce according to the instructions on the packet. Cool. Make the eggs up to ½ pt. with milk and stir the liquid into the cooked sauce. Adjust the seasoning and pour filling into the flan case. Bake in the oven for 30–35 min. at 375°F (mark 5) until set in the centre. Meanwhile grill the sausages from frozen for about 15 min. Place the sausages on the flan like the spokes of a wheel. Serve hot.

## Vegetarian vol-au-vents

6 3-in. frozen vol-au-vent cases
8 oz. frozen mixed vegetables
½ pt. frozen béchamel sauce (page 62)
1 level tsp. curry powder
½ level tsp. ground turmeric
salt and pepper
parsley sprigs

*Makes 6*

Bake the vol-au-vent cases in the oven at 400°F (mark 6) for about 15–20 min. Meanwhile cook the vegetables in boiling salted

*Onion cartwheel flan*

water until tender and heat the sauce in a saucepan until thawed. Add the curry powder, turmeric, salt and pepper to the sauce and cook for 1–2 min. Drain the vegetables and stir them into the sauce. Spoon the mixture into the pastry cases. Serve at once with a garnish of tiny parsley sprigs.

# Toreador pancakes

lard
2 oz. onion, skinned and chopped
8 oz. lean minced beef
1 level tsp. dried fines herbes
10½-oz. can condensed tomato soup
1 level tbsp. cornflour
1 tbsp. water
½ pt. pancake batter (i.e. made with
    4 oz. plain flour etc.—page 82)

*To complete dish from freezer*
1 oz. cashew nuts, chopped
grated Parmesan cheese

*Serves 4*

TO MAKE: In a frying pan, melt ½ oz. lard, add the onion and fry until soft and lightly coloured. Add the mince and continue to cook, stirring, for 5–7 min. until the mince changes colour. Sprinkle in the fines herbes. Make up the soup to half strength. Add to pan. Blend the cornflour with the water and stir into the pan. Bring to the boil, stirring, and simmer for 5 min. Cool. Make eight 7-in. pancakes in the usual way. Fold each into quarters and cool. Open up to form cornets and fill with meat mixture. Chill.

TO PACK AND FREEZE: Lay the pancakes in a single layer on a foil plate, cover with foil, seal, label and freeze.

TO USE: Unwrap, top the pancakes with the nuts and cheese, cover loosely with foil and reheat from frozen in the oven at 400°F (mark 6) for about 30 min.

# Sausage kebabs

(*see colour picture facing page 33*)

4 rashers streaky bacon, rinded
1 lb. small chipolata sausages
16 button mushrooms

*Serves 4*

TO MAKE: Cut each bacon rasher in half, stretch it with the back of a knife and roll up. Make 4 kebabs by putting sausages, bacon rolls and mushrooms alternately on skewers.

TO PACK AND FREEZE: Cushion the tip of each skewer with foil (to prevent it from piercing the package in the freezer). Wrap the kebabs individually in foil and then place in a polythene bag. Seal and label. Freeze.

TO USE: Unwrap kebabs and grill from frozen for 15–20 min. until the sausages and bacon are cooked through and well browned. Serve with rice or potato crisps.

# PLATE MEALS

# Ham and rice roulades

2 slices cooked ham
2 slices Gruyère cheese
2 oz. cooked long grain rice
2 oz. red pepper, blanched and chopped
salt and pepper
½ pt. cheese sauce (coating consistency)

*Serves 1–2*

TO MAKE: Lay the ham flat and top with slices of cheese. Mix the rice and pepper together, season to taste and divide the mixture between the ham and cheese slices. Roll up from the short ends. Place the rolls in a small freezer tray and cover with cheese sauce.

TO PACK AND FREEZE: See notes at beginning of chapter.

TO USE: Remove wrappings, cover the meal loosely with foil and reheat from frozen in the oven at 400°F (mark 6) for about 40 min. Serve with a green salad.

# Bolognese nests

(*see picture opposite*)

¼ oz. butter
2 oz. onion, skinned and finely chopped
4 oz. minced lean beef
1½ level tsps. flour
1½ level tsps. tomato paste
pinch mixed dried herbs
salt and pepper
4 tbsps. stock
creamed potato
3 oz. frozen diced carrots
3–4 oz. frozen Brussels sprouts, lightly cooked
1 oz. butter

*Serves 1*

TO MAKE: Melt the butter in a saucepan, add the onion and sauté gently until transparent and soft. Stir in the minced beef and brown well. Stir in the flour, tomato paste, herbs, seasoning and stock. Bring to the boil, then cool quickly. Pipe the potato to make two nests on a small freezer tray. Fill the centres with the cold mince mixture. Place the frozen carrots and cold cooked sprouts on the tray and dot with butter.

TO PACK AND FREEZE: See notes at beginning of chapter.

TO USE: Loosen the foil, but leave it in place. Reheat the meal from frozen in the oven at 400°F (mark 6) for 30–40 min.

# Stuffed marrow

2 thick slices marrow, peeled
1 oz. butter
2 oz. onion, skinned and chopped
4 oz. minced beef
2 oz. sausage meat
salt and pepper
3 oz. frozen Brussels sprouts, lightly cooked
1 tomato, sliced

*Serves 1*

TO MAKE: Remove the seeds from the marrow and steam the rings for about 10 min., until tender. Cool quickly. Melt a large knob of butter in a pan, sauté the onion, add the minced beef and sausage meat and cook uncovered for 10 min. Season to taste. Cool quickly. Place the marrow rings on a foil plate, stuff with the meat mixture. Put the vegetables with them on the plate and dot with butter.

TO PACK AND FREEZE: See notes at beginning of chapter.

TO USE: Remove wrappings, cover loosely with foil and place in the oven to reheat from frozen at 400°F (mark 6) for 40 min.

# Haddock à l'orange

6 oz. frozen haddock fillet
grated rind of ½ orange
3 oz. frozen green beans
4 oz. creamed potatoes
1 oz. butter

*Serves 1*

TO MAKE: Place the fish in a portioned freezer tray. Sprinkle with orange rind and place the vegetables in the other sections. Dot the meal with butter.

TO PACK AND FREEZE: See notes at beginning of chapter.

TO USE: Remove wrappings and cover lightly with foil. Bake from frozen in the oven at 400°F (mark 6) for 30 min.

# Sausage nut cakes

3 oz. sausage meat
3 oz. creamed potato
salt and pepper
1 oz. cashew nuts, chopped
1 oz. margarine
3 oz. frozen sweetcorn kernels
3–4 oz. frozen Brussels sprouts, lightly cooked
1 oz. butter

*Serves 1*

TO MAKE: Mix together the sausage meat, potato, salt and pepper. Form the mixture into 2 cakes and coat with the chopped nuts.

*Bolognese nests*

Fry the sausage cakes in the margarine for about 15 min. until golden on both sides; cool. Place sausage cakes and vegetables on a foil freezer plate. Dot the vegetables with butter.

**TO PACK AND FREEZE:** See notes at beginning of chapter.

**TO USE:** Remove wrapping and cover loosely with foil. Bake from frozen in the oven at 400°F (mark 6) for 40–45 min.

## Plaice goujons

4–6 oz. frozen plaice fillets, skinned
seasoned flour
beaten egg
fresh white breadcrumbs
deep fat for frying
2 oz. frozen peas
4 oz. mashed potato
1 oz. butter

*Serves 1*

**TO MAKE:** Cut the fish into strips with scissors. Dip in the flour, then the egg, and finally in the breadcrumbs, pressing them on well. Deep fat fry the goujons for 1–2 min., until crisp and golden. Drain and cool quickly. Put the goujons, frozen peas and potato dotted with butter into the sections of a portioned freezer tray.

**TO PACK AND FREEZE:** See notes at beginning of chapter.

**TO USE:** Remove wrappings and cover the meal loosely with foil. Place in the oven to reheat from frozen at 400°F (mark 6) for 35 min.

## Cauliflower bake

8–12 oz. frozen cauliflower
$\frac{3}{4}$ oz. butter
$\frac{3}{4}$ oz. flour
$\frac{1}{2}$ pt. milk
2 oz. lean cooked ham, chopped
2 oz. red pepper, blanched and chopped
salt and pepper
2 oz. mature Cheddar cheese, grated

*Serves 1*

**TO MAKE:** Cook the cauliflower in gently boiling salted water until just tender. Drain and cool. Melt the butter in a saucepan, stir in the flour and cook for 1–2 min. Add the milk, stirring all the time. Bring to the boil and simmer for a few min. Remove from the heat, add the ham and red pepper and adjust the seasoning. Cool. Place the cauliflower in a foil freezer tray. Cover with the sauce, sprinkle with cheese and chill.

**TO PACK AND FREEZE:** See notes at beginning of chapter.

**TO USE:** Remove wrappings and cover the meal loosely with foil. Reheat in the oven at 400°F (mark 6) for about 40 min. Remove the foil after 30 min. to allow the cheese to brown.

# DESSERTS

## Sweet cheese pancakes

(*see picture opposite*)

4 frozen 7-in. pancakes
4 oz. cottage cheese
1 egg yolk
1 level tbsp. caster sugar
1 level tsp. grated lemon rind
2 oz. sultanas
2 oz. butter
cinnamon sugar
soured cream

*Makes 4*

Unwrap the pancakes, spread them out separately and leave at room temperature for 20 min. Mix together all the remaining ingredients except the butter, cinnamon sugar, and soured cream. Place $\frac{1}{4}$ of the mixture in the centre of each pancake and fold each into an envelope. Melt the butter in a large frying pan and heat the pancakes through. Sprinkle with cinnamon sugar and serve at once with soured cream.

## Fraises cardinal

1 lb. frozen strawberries
$\frac{1}{2}$ lb. frozen raspberries
sugar to taste
2 oz. flaked almonds

*Serves 6–8*

Place the strawberries in the base of a glass serving dish. Place the raspberries and sugar in a basin and leave to stand at room temperature until just thawed. Sieve the raspberries, spread the purée over the strawberries and scatter the almonds on top.

## Apricot crunch

4 oz. fresh white breadcrumbs
2 oz. Demerara sugar
2 oz. butter
$\frac{1}{2}$–$\frac{3}{4}$ pt. frozen apricot purée, thawed
17-fl. oz. block dairy ice cream

*Serves 6*

Mix the breadcrumbs and sugar together. Melt the butter in a pan and fry the breadcrumb mixture until lightly golden, stirring occasionally. Turn out on to kitchen paper. Place alternate layers of this mixture and the apricot purée in a medium sized glass serving dish. Place 9–10 tbsps. ice cream over. Serve at once.

## Peach shortcakes

3 oz. butter or margarine
8 oz. self-raising flour
$\frac{1}{4}$ level tsp. salt
3 oz. caster sugar
1 egg, beaten
1–2 tbsps. milk

*To complete dish from freezer*
whipped cream
15$\frac{1}{2}$-oz. can peach slices, drained
icing sugar

*Serves 8*

*Sweet cheese pancakes*

**TO MAKE:** Grease an 8-in. sandwich tin. Rub the fat into the flour and salt with the fingertips until the mixture resembles fine breadcrumbs. Stir in the sugar. Add the egg a little at a time to the rubbed-in mixture together with a little milk to give a scone dough consistency. Knead the mixture lightly on a floured surface until smooth. Form into a round and roll out until it is 8 in. across. Press it evenly into the tin, mark into 8 portions and bake in the oven at 375°F (mark 5) for about 20 min., until golden and well risen. Turn out and cool on a wire rack. Break into 8 pieces.

**TO PACK AND FREEZE:** Wrap the scones in a polythene bag. Seal, label and freeze.

**TO USE:** Unwrap, place the scones on a baking sheet and cover with foil. Refresh in the oven at 375°F (mark 5) for 20 min. Cool on a wire rack. Split whilst still slightly warm and fill with cream and peaches. Dust with icing sugar and serve fresh.

# Strawberry swirl

8 oz. frozen strawberries, thawed
½ pt. natural yoghurt
sugar to taste

*Serves 3*

Put the strawberries and yoghurt into a blender and purée thoroughly. Add sugar to taste. Pour into stemmed glasses to serve.

# Strawberry vol-au-vents

6 3-in. frozen vol-au-vent cases
6 scoops strawberry ice cream
½ lb. small strawberries
2 tbsps. redcurrant jelly

*Serves 6*

Bake the vol-au-vent cases on a dampened baking sheet in the oven at 425°F (mark 7) for about 15–20 min. Cool on a wire rack. Place a scoop of strawberry ice cream in each vol-au-vent, top with whole strawberries and spoon warmed redcurrant jelly over. Serve with extra spoonsful of ice cream if liked. Serve at once.

# Beignets soufflés

2 oz. butter
¼ pt. water
2½ oz. plain flour, sifted
¼ oz. caster sugar
pinch salt
2 eggs, beaten

*To complete dish from freezer*
deep fat for frying
cinnamon sugar
whipped cream

*Serves 4*

**TO MAKE:** Melt the butter in the water and bring to the boil. Tip the flour, sugar and salt in all at once. Beat until the paste is smooth and forms a ball in the centre of the pan. Allow to cool slightly, then beat in the eggs. Place the pastry in a piping bag fitted with a ½-in. plain nozzle and pipe rounds about 1½-in. diameter on a greased baking sheet.

**TO PACK AND FREEZE:** Open freeze until solid. Place in a polythene bag, seal and label. Return to the freezer.

**TO USE:** Heat deep fat or oil to 360°F, or until a 1-in. cube of bread will brown in 60–70 sec. Unwrap the beignets, carefully lower them a few at a time into the hot fat and fry until golden brown and hollow inside. Drain and roll in cinnamon sugar. Serve immediately with whipped cream.

# Aebleskiver

½ oz. fresh yeast
milk
2 small eggs, separated
1 oz. caster sugar
¼ level tsp. crushed cardamom seeds
1 level tsp. grated lemon rind
2 oz. butter
6 oz. strong plain flour
a little unsalted butter

*Makes about 15*

**TO MAKE:** Cream the yeast with 2 tbsps. lukewarm milk. When thoroughly mixed, stir in the beaten egg yolks, sugar, cardamom and lemon rind. Melt the butter in ¼ pt. lukewarm milk and stir in the egg yolk mixture and flour alternately. Turn into a bowl, cover with a damp cloth and leave in a warm place for about 2 hr. until doubled in bulk.

Stiffly whisk the egg whites. Fold them into the risen batter just before cooking. Place a small knob of unsalted butter in the base of each depression of an aebleskiver pan (see below). Heat well, making sure the sides are coated by swirling the pan. When bubbling, add a spoonful of the mixture sufficient to fill each hollow two-thirds full. Cook until light brown, turning (off the heat) with a skewer before the mixture sets. Return to the heat and cook for 5–7 min., rotating the doughnuts frequently to ensure the centres are cooked. Test with a skewer. Cool quickly.

**TO PACK AND FREEZE:** Wrap in a polythene bag, seal and label. Freeze.

**TO USE:** Unwrap, place on a baking sheet and cover with foil. Reheat from frozen at 350°F (mark 4) for 30 min. Dust with icing sugar and serve.

*Note:* An aebleskiver pan is a special dimpled pan made of cast iron.

# Choux normande

8 oz. cooking apples, peeled, cored and roughly chopped
2 oz. sugar
1–2 oz. sultanas
2 oz. butter or margarine
¼ pt. water
2½ oz. plain flour
½ level tsp. ground cinnamon
2 eggs, beaten

*To complete dish from freezer*
icing sugar

*Makes 4*

**TO MAKE:** Place the apples and sugar in a saucepan. Heat gently until the sugar dissolves, add the sultanas and simmer gently for about 10 min. Cool. Melt the fat in the water and bring to the boil, remove from the heat and tip in the flour sifted with the cinnamon. Beat until the paste is smooth and forms a ball in the centre of the pan. Allow to cool slightly. Beat in the eggs gradually to give a piping consistency. Place the mixture in a forcing bag fitted with a plain ½-in. nozzle. Grease 4 individual 3½-in. soufflé dishes and pipe a single circle of paste inside each. Pipe a second circle on top of the first, leaving the centre free for the filling. Spoon the apple mixture into the hollows.

**TO PACK AND FREEZE:** Open freeze then place the dishes in a heavy duty polythene bag. Seal and label and return to the freezer.

**TO USE:** Unwrap and place the dishes on a baking sheet. Cover with buttered foil. Leave at room temperature for 30 min. before baking in the oven at 400°F (mark 6) for 30–40 min., until risen and golden. Dust with icing sugar and serve at once.

# Chocolate date nut Torte

6 oz. granulated sugar
2 level tbsps. plain flour
1 level tsp. baking powder
2 eggs
4 oz. shelled walnuts, coarsely chopped
8 oz. stoned dates, coarsely chopped
2 oz. plain chocolate, grated
¼ level tsp. ground cinnamon

*To complete Torte from freezer*
**whipped cream**

*Serves 8*

**TO MAKE:** Sift together 5 oz. sugar, the flour and baking powder into a bowl. Beat the eggs until light and fluffy. Add the dry ingredients and mix well. Fold in the nuts, dates and chocolate and turn the mixture into a well greased and floured 9-in. cake tin. Mix the remaining sugar and cinnamon and sprinkle over the Torte mixture. Bake in the oven at 300°F (mark 1–2) for 50 min.–1 hr.

until golden brown. Turn out and cool on a wire rack.

**TO PACK AND FREEZE:** See notes on page 218.

**TO USE:** Unwrap the Torte and allow to thaw at room temperature for 1 hr. or place, covered with foil, in the oven at 350°F (mark 4) for 20 min. Serve with whipped cream.

# Crêpes au mocha

4 oz. plain flour
1 level tbsp. caster sugar
pinch of salt
1 level tbsp. powdered chocolate
1 level tbsp. instant coffee powder
2 eggs
2 egg yolks
¾ pt. milk
2 tbsps. melted butter
butter for frying

*To complete dish from freezer*
**butter**
½ pt. double cream
1 tbsp. rum
sugar to taste

*Makes 8*

**TO MAKE:** Sift together the dry ingredients into a basin. Lightly beat the whole eggs and yolks and add to the dry ingredients. Beat in the milk and melted butter until smooth. Melt a knob of extra butter in a small frying pan; add about 2 tbsps. batter, tipping the pan to allow the batter to cover the base of the pan thinly. Cook crêpes for about 1 min. each side. Allow to cool.

**TO PACK AND FREEZE:** Interleave the pancakes with oiled greaseproof or non-stick paper. Wrap in foil, seal and label. Freeze.

**TO USE:** Unwrap and reheat each crêpe separately in a little hot butter in a frying pan. Whip the cream, rum and sugar together until stiff, place in the centre of each crêpe and roll up. Serve at once.

# Rhubarb, gooseberry and raisin dumplings

½ lb. raw rhubarb
2 oz. sugar
2 oz. seedless raisins
13½-oz. can gooseberries, drained
¼ level tsp. ground cinnamon
6 oz. shortcrust pastry (i.e. made with
    6 oz. flour etc.)

*Serves 4*

**TO MAKE:** Rinse and cut the rhubarb into 1-in. chunks. Place in a saucepan with the sugar. Heat gently until the sugar dissolves, then simmer the fruit gently until soft.

Remove from the heat, add the raisins, gooseberries and cinnamon. Divide the pastry into 4 and roll into circles about 6 in. in diameter. Place the fruit mixture in the centre of each circle. Damp the edges of the pastry with water, bring the edges together over the top, seal and flute.

**TO PACK AND FREEZE:** Open freeze on a baking sheet, then wrap in polythene, seal and label. Return to the freezer.

**TO USE:** Unwrap the dumplings, place on a baking sheet and bake from frozen in the oven at 375°F (mark 5) for about 40 min., until golden. Dredge with caster sugar. Serve with custard or cream.

# 3 Soups and Sauces in Store

Soups and sauces often take a while to prepare, so that we neglect them. If you cook in bulk and freeze them, though, you have them to hand whenever you want them. Basic béchamel and espagnole sauces are particularly useful to have in store since they can form the foundation of so many different sauces. Thawing and reheating are absolutely straightforward.

If you make a large quantity of sauce, pack and freeze it in $\frac{1}{2}$-pt. portions. This is the most likely quantity to be required at one time and it is a portion that is easy to thaw and reheat. It is also worth freezing a small quantity in an ice cube tray, so that you can take out a cube at a time if required. On the same principle, pack and freeze soups in 1–2 pt. portions. If more is needed you can always open 2 packs, but it is very difficult to divide a large one.

When freezing soups and sauces avoid over seasoning before freezing, as more can always be added when thawing or re-heating. This is also the time to add milk, cream or egg yolks to enrich a soup. Garlic is another item best added after freezing; after 2 weeks stored in a freezer the flavour of garlic tends to deteriorate.

## To pack and freeze
It is important to cool soups and sauces rapidly before freezing. First stand the pan in cold running water and then, when all traces of steam have gone, pour the liquid into the freezing container or another utensil and put it in the refrigerator to chill. To pack,
**Either** Line a rigid container with a heavy gauge polythene bag and pour in the liquid. Leave it to freeze until solid, then remove the bag from the pre-former, seal, label and return it to the freezer.
**Or** Pour the liquid into a waxed carton or rigid polythene container, leaving about $\frac{1}{2}$ in. headspace for expansion. Seal and label the container and freeze.

## Storage times
Soups and sauces may be stored for 2–3 months, but if highly seasoned they should be used within 2–3 weeks. Recipes containing chopped bacon should be used within 6 weeks.

## To use

Remove any outer wrappings, then run hot water over the outside of the polythene bag or rigid container to loosen the contents. Squeeze or ease out the block of soup or sauce and turn it into a pan just large enough to take the quantity used, plus 1–2 spoonsful milk or stock to hasten the start of the thawing. For thick sauces a double pan is best. Bring the liquid slowly to the boil, stirring. Adjust the seasoning and add any extra ingredients as specified in individual recipes.

**Note** If you have a sauce in one of the special 'boil-in-the-bag' freezer packs now available, the unopened bag can be put into a pan of boiling water to thaw and re-heat—allow extra heating time.

# SAVOURY SAUCES

## Espagnole sauce

4 oz. butter or good dripping
4 oz. green bacon, rinded and chopped
1 large onion, skinned and chopped
½ lb. carrots, peeled and finely sliced
2 stalks celery, finely sliced
6 mushrooms, or ¼ lb. stalks, sliced
  (optional)
3 oz. plain flour
3 pt. good meat stock
bouquet garni
4 level tbsps. tomato paste
salt and freshly ground black pepper

*Makes 4 ½-pt. portions*

This classic brown sauce is often used in French dishes and also makes the base for a number of other exciting sauces such as Madeira, Marengo, Bordelaise or Lyonnaise, but when made in small quantities for each dish it is time-consuming and—surprisingly enough—never tastes quite as good as when made in bulk.

**TO MAKE:** Melt the fat in a large pan and fry the bacon, onion, carrot and celery slowly until a deep golden brown, stirring occasionally. Add the mushrooms and cook for 5 min. more. Sprinkle with the flour and cook slowly, stirring often, until well browned but not burnt. Add 1 pt. boiling stock, stirring, then the bouquet garni. Bring to the boil and add a further 1 pt. hot stock, stirring. Continue to simmer—with just the occasional bubble rising—without the lid for about 1½ hr., stirring occasionally.

Skim off any fat as it rises. Add the tomato paste and cook for 15 min. more. Strain the mixture into a clean pan, add the remaining 1 pt. stock and continue to cook to reduce the sauce to about 2 pt. Adjust the seasoning at this point. Cool quickly.

**TO PACK AND FREEZE:** See notes at beginning of chapter.

**TO USE:** When using, say, for a Madeira sauce, add a fair quantity of extra liquid, then reduce the sauce by boiling to give the required consistency.

## Béchamel sauce

4 pt. milk
1 large carrot, peeled and roughly cut
1 onion, studded with 8 cloves
1 large sprig parsley
bouquet garni
1 level tsp. salt
¼ level tsp. freshly ground black pepper
8 oz. butter
8 oz. plain flour

*Makes 8 ½-pt. portions*

**TO MAKE:** Heat the milk with the carrot, clove-studded onion, parsley, bouquet garni, salt and pepper. Bring the milk almost to the boil, remove from the heat and leave to infuse for 30 min. Melt the butter in a large pan, but do not over-heat it. Remove from the heat, add the flour and blend well together. Cook the roux for 3 min., remove from the heat and gradually add the warm, strained milk. Bring slowly to the boil, stirring. Cool quickly.

TO PACK AND FREEZE: See notes at beginning of chapter.

TO USE: For thawing, see notes at beginning of chapter. Serve plain with vegetables or ham or add extra flavourings such as cheese, capers, sherry, tomato paste or anchovy.

## Cheese sauce

Add 2 oz. grated Cheddar cheese to $\frac{1}{2}$ pt. béchamel sauce on re-heating.

## Piquant gherkin sauce

$\frac{1}{2}$ oz. butter
1 shallot, skinned and finely chopped
1 level tbsp. flour
1 level tsp. tomato paste
$\frac{1}{2}$ pt. brown stock
1 oz. gherkins, finely chopped
salt and pepper
1 tbsp. lemon juice
1 tsp. chopped parsley

*Makes $\frac{1}{2}$ pt.*

TO MAKE: Melt the butter in a frying pan, add the shallot and brown lightly. Stir in the flour, cook for $\frac{1}{2}$ min., then add the tomato paste, stock and gherkins. Simmer the sauce gently for 10 min. Season with salt and pepper, add the lemon juice and parsley and cool.

TO PACK AND FREEZE: See notes at beginning of chapter.

TO USE: For thawing, see notes at beginning of chapter. Serve with grills or fish.

## Cumberland sauce

*(see picture page 66)*

thinly pared rind and juice of 1 lemon
    and 1 orange
1 level tbsp. caster sugar
3 tbsps. redcurrant jelly
6 fl. oz. port
1 level tbsp. arrowroot

*Makes about $\frac{1}{2}$ pt.*

TO MAKE: Cut half the lemon rind into very thin strips. Place the rest of the rinds, juices, sugar and jelly in a saucepan. Bring to the boil slowly. Simmer the sauce for 2–3 min. and strain. Return it to the pan, add the port and continue to simmer for 2 min. Blend the arrowroot with a little water and add to the saucepan, stirring, then add the lemon strips. Bring to the boil and cook until thickened and clear. Chill.

TO PACK AND FREEZE: See notes at beginning of chapter.

TO USE: For thawing, see notes. Serve with baked gammon or with cold meats.

## Spicy raisin sauce

3 oz. stoned raisins
2 cloves
$\frac{1}{2}$ pt. water
1 level tsp. cornflour
3 oz. Demerara sugar
salt and pepper
1 oz. butter
2 tsps. lemon juice

*Makes $\frac{1}{2}$ pt.*

TO MAKE: Put the raisins and cloves in a saucepan with the measured water. Bring to the boil and simmer for 10 min. Blend the cornflour with a little extra water, add it to the saucepan with the sugar and stir. Season well with salt and pepper and simmer until the sauce thickens. Add the butter and lemon juice. Cool.

TO PACK AND FREEZE: See notes at beginning of chapter.

TO USE: For thawing, see notes at beginning of chapter. Serve with gammon.

## Romesco sauce

8 oz. red or green peppers, seeded and
    sliced
8 oz. ripe tomatoes, sliced
2 cloves garlic, skinned and crushed
$\frac{1}{2}$ pt. olive oil
1 oz. almonds, toasted
2 tbsps. wine vinegar
salt and pepper

*Makes $\frac{3}{4}$ pt.*

**TO MAKE:** Place the peppers, tomatoes and garlic in an ovenproof dish, cover and bake in the oven at 400°F (mark 6) for about 30 min. or until soft. Remove from the oven and purée in a blender. Pass the purée through a sieve and return to the blender with the remaining ingredients. Blend well together. Chill.

**TO PACK AND FREEZE:** See notes.

**TO USE:** For thawing see notes at beginning of chapter. Serve with fish.

## Sauce provençale

2 oz. mushrooms, chopped
1 oz. shallots, skinned and chopped
1 clove garlic, skinned and chopped
2 tbsps. cooking oil
$\frac{1}{2}$ pt. well flavoured stock
$\frac{1}{4}$ pt. cider
1 level tsp. tomato paste
salt and freshly ground black pepper
bouquet garni

*Makes $\frac{3}{4}$ pt.*

**TO MAKE:** Fry the mushrooms, shallots and garlic in the oil until soft but not browned. Add the stock, cider, tomato paste, salt, pepper and the bouquet garni. Bring to the boil and simmer for 15–20 min. Chill.

**TO PACK AND FREEZE:** See notes at beginning of chapter.

**TO USE:** For thawing, see notes at beginning of chapter. Serve with fish.

## Sauce italienne
*(see picture page 66)*

2 oz. onions, skinned and finely chopped
1 tbsp. cooking oil
1 level tsp. flour
2 oz. mushrooms, chopped
salt and freshly ground black pepper
3 tbsps. white wine
$\frac{1}{2}$ pt. stock
1 level tbsp. tomato paste
1 level tbsp. dried mixed herbs

*Makes about $\frac{1}{2}$ pt.*

**TO MAKE:** Sauté the onion in the oil until soft but not coloured. Stir in the flour and mushrooms and mix well. Cook gently for 2–3 min., then add the salt, pepper, wine,

stock, tomato paste and herbs. Bring to the boil and simmer uncovered for 10 min. Chill.

**TO PACK AND FREEZE:** See notes at beginning of chapter.

**TO USE:** See notes. Serve with liver and grilled meats.

## Peanut sauce

2 oz. salted peanuts, chopped
2 oz. peanut butter
1 level tbsp. cornflour
$\frac{1}{2}$ pt. white stock
2 tbsps. lemon juice
$\frac{1}{4}$ level tsp. ground mace
$\frac{1}{4}$ level tsp. paprika pepper
3 tbsps. milk

*Makes about $\frac{1}{2}$ pt.*

**TO MAKE:** Place the peanuts with the peanut butter in a saucepan. Heat them gently and allow the butter to begin melting. Mix in the cornflour and gradually stir in the stock. Add the lemon juice, mace, paprika and milk. Bring to the boil and simmer for 2–3 min. Chill.

**TO PACK AND FREEZE:** See notes at beginning of chapter.

**TO USE:** See notes. Serve with chicken or turkey.

## Green pepper and tomato sauce

8 oz. green peppers, seeded and roughly
   chopped
$\frac{1}{2}$ pt. stock
1 lb. tomatoes, skinned and halved
1 oz. butter
2 level tbsps. flour
salt and freshly ground black pepper
2 tsps. lemon juice
1 level tsp. sugar

*Makes about 1 pt.*

**TO MAKE:** Cook the peppers in a covered pan with the stock until soft. Place the tomatoes in a blender, add the peppers and stock and blend until smooth. Melt the butter, add the flour and cook for 1 min. Gradually, add the tomato and pepper purée, then season with salt, pepper, lemon

*Crème vichyssoise* (page 69); *Beetroot soup* (page 71);
*Lentil and bacon soup* (page 70)

juice and sugar. Bring to the boil and simmer for 2–3 min., stirring. Chill.

**TO PACK AND FREEZE:** See notes at beginning of chapter.

**TO USE:** See notes.

# Sauce Bercy

1 oz. butter
1 tbsp. chopped shallot
¼ pt. dry white wine
½ pt. stock
salt and freshly ground black pepper
2 tbsps. lemon juice
2 tsps. chopped parsley
2 level tbsps. cornflour

*Makes about ¾ pt.*

**TO MAKE:** Melt the butter in a pan and sauté the shallot until soft but not coloured. Add the wine, bring to the boil and boil until reduced by half. Add the stock, salt, pepper, lemon juice and parsley. Blend the cornflour with a little water and add to the pan, stirring. Bring the sauce to the boil and simmer gently for 2–3 min. Chill.

**TO PACK AND FREEZE:** See notes at beginning of chapter.

**TO USE:** See notes. Serve with a white fish such as haddock.

# Celery sauce

1 oz. butter
6 oz. celery, trimmed and roughly
    chopped
2 oz. onions, skinned and chopped
2 cloves garlic, skinned and crushed
½ pt. water
pinch each ground coriander, paprika
    pepper, turmeric and cinnamon
2 level tbsps. tomato paste

*Makes about ¾ pt.*

**TO MAKE:** Melt the butter in a pan and sauté the celery and onions until soft; add the garlic and cover with the water. Add the remaining ingredients, bring to the boil, cover and simmer for 20 min. Purée the mixture in a blender and then sieve. Chill.

**TO PACK AND FREEZE:** See notes at beginning of chapter.

**TO USE:** Reheat according to notes and serve with chicken, lamb or pork.

# Sauce normande

3 oz. butter
3 oz. onions, skinned and finely chopped
1 oz. flour
½ pt. cider
salt and freshly ground black pepper
½ level tsp. grated nutmeg
1 tbsp. lemon juice

*To complete sauce from freezer*
2 tbsps. cream

*Makes about ½ pt.*

**TO MAKE:** Melt the butter in a pan and sauté the onion. Stir in the flour and cook gently for a few min., then add the cider and remaining ingredients and simmer for a few min. more. Cool quickly and chill.

**TO PACK AND FREEZE:** See notes at beginning of chapter.

**TO USE:** Reheat the sauce as described in the notes and stir in the cream just before serving; do not boil after adding the cream. Serve with sausages and grills.

# Apricot and currant sauce

*(see picture overleaf)*

4 oz. dried apricots, chopped
2 oz. currants
1½ pt. water
½ oz. butter
2 level tbsps. cornflour
1 tbsp. wine vinegar
¼ level tsp. salt
1 level tsp. sugar
freshly ground black pepper
1 tbsp. lemon juice

*Makes 3 ½-pt. portions*

**TO MAKE:** Put the apricots and currants in an open pan with the water and cook until soft. Drain but reserve the juice. Melt the butter in a saucepan, add the cornflour and blend well. Add the juice from the fruit and all the remaining ingredients. Bring to the boil and simmer for 10 min. Chill.

**TO PACK AND FREEZE:** See notes at beginning of chapter.

*Apricot and currant sauce, Sauce italienne, Chocolate sauce, Walnut and butterscotch sauce, Cumberland sauce*

**TO USE:** Reheat according to notes and serve with grilled lamb chops or roast lamb.

## Apricot and horseradish sauce

15-oz. can apricots
½ oz. horseradish, grated
1 tbsp. lemon juice
½ level tsp. dried tarragon

*Makes ¾ pt.*

**TO MAKE:** Purée all the ingredients in a blender or pass them through a sieve.

**TO PACK AND FREEZE:** See notes at beginning of chapter.

**TO USE:** This sauce may be served hot or cold with cold meats or poultry. If serving cold, leave to thaw at room temperature for 4 hr. To serve hot, see notes.

## Hot cucumber sauce

2 cucumbers, peeled and chopped
¾ pt. white stock
1 oz. butter
1 oz. flour
2 tbsps. orange juice
salt and freshly ground black pepper

*Makes 1¼ pt.*

**TO MAKE:** Cook the cucumber in the stock until soft, then purée it in a blender. Melt the butter in a saucepan, add the flour and stir well. Cook for 1 min. Add the purée gradually and bring to the boil. Add the orange juice and adjust the seasoning with salt and pepper. Chill.

**TO PACK AND FREEZE:** See notes at beginning of chapter.

**TO USE:** Reheat according to notes and serve with salmon steaks or lamb chops.

# Tomato sauce

1 tbsp. cooking oil
1 small onion, skinned and chopped
1 medium carrot, peeled and chopped
1 level tbsp. flour
$\frac{3}{4}$ pt. stock
14-oz. can tomatoes
$\frac{1}{2}$ level tsp. dried basil
1 level tbsp. tomato paste
pinch ground ginger
salt and freshly ground black pepper

*Makes 1$\frac{1}{4}$ pt.*

**TO MAKE:** Heat the oil and gently sauté the onion and carrot until the onion is transparent. Remove from the heat, stir in the flour and gradually blend in the stock and tomatoes, with their juice. Add the basil, tomato paste, ground ginger, salt and pepper. Bring to the boil and simmer without a lid for about 10 min. Allow to cool a little and then purée in a blender. Chill.

**TO PACK AND FREEZE:** See notes at beginning of chapter.

**TO USE:** Reheat according to notes and serve with beef or lamb dishes, pasta and made-up dishes such as fish cakes or rissoles.

# Pesto dressing

2 level tbsps. dried basil
2 level tbsps. dried parsley
2 cloves garlic, skinned and crushed
1 tbsp. chopped walnuts
1 oz. Parmesan cheese, grated
$\frac{1}{4}$ pt. olive oil
2 tbsps. lemon juice
1 tbsp. wine vinegar
freshly ground black pepper

**TO MAKE:** Purée all the ingredients in a blender.

**TO PACK AND FREEZE:** See notes at beginning of chapter.

**TO USE:** Thaw at room temperature for about 2 hr. Serve as a dressing for tomato and onion salads.

# SWEET SAUCES

## Walnut butterscotch sauce

(*see picture opposite*)

$\frac{1}{2}$ oz. butter
1 level tbsp. Demerara sugar
2 oz. golden syrup
$\frac{1}{4}$ pt. water
2 level tsps. custard powder
2 tbsps. lemon juice
1 oz. shelled walnuts, chopped

*Makes $\frac{1}{4}$ pt.*

**TO MAKE:** Melt the butter in a pan and add the sugar, syrup and water. Dissolve the custard powder in 1 tsp. water and stir it into the syrup with the lemon juice and chopped walnuts. Bring the sauce to the boil and cook for 1 min., stirring. Chill.

**TO PACK AND FREEZE:** See notes at beginning of chapter.

**TO USE:** Reheat according to notes and serve with ice cream or sponge puddings.

## Suchard sauce

4 oz. granulated sugar
$\frac{1}{2}$ pt. water
vanilla pod
6 oz. plain chocolate

*Makes $\frac{3}{4}$ pt.*

**TO MAKE:** Dissolve the sugar in the water in a saucepan. Add the vanilla pod, bring to the boil and simmer for 5 min. Remove the pod, add the chocolate and allow it to melt. Beat well and bring to the boil. Simmer until slightly syrupy. Chill.

**TO PACK AND FREEZE:** See notes at beginning of chapter. As this sauce does not freeze absolutely firm, use a rigid container.

**TO USE:** This sauce may be served hot or cold. If serving cold, simply leave it to thaw at room temperature for 2 hr. If serving hot, reheat according to notes. Serve with fruit and ice cream, profiteroles or beignets.

## Red wine sauce

2 level tbsps. granulated sugar
½ pt. water
thinly pared rind of ½ lemon
¼ pt. red wine
4 tbsps. seedless bramble jam
4 level tsps. arrowroot
2 tbsps. lemon juice

*Makes ¾ pt.*

**TO MAKE:** Place the sugar, water, lemon rind, wine and jam in a saucepan. Bring slowly to the boil and simmer for 5 min. Remove the lemon rind. Blend the arrowroot with the lemon juice and stir it into the sauce. Bring to the boil, still stirring, reduce the heat and cook until thickened. Chill.

**TO PACK AND FREEZE:** See notes at beginning of chapter.

**TO USE:** Thaw overnight in the refrigerator and serve chilled, or reheat according to notes. Serve with fruit, ice cream or sponge puddings.

## Fudge sauce

½ pt. milk
½ vanilla pod, split
2 oz. butter
3 oz. light soft brown sugar
1 tbsp. golden syrup
1 level tsp. arrowroot

*Makes ¾ pt.*

**TO MAKE:** Heat the milk in a pan with the vanilla pod and butter. Remove the pan from the heat and leave the milk to infuse for about 10 min., then remove the vanilla pod. Put the sugar and golden syrup in a saucepan over gentle heat to dissolve. Raise the temperature until the sugar and syrup caramellise, then stir in the flavoured milk and leave over a very low heat to dissolve the caramel. Blend the arrowroot with a little water and stir it into the milk mixture. Bring the sauce to the boil, stirring, and boil for 1 min. Chill.

**TO PACK AND FREEZE:** See notes at beginning of chapter.

**TO USE:** Reheat according to notes and serve with ice cream or sponge puddings.

# SOUPS

## Carrot and courgette soup

2 oz. butter
½ lb. carrots, peeled and sliced
½ lb. courgettes, trimmed and sliced
1 pt. well flavoured stock
1 bayleaf
1 level tbsp. tomato paste
1 level tbsp. caster sugar
salt and freshly ground black pepper

*Serves 4*

**TO MAKE:** Melt the butter in a saucepan, add the carrots and courgettes, cover and cook gently for about 10 min., until soft. Pour in the stock and remaining ingredients. Bring the soup to the boil and simmer for 30 min., remove the bayleaf then pour it into a blender and blend for a few sec.; the soup should not be completely smooth. Chill.

**TO PACK AND FREEZE:** See notes at beginning of chapter.

**TO USE:** See notes.

## Iced apple and apricot soup

2 lb. apples, peeled, cored and chopped
8 oz. dried apricots, chopped
2 pt. light stock
¼ pt. dry white wine
¼ pt. soured cream

*Serves 8*

**TO MAKE:** Cook the apples and apricots in the stock until soft, then purée them in a blender. Add the wine and soured cream and chill.

**TO PACK AND FREEZE:** See notes at beginning of chapter.

**TO USE:** Thaw in the refrigerator overnight and serve chilled.

# Crème vichyssoise

(*see colour picture facing page 64*)

4 leeks, trimmed, finely sliced and washed
2 oz. butter
1 onion, skinned and chopped
2 pt. chicken stock
2 potatoes, peeled and thinly sliced
salt and freshly ground black pepper

*To complete dish from freezer*
⅓ pt. single cream
chopped chives

*Serves 12*

**TO MAKE:** Sauté the well drained leeks gently in the butter in a large saucepan for about 10 min., with the onion. Do not allow to brown. Add the stock and potatoes and cook until the vegetables are tender. Cool and pass through a sieve. Adjust the seasonings, and chill.

**TO PACK AND FREEZE:** See notes at beginning of chapter.

**TO USE:** Unwrap and reheat gently in a saucepan without boiling. Cool, stir in the cream, turn into a serving dish and sprinkle with chopped chives.

# Chicken and horseradish soup

1 large chicken portion (10–12 oz.)
½ pt. chicken stock
1 oz. horseradish, grated
½ pt. milk
1 oz. butter
2 level tbsps. flour
salt and freshly ground black pepper

*Serves 2–3*

**TO MAKE:** Place the chicken portion and stock in a pan, bring to the boil and simmer until tender. Put the horseradish and milk in a small pan, bring to the boil, remove from the heat and leave to infuse for a few min. Melt the butter in a pan and stir in the flour and cook for 1 min. Gradually stir in the milk, bring to the boil and cook for 2 min. Drain the stock from the chicken and add the stock to the soup. Discard the skin and bones from the chicken, chop the flesh and add it to the soup. Adjust seasoning and chill.

**TO PACK AND FREEZE:** See notes at beginning of chapter.

**TO USE:** Reheat according to notes.

# German spinach soup

(*see picture below*)

1 oz. butter
3 oz. onion, skinned and chopped
4 oz. bacon, rinded and chopped
1 lb. frozen leaf spinach
3 level tbsps. flour
1½ pt. stock
salt and freshly ground black pepper
½ level tsp. grated nutmeg

*Serves 6–8*

**TO MAKE:** Melt the butter in a saucepan and sauté the onion and bacon. Add the frozen spinach and allow to thaw slowly in the pan. Add the flour and mix well. Stir in the stock, salt, pepper and nutmeg. Bring the soup to the boil, stirring, cover and simmer for 15–20 min. then purée in a blender and chill.

**TO PACK AND FREEZE:** See notes at beginning of chapter.

*German spinach soup*

TO USE: Reheat according to notes and serve hot, sprinkled with grated Gruyère cheese or accompanied by cheese straws.

# Kartoffelsuppe

(*see picture below*)

4 oz. smoked bacon rashers, rinded and diced
1½ lb. potatoes, peeled and diced
2 carrots, peeled and sliced
1 stick celery, sliced
1 leek, trimmed, sliced and washed
4 oz. onions, skinned and sliced
2½ pt. water
1 level tsp. salt
freshly ground black pepper
¼ level tsp. dried marjoram
½ level tsp. dried mixed herbs
1 sprig parsley with stem

*To complete soup from freezer*
1 tsp. chopped parsley
4 oz. German sausage, chopped
croûtons

*Serves 6–8*

TO MAKE: In a saucepan, fry the bacon over a low heat to extract the fat. Add the prepared vegetables and sauté for 2–3 min. Cover with the water, add the salt, pepper, marjoram and mixed herbs. Bring to the boil, add the parsley, cover and then simmer for about 45 min. until all the vegetables are well cooked. Purée the soup in a blender or pass it through a sieve. Cool rapidly and chill.

TO PACK AND FREEZE: See notes at beginning of chapter.

*Kartoffelsuppe*

TO USE: Reheat according to notes. When half the soup becomes liquid, add the chopped parsley and sausage, then continue to heat to serving temperature. Serve garnished with croûtons.

# Watercress soup

4 oz. butter
2 oz. plain flour
1¼ pt. chicken or veal stock
½ pt. milk
salt and freshly ground black pepper
3 oz. onions, skinned and chopped
2 bunches watercress

*Serves 6*

TO MAKE: Melt 3 oz. butter in a pan and stir in the flour. Cook over a gentle heat for 1–2 min. Remove from the heat and stir in all the stock and milk. Return to the heat and bring to the boil, stirring continuously. Simmer gently for 3 min. Season with salt and pepper. Sauté the onion in the remaining 1 oz. butter until soft. Wash the watercress and trim it, leaving some of the stem, then roughly chop it and add it to the onion. Cover with a lid and sauté for a further 4 min. Stir the sauté vegetables into the sauce. Purée in a blender or pass through a sieve. Chill.

TO PACK AND FREEZE: See notes at beginning of chapter.

TO USE: Reheat according to notes.

# Lentil and bacon soup

(*see colour picture facing page 64*)

6 oz. lentils
2½ pt. stock
1 clove garlic, skinned and crushed
1 clove
salt and pepper
½ lb. lean bacon rashers, rinded and diced
8-oz. can tomatoes
4 oz. onions, skinned and chopped
1 lb. potatoes, peeled and diced
2 tbsps. lemon juice

*To complete soup from freezer*
crispy fried bacon rolls, or chopped parsley, or grated cheese, or croûtons

*Serves 6*

TO MAKE: Wash the lentils and place them in a saucepan with the stock. Add the garlic, clove, salt, pepper, bacon, tomatoes and onion. Bring to the boil, cover and simmer for 1 hr. until the lentils and bacon are soft. Add the potatoes and cook for a further 20 min. Remove the clove, pour the soup into a blender and purée until smooth. Add the lemon juice and stir. Chill.

TO PACK AND FREEZE: See notes at beginning of chapter.

TO USE: Reheat the soup according to notes and garnish with bacon rolls, chopped parsley, grated cheese or croûtons.

*Maximum storage time:* 6 weeks.

# Beetroot soup

(*see colour picture facing page 64*)

1¼ lb. cooked beetroot, peeled and chopped
4 oz. onions, skinned and chopped
3 pt. beef stock
2 tbsps. wine vinegar
salt and pepper

*Serves 8–12*

TO MAKE: Put the beetroot, onion and stock in a large pan, cover and simmer for 30 min. Purée in a blender, then add the wine vinegar, salt and pepper. Chill.

TO PACK AND FREEZE: See notes at beginning of chapter.

TO USE: See notes.

# Almond, celery and grape soup

1 onion, skinned and finely chopped
1 small head celery, washed and chopped
1½ oz. butter
2 level tbsps. flour
1½ pt. white stock
½ pt. milk
2 oz. nibbed almonds

*To complete soup from freezer*
salt and freshly ground black pepper
1 egg yolk blended with 2–3 tbsps. double cream
4 oz. white grapes, peeled and pipped

*Serves 6–8*

TO MAKE: Sauté the onion and celery in the butter until soft. Stir in the flour and cook for 1 min. Gradually stir in the stock, add the milk and nuts and simmer gently for 30 min. Chill.

TO PACK AND FREEZE: See notes at beginning of chapter.

TO USE: Reheat according to notes. Adjust seasoning and add the egg yolk, cream and grapes just before serving.

# Kipper and lemon soup

3 pairs frozen kippers
pared rind and juice of 1 lemon
water
1 oz. butter
2 level tbsps. flour
1 pt. white stock
black pepper

*Serves 2–3*

TO MAKE: Place the frozen kippers in a pan with the lemon rind and juice and enough water to just cover. Cover with a lid and cook for 20–25 min. until soft. Drain the fish, reserving the stock, and skin and bone the kippers. Purée the fish in a blender. Melt the butter, stir in the flour and cook for 1 min. Gradually add the fish stock, then the 1 pt. white stock; cook for 2 min. Season with pepper, add the kipper and mix well. Chill.

TO PACK AND FREEZE: See notes.

TO USE: See notes.

# Walnut soup

1 oz. butter
2 oz. onion, skinned and finely chopped
1 oz. flour
1½ pt. chicken stock
bayleaf
salt and pepper
2 oz. shelled walnuts, grated with a Mouli grater
½ pt. milk

*To complete soup from freezer*
1 egg yolk blended with 4 tbsps. double cream

*Serves 6*

TO MAKE: Melt the butter, add the onion and cook gently until beginning to soften. Stir in the flour and cook for 1–2 min. Add the stock gradually, stirring all the time. Add the bayleaf, salt, pepper, nuts and milk. Bring the soup to the boil, cover and simmer for 30 min. Remove bayleaf and chill.

TO PACK AND FREEZE: See notes at beginning of chapter.

TO USE: Reheat according to notes. Add the egg yolk and cream just before serving, taking care not to boil the soup.

# Curried cucumber soup

1 oz. butter
5 oz. onions, skinned and finely chopped
1 cucumber, peeled and grated
½–1 level tsp. curry powder
1 level tbsp. flour
1½ pt. chicken stock

*To complete soup from freezer*
¼ pt. milk
croûtons

*Serves 4–6*

TO MAKE: Melt the butter in a saucepan, add the onion and fry without colouring for 10 min. Add the cucumber and fry for a further 5 min. Stir in the curry powder and flour and cook for 2–3 min., then gradually stir in the stock and bring to the boil. Cover, reduce the heat and simmer for 30 min. Chill.

TO PACK AND FREEZE: See notes at beginning of chapter.

TO USE: Reheat according to notes, add the milk and heat through again, without boiling. Serve the soup garnished with croûtons.

# Chinese spring soup

2 oz. spring onions, trimmed and chopped
2 oz. new carrots, scraped and diced
2 oz. frozen peas
1¼ pt. chicken stock
1 oz. butter
1 oz. flour
¼ pt. milk
salt and freshly ground black pepper

*Serves 6*

TO MAKE: Cook the prepared vegetables in the stock for about 6 min. until soft. Melt the butter in a saucepan, add the flour and stir well. Cook for 1 min. then gradually add the milk, stirring all the time, and bring to the boil. Cook for 1–2 min. Add the vegetables and stock and adjust the seasoning. Chill.

TO PACK AND FREEZE: See notes at beginning of chapter.

TO USE: See notes.

# 4 Ways With Fish

Fish for freezing must be absolutely fresh, within 24 hr. of the catch. The fish should be kept iced or refrigerated during any short journey home and frozen quickly in a freezer with a fast-freeze switch. Fish must be in prime condition, and it will then be satisfactory to freeze it raw, with good results.

Thin pieces of fish freeze the fastest, so it is best to divide fish into fillets, cutlets or portions and to freeze these in shallow packages. Small fish such as mackerel, trout, mullet or herring can be frozen whole. Wash the fish and remove the scales by scraping from tail to head with the back of a knife. Gut it, wash thoroughly under running water, drain and dry it on a clean cloth or absorbent paper. For special occasions and short term storage, fish such as salmon and salmon trout may be frozen whole, provided they are not too large. If wished, the belly cavity may be stuffed with paper to keep it a good shape.

### Packing raw fish for the freezer
It is essential that the fish be protected from the drying effect of the cold air in the freezer. The best packaging is material in roll form, such as foil or polythene, which can be moulded to the shape required. Overwrap if you wish. To package fillets or portions it is easiest to place separators of small pieces of thin plastic film between each piece of fish, so that the portions can be frozen separately but packaged together.

### To freeze raw fish in ice
For best results, place whole fish unwrapped in the freezer until solid. Remove the fish and dip it in cold fresh water. This forms a thin layer of ice over the fish. Return it to the freezer and repeat the process until the ice glaze is $\frac{1}{4}$ in. thick. Wrap in heavy duty polythene, seal and label; overwrap with mutton cloth or stockinette to protect it. Store carefully as ice may crack.

### Storage times
Salmon and oily fish such as halibut, mackerel and herring, 2 months; white fish such as cod, whiting, plaice and sole, 3 months; shellfish, 1 month.

## Cooking frozen fish

There is no need to thaw small portions of fish before cooking. Simply separate the small whole fish or fillets and cook them from frozen. When thawing large whole fish or large portions for deep fat frying, allow 5–6 hr. per lb. in a refrigerator or 3–4 hr. at room temperature. To thaw whole fish quickly, about 1 hr. submerged in cold water will do the trick; but this method of thawing is apt to result in some loss of flavour and texture and is not really to be recommended. Once thawed, the fish should be used promptly.

When cooking frozen fish, allow a little longer cooking time at a lower temperature than you would use for fresh fish. When the flesh flakes easily in the centre of the fish it is cooked through. Frozen fish can of course be cooked and re-frozen as a finished dish.

## Bulk buying

If really fresh fish cannot be obtained for freezing, buy packs of commercially frozen fish interleaved with plastic film dividers. Bulk packs, known as 'shatter-packs', are now being produced for freezer owners. These are individually frozen fillets packed in bags so that the exact number of fillets required can be removed at any one time, the remainder being left in the freezer.

## Freezing cooked fish dishes

There are many cooked fish dishes that can be frozen to advantage. These dishes should be cooled as quickly as possible after cooking, then chilled in the refrigerator, packed and frozen promptly. Shallow single layer dishes will freeze and reheat more quickly than deeper dishes. The maximum storage life of these dishes is 2 months.

## Packaging cooked fish dishes for the freezer

Allow the food to cool thoroughly before packing in one of the following ways:

### Either

Line a rigid container, such as a casserole, ovenproof dish or even a saucepan, with foil and spoon in the prepared food. Freeze until firm, remove the pre-formed package, fold the foil over and overwrap with a polythene bag. Seal, label and return to the freezer. When required, unwrap and return the frozen food to the original container and follow the recipe instructions.

### Or

Pack in a rigid foil container leaving $\frac{1}{2}$ in. headspace if there is liquid. Put the lid in place or cover with foil; seal, label and freeze. When required, remove top foil or lid and re-cover with foil if necessary.

### Or

Open freeze on a baking sheet. This method is suitable for quiches, pizza,

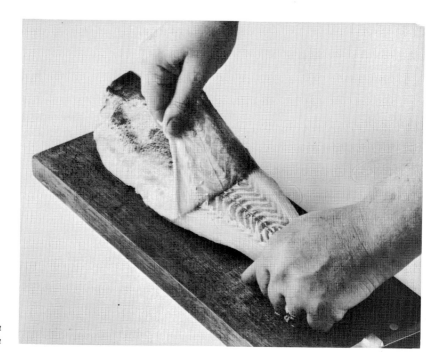

*Skinning a
frozen fish*

pancakes etc. Freeze the food uncovered and when it is solid, wrap in poly-thene, seal, label and return it to the freezer. If a flan or pastry dish is likely to be fragile, open freeze and wrap in heavy duty foil before overwrapping with polythene.

# Pilchard canneloni

2 sticks celery, trimmed
3 oz. mushrooms, wiped
1 oz. butter
1-lb. can pilchards in tomato sauce
freshly ground black pepper
1 tbsp. lemon juice
12 canneloni

*For sauce*
1½ oz. butter
1½ oz. flour
1 pt. milk
salt and pepper

*Serves 4*

**TO MAKE:** Finely chop the celery and mushrooms, melt the butter and sauté the vegetables until tender. Mash the contents of the can of pilchards, add the vegetables,

pepper and lemon juice. Meanwhile cook the pasta in boiling salted water until just tender; drain and rinse with cold water. Stuff the canneloni with the fish filling and arrange in a foil-lined baking dish or rigid foil container in a single layer.

Melt the butter for the sauce; remove from the heat, stir in the flour and gradually blend in the milk. Return the pan to the heat, bring the sauce to the boil, stirring all the time, boil for 1 min. until thickened, season to taste with salt and pepper and pour over the canneloni, to cover completely. Cool rapidly.

**TO PACK AND FREEZE:** See notes at beginning of chapter.

**TO USE:** Remove wrappings and reheat from frozen in the oven at 400°F (mark 6) for 45 min.

# Haddock provençale

1½ lb. haddock fillet, skinned
1 oz. seasoned flour
4 tbsps. corn oil
2 onions, skinned and sliced
1 clove garlic, skinned and crushed
2 tsps. chopped parsley
¼ pt. wine vinegar
3 level tsps. tomato paste
¼ pt. water

*To complete dish from freezer*
4 oz. shelled prawns

*Serves 4*

**TO MAKE:** Cut the fish into 2-in. squares and coat with seasoned flour. Heat the oil in a frying pan and fry the fish quickly until brown all over. Drain and place in a rigid plastic container. Fry the onion in the same oil until golden. Add to the fish with the garlic and parsley. Blend the vinegar, tomato paste and water and pour over the fish.

**TO PACK AND FREEZE:** See notes at beginning of chapter.

**TO USE:** Unwrap, place in a casserole, cover and reheat from frozen in the oven at 425°F (mark 7) for about 30 min. Add the prawns and heat for a further 10 min. Serve with pasta or rice.

# Curried fish

1 oz. butter
8 oz. onions, skinned and chopped
1 level tbsp. curry powder
4 level tbsps. flour
1 pt. stock
1 lb. cod fillet, skinned and cubed
2 tomatoes, skinned and quartered
1 large banana, peeled and sliced
2 oz. sultanas
salt

*Serves 4*

**TO MAKE:** Melt the butter and fry the onion until transparent. Stir in the curry powder and flour and cook for a few min. Remove the pan from the heat and gradually add the stock. Add the cod and remainder of the ingredients, return the pan to the heat and simmer the curried fish for 10–15 min. Cool.

**TO PACK AND FREEZE:** See notes at beginning of chapter.

**TO USE:** Remove wrappings and thaw overnight in the refrigerator. Reheat in a saucepan, simmering gently for a few min. until heated through. Stir gently from time to time but take care not to break up the fish.

# Kipper pizza

*For scone base*
8 oz. self-raising flour
½ level tsp. salt
2 oz. margarine
¼ pt. milk

*For topping*
2 oz. butter
2 oz. flour
½ pt. milk
3 oz. Cheddar cheese, grated
salt and pepper
pinch of dry mustard
12 oz. frozen kipper fillets
3 tomatoes

*Serves 4*

**TO MAKE:** First make the scone base. Sift the flour and salt together into a bowl, rub in the margarine with the fingertips until the mixture resembles breadcrumbs. Bind together with the milk. Knead the dough lightly on a floured board and roll to a circle about ¼ in. thick. Place on a baking sheet.

For the topping, melt the butter in a saucepan, stir in the flour and cook for 1–2 min. Remove the pan from the heat, blend in the milk and cook until thickened; add the cheese, salt, pepper and mustard. Spread the sauce on to the scone base. Cook the kippers according to the instructions on the packet, skin them and cut into strips. Slice the tomatoes and arrange a lattice pattern of kipper strips on the base, placing the tomatoes in between. Bake at 375°F (mark 5) for 30 min. Cool quickly.

**TO PACK AND FREEZE:** Transfer the pizza to a large piece of foil placed on a baking sheet. Open freeze, and when firm wrap the foil around the pizza, overwrap, seal and label.

**TO USE:** Unwrap pizza, brush with a little oil and cover loosely with foil again. Place on a baking sheet and reheat from frozen in the oven at 375°F (mark 5) for 1 hr.

# Greek fish pâté

6 oz. butter, plus a little extra for topping
a few grains cayenne pepper
1 level tsp. garlic salt
2 tbsps. chopped parsley
2 tbsps. double cream
8 oz. pressed cods' roe
1 tbsp. lemon juice

*To complete dish from freezer*
1 lemon, sliced, or sprigs of parsley

*Serves 4*

**TO MAKE:** Cream 6 oz. butter with the cayenne, garlic salt, parsley and cream (use the cayenne very sparingly). Mash the roe and lemon juice and add to the flavoured butter. Spoon into 4 small pots or ramekins. Top with melted butter.

**TO PACK AND FREEZE:** Open freeze until firm, wrap each pot in foil, overwrap in polythene, seal and label.

**TO USE:** Leave wrapped in the refrigerator for about 8 hr., or at room temperature for 4 hr. Unwrap and garnish with twists of lemon or sprigs of parsley. Serve with freshly made toast, wrapped in a napkin to keep it warm.

# Fish chowder

½ oz. butter
2 rashers bacon, rinded and chopped
1 onion, skinned and sliced
3 potatoes, peeled and sliced
15-oz. can tomatoes, drained
1 pt. light stock
salt and pepper
1 bayleaf
2 cloves

*To complete dish from freezer*
1 lb. frozen haddock, skinned and cubed
parsley for garnish

*Serves 4*

**TO MAKE:** Melt the butter and lightly fry

the bacon and onion until soft but not coloured. Add the potatoes. Beat the tomatoes to a thick purée, add the stock and mix into the potato and onion; add salt, pepper, bayleaf and cloves. Simmer for about 15 min., until the potatoes are almost cooked. Remove bayleaf and cloves. Cool rapidly.

**TO PACK AND FREEZE:** See notes at beginning of chapter.

**TO USE:** Unwrap and turn the frozen chowder into a saucepan. Thaw over a gentle heat, add fish and simmer very gently until the fish is soft but retaining its shape. Serve garnished with parsley.

# Plaice with bananas and lemon and wine sauce

4 plaice fillets
2 bananas, peeled
grated rind and juice of ½ lemon
¼ pt. fish stock or water
dry white wine
1 oz. butter
2 level tbsps. flour
salt and pepper
1 egg yolk blended with 1 tbsp. double
  cream

*Serves 4*

**TO MAKE:** Skin the plaice fillets. Cut the bananas in half crosswise and dip them in lemon juice. Roll the fish fillets round the banana halves. Place them in a shallow foil container or foil lined baking dish, cover with the stock or water and poach, uncovered, in the oven at 375°F (mark 5) for 10 min. Drain off the cooking liquid and make up to ½ pt. with the wine.

Melt the butter in a pan, remove from the heat and stir in the flour; gradually add the ½ pt. stock and wine, salt, pepper and lemon rind. Bring to the boil, stirring, and simmer for a few min. Remove from the heat and cool slightly before mixing in the egg yolk and cream. Spoon the sauce over the fish and cool.

**TO PACK AND FREEZE:** See notes at beginning of chapter.

**TO USE:** Unwrap and reheat, covered, from frozen in the oven at 375°F (mark 5) for 45 min.

# Fish goulash

1 oz. butter
½ large onion, skinned and sliced
½ green pepper, seeded and sliced
1 level tbsp. flour
1 level tbsp. paprika pepper
½ pt. water
juice of ½ lemon
2 8-oz. cod fillets
2 tbsps. double cream

*Serves 2*

**TO MAKE:** Melt the butter and sauté the onion and pepper, add the flour and paprika. Mix well. Stir in the water and lemon juice, bring to the boil, then lower the heat. Lay the fish on top of the vegetables and simmer gently for about 10 min., until just cooked.

**TO PACK AND FREEZE:** Lay the fish in a rigid foil or plastic container. Drain the vegetables and spoon them over the fish, stir the double cream into the pan juices and pour this sauce over the fish. Cool, freeze uncovered, then wrap, overwrap, seal and label.

**TO USE:** Unwrap and turn into a casserole or leave in the foil container. Cover and heat from frozen in the oven at 375°F (mark 5) for 45 min.

# Thatched cod

1½ oz. margarine or butter
4 frozen fillets haddock or cod, 4–6 oz. each
4 oz. onion, skinned and chopped
2 oz. fresh white breadcrumbs
2 oz. mature Cheddar cheese, grated
3 firm tomatoes, skinned and roughly chopped
¼ level tsp. salt
freshly ground black pepper
1 tbsp. chopped parsley
grated rind and juice of 1 lemon

*Serves 4*

**TO MAKE:** Melt ½ oz. of the fat in a small saucepan. Use a little to grease the inside of

a shallow, foil lined ovenproof serving dish or rigid foil container. Arrange the fillets in a single layer and brush with the rest of the melted fat. Fry the onion in the remaining 1 oz. fat, without colouring it. Combine it with the breadcrumbs, cheese, tomatoes, salt, pepper, parsley, lemon rind and juice. Spoon the mixture evenly over the fillets.

**TO PACK AND FREEZE:** See notes at beginning of chapter.

**TO USE:** Unwrap, cover and cook from frozen in the oven at 375°F (mark 5), for 30 min; uncover and cook for a further 15 min.

# Kulebayaka

(*see colour picture facing page 65*)

1½ oz. butter
½ lb. frozen leaf spinach
salt and freshly ground black pepper
1 egg, beaten
2 oz. long grain rice
4 oz. onion, skinned and chopped
½ lb. tomatoes, skinned and seeded
¼ lb. button mushrooms, trimmed and sliced
13-oz. pkt. frozen puff pastry, thawed
7½-oz. can red salmon, drained and flaked

*To complete dish from freezer*
1 egg, beaten
melted butter
lemon wedges

*Serves 6–8*

**TO MAKE:** Melt ½ oz. butter in a pan, add the frozen spinach, cover and cook gently, turning frequently. Drain well, chop roughly and season. When cool, add half the egg. Cook the rice in boiling salted water for about 12 min., until tender; drain. Sauté the onion in ½ oz. butter until golden. Dice the tomatoes and sauté them with the mushrooms in the remaining ½ oz. butter. Season well and leave to one side.

Roll out the pastry to a rectangle 11 in. by 16 in. Brush it with some of the remaining beaten egg. In the centre of the pastry, layer up the onion, tomatoes, mushrooms, rice, spinach and salmon.

Adjust seasoning. Leave a border of pastry round the edges; brush these with beaten egg, fold both long sides over $\frac{1}{2}$ in. and roll up like a Swiss roll. Tuck the ends under the roll.

**TO PACK AND FREEZE:** Place the kule-bayaka on a baking sheet and open freeze—see notes at beginning of chapter.

**TO USE:** Unwrap, place on a baking sheet and cook from frozen in the oven at 400°F (mark 6), for $1\frac{1}{2}$ hr. After 30 min. of cooking, slash across the top with a knife 3–4 times and brush with beaten egg. To serve, pour melted butter down the slash marks and accompany with lemon wedges.

# Fish 'n chip balls

8 oz. smoked haddock fillet
2 slices lemon
parsley stalks
6 oz. full fat soft cheese
2 tsps. lemon juice
1 tbsp. chopped parsley
freshly ground black pepper
2 1-oz. pkts. plain potato crisps, crushed

*Makes about 30*

**TO MAKE:** Place the fish with the slices of lemon and parsley stalks in a pan with water to cover. Bring to the boil, reduce heat and simmer until cooked. Drain well. Remove any skin, membrane and bones. Allow to cool and flake the flesh finely. Cream the cheese with the lemon juice until fully blended. Add the chopped parsley and freshly ground black pepper. Beat the cheese with the flaked fish until smooth.

Using lightly floured hands, take small walnut-size portions of the mixture and roll into balls. Roll each ball in crushed crisps, pressing the crisps well into the mixture.

**TO PACK AND FREEZE:** Place in a single layer on a baking sheet and open freeze. Pack in a heavy duty polythene bag.

**TO USE:** Remove the balls from the bag and lay them on a baking sheet. Leave to thaw for about 3 hr. at room temperature. Serve on cocktail sticks as a savoury with drinks.

# Fish envelopes

$7\frac{1}{2}$-oz. pkt. 'boil-in-the-bag' frozen haddock fillets
milk
1 oz. butter or margarine
1 oz. flour
salt and freshly ground black pepper
little grated nutmeg
grated rind and juice of $\frac{1}{2}$ lemon
1 egg, hard-boiled and chopped
1 tbsp. chopped parsley
$7\frac{1}{2}$-oz. pkt. frozen puff pastry, thawed

*Makes 4*

**TO MAKE:** Cook the haddock as directed on the packet. Open, pour out the juices and make up to $\frac{1}{2}$ pt. with milk. Melt the fat in a pan, stir in the flour and cook for a few min. Remove from the heat and stir in the milk. Return the pan to the heat and bring to the boil, stirring. Season well with salt, pepper and nutmeg, stir in the lemon rind, juice, chopped hard-boiled egg and parsley. Cool.

Roll out the pastry to a 12-in. square. Cut into quarters. Place an equal amount of filling in the centre of each square, brush the edges with milk and seal envelope-fashion. Crimp the edges together and glaze with milk.

**TO PACK AND FREEZE:** Open freeze—see notes at beginning of chapter.

**TO USE:** Unwrap, place on a baking sheet and cook from frozen in the oven at 400°F (mark 6) for 45 min.

# Seafood pie

$7\frac{1}{2}$-oz. pkt. frozen cod fillets
$\frac{1}{2}$ pt. milk
salt and pepper
$1\frac{1}{2}$ oz. butter
1 oz. flour
juice of $\frac{1}{2}$ lemon
a little grated nutmeg
4 oz. frozen shrimps
1 lb. old potatoes

*Serves 2*

**TO MAKE:** Place the frozen cod in a pan, pour over the milk and season lightly with salt and pepper. Bring to the boil, cover, reduce the heat and simmer for 5 min. Melt the butter in a saucepan, stir in the flour and cook for 1 min. without browning. Drain the fish and add the cooking liquid to the roux in the saucepan. Whisk well and bring to the boil, stirring. Add the lemon juice, nutmeg and shrimps. Cool.

Remove any dark skin from the cod, flake the flesh, using 2 forks, cool and turn it into a foil lined 1-pt. baking dish or a rigid foil container. Pour the shrimp sauce over. Meanwhile boil the potatoes, drain and cream them until light and fluffy. Use the creamed potato to cover the fish filling. Level it with a knife and run a fork across to make a lattice pattern. Cool rapidly.

**TO PACK AND FREEZE:** See notes at beginning of chapter.

**TO USE:** Remove wrappings and reheat, uncovered, from frozen in the oven at 375°F (mark 5) for 1 hr.

# Codfish pie

4 medium-size potatoes
salt
2 lb. cod fillet, skinned
6 tbsps. olive oil
3 onions, skinned and sliced
2 cloves garlic, skinned and crushed with a
    little salt
2 oz. stoned black olives, chopped
freshly ground black pepper
5 tbsps. dry white wine

*To complete dish from freezer*
chopped parsley
white wine sauce (sauce Bercy)—see
    page 65

*Serves 4–6*

**TO MAKE:** Cook the potatoes in their skins in boiling salted water until tender. Drain, skin and cut them into ¼-in. slices; cut each slice in half. Poach the cod until tender in lightly salted water. Drain and flake the fish.

Heat the oil in a large frying pan, add the onion and potatoes and brown well for about 10 min. Add the garlic, olives, salt and pepper.

**TO PACK AND FREEZE:** Arrange half the vegetables in the bottom of a foil lined,

shallow baking dish or large foil container. Lay the cod on top, season to taste and cover with the remainder of the vegetables. Pour the wine over and cool quickly. Freeze—see notes at beginning of chapter.

**TO USE:** Unwrap and reheat covered from frozen in the oven at 350°F (mark 4) for 1½ hr. Serve sprinkled with chopped parsley and accompanied by hot Bercy sauce.

# Salmon mousse

2 7½-oz. cans red or pink salmon
milk
½ oz. powdered gelatine
4 tbsps. water
1 oz. butter
3 level tbsps. flour
2 eggs, separated
¼ pt. single cream
1 level tbsp. tomato paste
1 tbsp. lemon juice
salt and freshly ground black pepper
edible pink colouring, optional

*To complete dish from freezer*
½ large cucumber
2 tbsps. distilled white vinegar
1 bunch watercress

*Serves 10*

**TO MAKE:** Drain the juice from the salmon and make up to ½ pt. with the milk. Remove the skin and bones from the salmon and mash the flesh until smooth. Soften the gelatine in the water. Melt the butter in a saucepan, stir in the flour and cook for 2–3 min. Remove from the heat and gradually add the salmon liquid and milk. Bring to the boil, stirring constantly. Remove from the heat, add the soaked gelatine and stir to dissolve. Beat in the yolks, allow to cool slightly and stir in the cream, tomato paste, lemon juice and the salmon. Season to taste with salt and pepper. Tint pink if wished.

Whisk the egg whites stiffly and fold into the mixture. Pour into a 2-pt. ring mould and leave to set.

**TO PACK AND FREEZE:** Unmould the mousse on to a foil plate and open freeze— see notes at beginning of chapter.

**TO USE:** Unwrap the mousse, transfer to a serving plate and thaw overnight in the refrigerator. Cut the unpeeled cucumber into julienne strips and toss them in vinegar. Fill the centre of the salmon ring with cucumber and serve garnished with watercress.

# Moules à la bruxelloise

(*see colour picture facing page 65*)

1 oz. butter
3 sticks celery, finely sliced
4 oz. onion, skinned and chopped
1 clove garlic, skinned and crushed
2 pt. mussels
2 bayleaves
2 blades mace
1 level tsp. salt
$\frac{1}{2}$ pt. mussel stock (see method)
$\frac{1}{4}$ pt. dry white wine

*To complete dish from freezer*
chopped parsley
crusty French bread

*Serves 4*

**TO MAKE:** Melt the butter in a medium-sized pan. Fry the celery and onion until transparent, remove from the heat and stir in the garlic. Wash the mussels very thoroughly in cold running water; scrub the shells well, removing any barnacles and discard any open ones that do not close if given a sharp tap with the back of the brush. Put them in a large pan, cover with $\frac{3}{4}$ pt. water, add the bayleaves, mace and salt. Bring to the boil and cook for a few min. with the lid on the pan, until the shells open. Any that remain closed should be thrown away. Strain off the stock and add it with the wine and mussels to the sauté vegetables. Cover the pan and boil rapidly for 7 min. Remove the mussels and discard the top half of each shell. Cool quickly.

**TO PACK AND FREEZE:** Spread the mussels on a baking sheet and open freeze. Place the frozen mussels and still liquid sauce in a rigid container or a double polythene bag and freeze. Store for up to 1 month only.

**TO USE:** Allow to thaw overnight in the refrigerator, then turn into a saucepan and heat gently over a low heat, on top of the stove. Serve with chopped parsley, accompanied by crusty bread to soak up the sauce.

# Kipper meringue flan

6 oz. shortcrust pastry (i.e. made with 6 oz. flour etc.)
$7\frac{1}{2}$-oz. pkt. 'boil-in-the-bag' kipper fillets
3 oz. Cheddar cheese, grated
2 eggs, separated
2 tbsps. milk
salt and freshly ground black pepper

*Serves 4–6*

**TO MAKE:** Roll out the pastry and use it to line a $7\frac{1}{2}$-in. flan ring placed on a baking sheet. Bake blind at 400°F (mark 6) for about 25 min. Cook the kippers as directed on the packet, drain and skin. Mash the kippers with 2 oz. cheese, egg yolks and milk. Place this mixture in the base of the flan. Season with salt and pepper (use salt sparingly as the kippers may already be salty).

**TO FREEZE AND PACK:** Open freeze the flan and freeze the egg whites and 1 oz. grated cheese separately.

**TO USE:** Thaw the flan and egg whites at room temperature for 2 hr. Remove the wrappings and place the flan on a baking sheet. Cook in the oven at 350°F (mark 4) for 20–25 min. Beat the egg whites to a stiff meringue, spoon this over the top of the flan and sprinkle with the remaining 1 oz. cheese. Turn the oven to 400°F (mark 6) and quickly brown the meringue. Serve hot.

# Sardine quiche

6 oz. shortcrust pastry (i.e. made with 6 oz. flour etc.)
1 oz. butter
8 oz. onions, skinned and thinly sliced
salt and freshly ground black pepper
$4\frac{1}{2}$-oz. can sardines in olive oil, drained
2 large eggs
milk
lemon wedges

*Serves 4–6*

**TO MAKE:** Roll out the pastry thinly and use it to line an $8\frac{1}{2}$-in. loose-bottomed, French fluted flan tin. Melt the butter in a pan and fry the onions until soft but not brown. Season with salt and pepper and cool. Cut off and discard the tails of the sardines. Place the onions in the flan base and lay the sardines on top, as the spokes of a wheel; build up the onion a little under each sardine.

Beat the eggs, season and make up to $\frac{1}{2}$ pt. with the milk. Carefully spoon the custard round the sardines. Bake in the oven at 400°F (mark 6) for 20 min. Reduce the temperature to 325°F (mark 3) and cook for about a further 20 min. Cool rapidly.

TO PACK AND FREEZE: Open freeze—see notes at beginning of chapter.

TO USE: Unwrap and thaw at room temperature for 2 hr. Place on a baking sheet, cover with foil and reheat in the oven at 350°F (mark 4) for about 20–25 min. Serve hot or cold.

# Tuna-stuffed pancakes

7$\frac{1}{2}$-oz. can tuna, drained
2 oz. margarine
4 level tbsps. plain flour
$\frac{1}{2}$ pt. milk
1 level tsp. French mustard
1 tbsp. lemon juice
2 oz. mature Cheddar cheese
salt and freshly ground black pepper

*For pancake batter*
4 oz. flour
pinch salt
1 large egg, beaten
$\frac{1}{2}$ pt. milk
butter for frying

*To complete dish from freezer*
a little butter or margarine
chopped parsley

*Serves 4*

TO MAKE: Flake the tuna. Melt the margarine in a saucepan, stir in the flour and cook for a few min. Remove from the heat and gradually beat in the milk. Bring to the boil, stirring. Add the mustard, lemon juice, cheese and tuna. Adjust seasoning with salt and pepper.

To make the pancakes, sift the flour and salt into a bowl, make a well in the centre and add the egg and half the milk. Beat well until smooth. Stir in the remaining milk until well mixed. Melt a little butter in a 7-in.

frying pan and when hot, pour in a spoonful of batter. Cook until brown then turn the pancake and cook for 10–15 sec. on the other side. Make 8 pancakes in all.

Fold the pancakes into quarters, open out each one to make a pocket and fill with the tuna sauce. Arrange in a shallow ovenproof dish or on an 8-in. foil plate.

TO PACK AND FREEZE: Wrap and overwrap, seal, label and freeze.

TO USE: Unwrap the pancakes and dot them with a little butter. Cover with foil. Reheat from frozen in the dish in which they were frozen, for about 1$\frac{1}{4}$ hr. in the oven at 350°F (mark 4). Sprinkle with chopped parsley and serve.

# Devilled salmon pâté

(*see picture opposite*)

7-oz. can salmon
6 tbsps. milk
bouquet garni
a few black peppercorns
$\frac{1}{2}$ oz. butter
1 oz. flour
2 tbsps. double cream
juice of $\frac{1}{2}$ lemon
a few drops Tabasco sauce
1 level tsp. French mustard
salt and freshly ground black pepper

*To complete dish from freezer*
1 lemon
parsley sprigs

*Serves 4–6*

TO MAKE: Mix the juice drained from the salmon with the milk. Add the bouquet garni and peppercorns. Leave covered for 10 min., then strain. Melt the butter in a small pan, add the flour and cook for 1 min. Remove from the heat and stir in the flavoured milk. Bring to the boil, stirring. Flake the salmon and add it to the sauce with the remainder of the ingredients. Mix well or purée in an electric blender.

TO PACK AND FREEZE: See notes at beginning of chapter.

TO USE: Leave wrappings in place and thaw in the refrigerator for 8–12 hr., or at room temperature for 4–6 hr. Unwrap and garnish

*Devilled salmon pâté*

with wedges or twists of lemon and sprigs of parsley. Serve freshly made toast as an accompaniment.

# Mackerel à la Kiev

(*see colour picture facing page 65*)

4 mackerel, filleted
2 eggs
fresh white breadcrumbs
a little flour
4 oz. butter
2 level tbsps. dry mustard

*Makes 6*

TO MAKE: Place the mackerel in a large frying pan, cover with water and simmer gently for about 5 min. or until cooked. Drain the fish, remove the skin and bones and flake the flesh. Add 1 beaten egg and a few breadcrumbs to bind and divide the mixture into 6. Using floured hands, flatten the portions into 6 cakes. Refrigerate until ready to use. Soften the butter, mix with the mustard and roll it in foil. Chill.

Divide the flavoured butter into 6, put 1 portion in the centre of each fish cake then form them into oval balls encasing the butter. Dip in beaten egg and coat with breadcrumbs.

TO PACK AND FREEZE: Open freeze—see notes at beginning of chapter.

TO USE: Unwrap and thaw loosely covered for 2½ hr. at room temperature. Deep fry slowly for about 15 min. until golden brown. Serve at once, while piping hot.

# Waterzooi de poisson

1½ lb. cod fillet, skinned
3 sticks celery, roughly chopped
1 oz. butter
1 level tsp. salt
freshly ground black pepper

*To complete dish from freezer*
3 egg yolks
2 fl. oz. double cream
2 level tsps. cornflour
chopped parsley
rye bread

*Serves 4–6*

**TO MAKE:** Cut the fish into large pieces about $1\frac{1}{4}$ in. square. Place in a saucepan and cover with cold water. Add the celery, butter, salt and pepper. Poach gently for about 8–10 min., until the flesh flakes fairly easily when pierced with a fork. Carefully remove the fish, using a draining spoon. Place it in a $2–2\frac{1}{2}$-pt. foil-lined, flameproof casserole and scatter the celery over. Strain the fish stock then rapidly boil it down to about 1 pt. Cool and pour it over the fish.

**TO PACK AND FREEZE:** See notes at beginning of chapter.

**TO USE:** Unwrap the fish and put it in a flameproof casserole. Cover and reheat from frozen in the oven at 400°F (mark 6) for $1\frac{1}{2}$hr. Beat the egg yolks with the cream and corn-flour until smooth, stir in a little hot fish stock, then add to the casserole. Stir continuously over a gentle heat until thickened. Adjust seasoning. Garnish with chopped parsley and serve accompanied by rye bread.

# Smoked fish flan

10-oz. pkt. frozen buttered kipper fillets
6 oz. rich shortcrust pastry (i.e. made with
  6 oz. flour etc.)

*For choux pastry*
$1\frac{1}{2}$ oz. butter
$\frac{1}{4}$ pt. water
$2\frac{1}{2}$ oz. plain flour
salt
2 large eggs, beaten

*For sauce*
2 oz. butter
2 level tbsps. flour
milk
2 tbsps. single cream
4 oz. Cheddar cheese
salt and freshly ground black pepper
grated Parmesan cheese

*To complete dish from freezer*
watercress sprigs
lemon wedges

*Serves 6*

**TO MAKE:** Cook the kipper fillets as directed on the packet. Drain the cooking liquid into

a measuring jug. Use a small palette knife and a fork to remove the dark skin from the fish, and discard it. Cut each fillet roughly into three and leave to cool.

Roll out the shortcrust pastry and use it to line a 9-in. plain flan ring placed on a lightly greased baking sheet. Prick the base and chill in the refrigerator. To make the choux pastry, melt the butter in the water, sift the flour and salt on to a sheet of grease-proof paper. When it boils, remove the water from the heat and tip the flour into the pan all at once. Beat the mixture just enough to take it away from the pan sides. Add the eggs slowly, beating hard between each addition. Keep back just enough egg to brush the base and edge of the flan. Leave the choux paste to cool.

To make the sauce, melt the butter in a saucepan. Sprinkle in the flour and blend well. Make up the cooking liquid from the kippers to $\frac{1}{2}$ pt. with milk, remove the pan from the heat and whisk in the liquid. Return it to the heat, whisking gently, and stir in the cream. Grate the Cheddar cheese and add to the sauce. Season with salt and pepper. Cover with a round of buttered paper to prevent a skin forming and leave to cool.

Brush the base and edge of the shortcrust flan with beaten egg. Spoon the choux paste into a piping bag fitted with a $\frac{1}{2}$-in. plain nozzle and pipe about half the choux round the edge of the flan case. Lightly draw the choux up the sides of the flan case with a knife. Turn the cold kipper fillets into the centre of the flan and spread them out evenly. Spoon the sauce over the kippers and pipe the remaining choux paste over the top. Brush with a little beaten egg, dust lightly with grated Parmesan and bake at 425°F (mark 7) for about 45 min. Cool rapidly. Remove flan ring.

**TO PACK AND FREEZE:** Open freeze—see notes at beginning of chapter.

**TO USE:** Leave the wrappings in place and thaw for 1 hr. at room temperature. Unwrap, place the flan on a baking sheet and warm through in the oven at 325°F (mark 3) for about 1 hr. Garnish with watercress and lemon wedges.

*Seafood timbale*

# Seafood timbale

(*see picture above*)

¾ lb. haddock fillet, skinned
1½ level tsps. salt
pinch of pepper
1½ oz. butter
1½ oz. flour
5 tbsps. cider
2 eggs, separated
5 tbsps. single cream

*For sauce*
¾ oz. butter
¾ oz. (2 level tbsps.) flour
¾ pt. light stock

*To complete sauce from freezer*
3 tbsps. single cream
a few drops lemon juice
salt and pepper
4 oz. frozen or fresh peeled shrimps or
    prawns

*Serves 4–6*

**TO MAKE:** Mince the uncooked fish with
the salt and pepper. Melt the butter, remove
from the heat and stir in the flour, add the
cider, return to the heat and cook gently,
stirring. Add the fish, egg yolks and cream.
Whisk the egg whites stiffly and fold into
the fish mixture. Transfer it to a greased
1½-pt. ring mould. Place the mould in a
baking tin, with enough water to come
about halfway up the mould. Cook the
timbale in the oven at 350°F (mark 4) for
45 min. Cool rapidly.

Meanwhile, melt the butter for the sauce
in a pan, stir in the flour and cook for 1–2
min. Remove from the heat, blend in the
stock and cook for a further 2–3 min.,
stirring constantly.

**TO PACK AND FREEZE:** Unmould the
timbale, open freeze, then wrap as usual.
Freeze the sauce separately in a waxed
container or preformed polythene bag.

**TO USE:** Remove the polythene but keep
the timbale loosely wrapped in foil. Reheat
it from frozen in the oven at 400°F (mark 6)
for 1 hr. Turn the frozen sauce into a

saucepan and thaw over gentle heat, beating occasionally. Remove from the heat and add the cream, lemon juice, salt, pepper and prawns. Serve with the timbale.

## Haddock cutlets with sultana sauce

1 lb. haddock fillets, skinned
½ pt. milk
3 oz. fresh white breadcrumbs
1 egg
½ level tsp. grated nutmeg
salt and freshly ground black pepper
2 oz. butter

*For sultana sauce*
½ pt. soured cream
¼ level tsp. grated nutmeg
2 oz. sultanas
4 tbsps. milk
1 tbsp. lemon juice
1 tbsp. white wine

*Makes 12*

**TO MAKE:** Purée the uncooked fish with the milk in a blender, or mince it and then mix with the milk. Add the breadcrumbs, egg, nutmeg, salt and pepper. Beat well to blend all the ingredients together. Shape the mixture into 12 compact rounds. Melt the butter and fry the cutlets until they begin to brown then drain well on absorbent paper. Make the sauce by gently heating all the ingredients together. Cool.

**TO PACK AND FREEZE:** Freeze the sauce in a waxed container or pre-formed poly-thene bag. Place the cutlets in a single layer on a baking sheet and open freeze—see notes at beginning of chapter.

**TO USE:** Unwrap the cutlets, place them on a baking sheet and cover loosely with foil. Reheat from frozen in the oven at 375°F (mark 5) for 45 min. Turn the frozen sauce into a pan and reheat on top of the stove, beating it occasionally; do not boil. Serve separately.

# RECIPES USING
# FISH FROM THE FREEZER

## Casseroled cod

1 lb. frozen cod fillet, thawed and skinned
4 oz. onions, skinned and sliced
6 oz. mushrooms, sliced
2 oz. butter
1 level tbsp. flour
½ level tsp. made mustard
¼ pt. stock
½ tsp. anchovy essence
salt and pepper
4 oz. frozen peas
4 oz. frozen sweetcorn kernels
¼ pt. soured cream

*Serves 4*

Cut the fish into cubes. Sauté the onions and mushrooms in the butter, stir in the flour and mustard and cook for a few min. Add the stock and anchovy essence, season with salt and pepper (be sparing with salt as the

anchovy essence is very salty). Add the fish to the pan with the frozen peas and sweet-corn. Simmer very gently for 8–10 min. until cooked. Gently stir in the soured cream and reheat without boiling. Serve at once.

## Haddock with Dubrovnik sauce

½ oz. butter
1 oz. onion, skinned and chopped
½ lb. tomatoes, skinned and sliced
7 fl. oz. cider
4 frozen haddock fillets
2 oz. Cheddar cheese, grated

*Serves 4*

Melt the butter and fry the onion until transparent; add the tomatoes and fry gently for a few more min. Add the cider and simmer uncovered for 5 min. Arrange the fish in an ovenproof dish, pour over the tomato sauce and sprinkle with the cheese. Bake uncovered in the oven at 400°F (mark 6) for 20–30 min., until the fish is tender and the flesh flakes easily.

# Prawn provençale

1 oz. butter
1 onion, skinned and finely chopped
1 clove garlic, skinned and crushed
15-oz. can tomatoes, drained
4 tbsps. dry white wine
salt and freshly ground black pepper
pinch sugar
1 tbsp. chopped parsley
8 oz. frozen prawns

*Serves 2*

Melt the butter and fry the onion and garlic gently for about 5 min., until soft but not coloured. Add the tomatoes, wine, salt, pepper, sugar and parsley. Stir well and simmer gently for about 10 min. Add the prawns to the sauce and continue simmering for a further 10 min., or until they are just heated through. Serve at once with crusty French bread or boiled rice.

# Crab thermidor

2 oz. butter
1 tbsp. finely chopped shallot
2 tsps. chopped parsley
1–2 tsps. chopped fresh tarragon
4 tbsps. dry white wine
½ pt. frozen béchamel sauce (page 62)
½ lb. frozen crab meat, thawed
2 tbsps. grated Parmesan cheese
pinch dry mustard
salt
paprika pepper

*Serves 6*

Melt 1 oz. butter in a saucepan and gently sauté the shallot with the parsley and tarragon. After a few min. add the wine and simmer for 5 min. Add the frozen béchamel sauce and simmer until the sauce is thawed and of a creamy consistency, stirring frequently. Add the crab meat to the sauce with 1½ tbsps. Parmesan, the remaining

butter, mustard, salt and paprika. Arrange the mixture in 6 shells or individual cocottes and sprinkle with the remaining Parmesan. Put under a pre-heated grill to brown and serve at once.

# Devilled prawn appetiser

1 tbsp. Worcestershire sauce
8 drops Tabasco sauce
3 tsps. lemon juice
¼ tsp. anchovy essence
¼ pt. double cream
8 oz. frozen prawns, thawed

*Serves 6*

Add all the seasonings to the cream and whip until thick. Divide the prawns between 6 scallop shells. Pile the cream on to the prawns and place in a very hot oven at 450°F (mark 8) or under a preheated grill to brown. Serve at once.

# Gebakken schol fillet vit de pan

5 6-oz. frozen plaice fillets
1½ level tsps. salt
juice of 1 lemon
1½ oz. butter
5 rashers streaky bacon, rinded
1 oz. plain flour
¼ level tsp. dill seeds
¼ level tsp. grated nutmeg
freshly ground black pepper
2 oz. Gouda cheese, ⎞ both these may
  grated ⎟ be used straight
3 oz. fresh white ⎬ from the
  breadcrumbs ⎠ freezer
½ oz. nibbed almonds
2 oz. butter

*Serves 5*

Skin the fillets whilst still frozen. Lay them in a shallow dish, season with salt and lemon juice and leave for about 30 min., or until thawed. Melt the butter in a wide, shallow flameproof serving dish large enough to hold the fish in a single layer. Tilt the dish to coat the sides with butter. Fry the bacon gently in the butter until lightly browned. Drain well and keep to one side. Pat the fish

Cod bubble

dry with absorbent kitchen paper and fold in half lengthwise. Roll the fillets in flour, sprinkle with dill seeds and nutmeg and arrange side by side in the buttered dish. Top with the bacon and season with black pepper.

Mix the cheese, breadcrumbs and almonds together. Scatter the mixture over the fish and dot with butter. Cook, covered, on top of the stove over a gentle heat for 15 min. Uncover and flash under a pre-heated grill to brown and crisp the topping.

## Cod bubble

(*see picture above*)

4 frozen cod steaks, 1 in. thick
2 oz. butter
4 tomatoes, skinned and sliced
salt and pepper
4 oz. Cheddar or Lancashire cheese,
    grated

*Serves 4*

Line the grill pan with buttered foil. Place the frozen cod steaks in the pan and dot with half the butter. Place under a medium grill and cook for 7 min., basting the fish from time to time. Turn the cod, dot with the remaining butter and cook for a further 7 min. Place the sliced tomatoes on top of the steaks, season with salt and pepper and cover with cheese. Return to the grill and cook for about 5 min., until the cheese is bubbly and golden brown. Serve at once.

## Smoked fish and egg slice

(*see picture opposite*)

13-oz. pkt. frozen puff pastry, thawed
1 egg, beaten
6 eggs, hard-boiled
6 oz. full fat soft cream cheese
$\frac{1}{4}$ pt. soured cream
1 oz. finely chopped onion
1 tbsp. lemon juice
black pepper
12 oz. frozen smoked haddock fillets,
    thawed
parsley

*Serves 6–8*

*Smoked fish and egg slice*

Roll out the pastry to a rectangle about 8 in. by 12 in. Make 1-in. cuts, 1 in. apart on the long sides. Brush the edges with a little beaten egg and fold back the 'tabs' diagonally to form triangles. Prick the base well and brush again with the beaten egg. Leave on a wetted baking sheet in a cool place for 30 min. Bake at 425°F (mark 7) for about 30 min., until golden. Cool on a wire rack.

Roughly chop 4 of the hard-boiled eggs and slice the other 2 lengthwise for decoration. Beat together the chopped eggs, cheese, cream, onion, lemon juice and pepper. Cut the fish fillets in 3 by cutting off the tail end and halving the remaining piece down the centre. Roll the pieces and secure with cocktail sticks. Place them in a pan, cover with a little water or milk and poach gently until tender. Remove the skewers and halve each roll. Any flaked pieces may be added to the egg and cheese mixture. Spread the egg filling on the pastry base and decorate with the fish wheels, egg slices and parsley.

# Malaysian fish stew

1 lb. frozen firm white fish, e.g. cod or
 haddock
15½-oz. can pineapple chunks
milk
2 oz. butter
2 oz. flour
2 level tsps. curry powder
2 tbsps. mango chutney
2 oz. long grain rice, cooked
salt and freshly ground black pepper
1 tsp. lemon juice

*Serves 4*

Skin the fish and poach it in the juice from the can of pineapple for about 20 min., until tender. Drain and make the cooking liquid up to 1 pt. with milk. Melt the butter, remove from the heat and stir in the flour and curry powder. Blend in the cooking liquid and milk, add the mango chutney, pineapple chunks, rice, salt, pepper, lemon juice and fish. Heat through gently before serving.

# Trout en papillote

(*see colour picture facing page 32*)

4 frozen trout
2 oz. butter
4 sprigs parsley
4 shallots, skinned and finely chopped
pared rind of ½ lemon
salt and freshly ground black pepper

*Serves 4*

Cut 4 pieces of foil large enough to wrap the trout in. Butter the foil with half the butter. Lay the frozen trout on the foil and add the remaining ingredients, dividing them equally between the fish and finishing with the remaining 1 oz. butter dotted over the top. Close the foil round the fish, place them on a baking sheet and bake at 325°F (mark 3) for about 1 hr. Take the fish to the table still wrapped in the foil.

# 5 Ways With Meat

**Packing meat for the freezer**

Whether commercially deep frozen or fresh, joints should be of a manageable size, and packed, for preference without the bone, in a parcel type wrap, in freezer paper or foil, overwrapped in a heavy gauge polythene bag, sealed and labelled. Freeze as soon as possible after purchase.

Any small cut of meat, such as steaks or chops which can be cooked from frozen, should be interleaved with double cellophane or polythene film to enable them to be separated easily.

Any meat to be used for stews, etc. may be boned and cut into pieces of the required size: pack tightly in a polythene bag or rigid container, excluding as much air as possible and freeze rapidly. Keep offal and fresh mince in small portions for easy thawing; freeze rapidly.

**Storage times for meat:**

| | | | | | | | | |
|---|---|---|---|---|---|---|---|---|
| Beef | .. | .. | .. 8 months | Mince | .. | .. | .. | 3 months |
| Lamb | .. | .. | .. 6 months | Offal | .. | .. | .. | 3 months |
| Pork and veal | .. | | .. 6 months | Sausages | | .. | .. | 3 months |

When cooking small cuts of meat, such as steaks, chops, escalopes, etc., from frozen it is best to start cooking over a low heat until the meat is thawed, and then increase the heat at the end of the time to brown the meat, if necessary.

It is advisable to thaw very large joints of meat—allow about 6 hr. per lb. in the refrigerator, or 3 hr. per lb. at room temperature. Slow (refrigerator) thawing is best as there is less loss of meat juices. Mince and stewing steak are best thawed prior to cooking. Once thawed, the meat must be cooked at once.

Sausages may also be cooked from frozen; they are best baked or grilled, as in frying they tend to split; turn occasionally during cooking.

**Large sausages**

    baked    $1-1\frac{1}{4}$ hr. 350°F (mark 4)

    grilled    10 min. starting at a low temperature

**Small chipolata sausages**
    baked     45 min. at 375°F (mark 5)
    grilled    12–15 min.
    fried      12–15 min.

**Cooking sliced liver from frozen**
    fried          20 min.
    baked in sauce   1 hr. at 350°F
                         (mark 4)

### Cooking joints of meat from frozen

Joints of meat cooked from frozen often have a very much better flavour than those which have been thawed prior to cooking, although the thawed joints may often be a little more tender.

When cooking joints from frozen it is essential that the internal temperature is checked with a meat thermometer. About 30 min. before the end of cooking time insert the thermometer into the centre of the joint, taking care that it does not touch the bone. Cooking time may vary according to the shape and composition of the joint, and individual preferences.

**Beef**—seal the joint in hot fat in a pre-heated oven at 450°F (mark 8) uncovered, for 20 min. Reduce the oven temperature to 350°F (mark 4), cover the meat and cook 50 min. per lb., basting frequently.

Temperature of slightly underdone meat using a meat thermometer—160°F.

**Lamb**—seal the joint in hot fat in a pre-heated oven at 450°F (mark 8) uncovered, for 20 min. turning once. Reduce the oven temperature to 350°F (mark 4), cover the meat and cook 60 min. per lb. basting frequently.

Temperature of meat using a meat thermometer—180°F.

**Pork**—seal the joint in hot fat in a pre-heated oven at 450°F (mark 8) uncovered, for 20 min. turning once. Reduce the oven temperature to 350°F (mark 4). Cover the meat and cook 60 min. per lb., basting frequently.

Temperature of meat using a meat thermometer—190°F.

Roasting bags may be used to advantage in roasting joints from frozen. The cooking times are the same per lb. as the open method of roasting, omitting the initial 20 min. to seal the meat. Therefore, less cooking time is involved and the oven is kept clean. If a very well browned joint is preferred or very crisp pork crackling, the roasting bag may be slit 20–30 min. before the end of cooking time.

### Packing cooked meat dishes for the freezer

*Casseroles*

Cool the dish quickly after cooking. A casserole may be frozen as a pre-formed pack or in a rigid foil container with a lid. **Either** Line a casserole dish with foil, fill with the meat mixture and freeze until firm; remove the foil covered block from the casserole, fold the foil over and overwrap with polythene. Seal, label and return to the freezer. To use, unwrap the frozen meat, return it to the original casserole or to a saucepan and thaw or reheat from frozen according to the recipe. Thaw and cook covered unless the recipe specifies otherwise. **Or** Turn the cool, cooked meat into a rigid foil container and cover with a lid, leaving $\frac{1}{2}$ in. headspace for expansion. Make sure that all chunks of meat are covered by the sauce. Overwrap, seal and label the pack and freeze. To use, remove any overwrappings and the lid if it is cardboard. Then either cover with foil and thaw or reheat from frozen in the foil container, or turn out the meat into a saucepan or casserole if you prefer. Again, thaw and cook covered unless the recipe specifies otherwise.

## Moulds and pâtés

These are usually frozen in the tin in which they are cooked then turned out and wrapped in foil, overwrapped in polythene and sealed and labelled for returning to the freezer. To use, remove overwrappings, loosen the foil and thaw or reheat according to the recipe. Maximum storage time for pâtés and meat moulds is 1 month.

## Meat pies

Freeze cooked or uncooked as you prefer. Open freeze in the pie dish, then turn out and wrap in foil, overwrap with polythene, seal and label. To use, remove wrappings, return to the pie dish and bake or reheat according to recipe instructions.

# MEAT RECIPES FOR THE FREEZER

## LAMB

## Basic lamb casserole

6 lb. stewing lamb (weight of prepared
  meat)
$1\frac{1}{2}$ oz. lard
$1\frac{1}{2}$ lb. onions, skinned and chopped
$1\frac{1}{2}$ pt. stock
salt and pepper

*Makes 4 casseroles, each to serve 4*

**TO MAKE:** Cut the lamb into $\frac{1}{2}$-in. cubes. Heat the lard in a large frying pan and brown the meat evenly all over, frying about a quarter of the meat at a time. As each portion is browned, take it from the pan, set aside for

the moment and reheat the fat. Lastly, fry the onions and cook until transparent, stirring. Transfer the meat and onion to a casserole or casseroles, pour over the boiling stock, season, cover and cook in the oven at 350°F (mark 4) for $1\frac{1}{2}$ hr. Cool rapidly.

**TO PACK AND FREEZE:** Divide into 4 and pack according to notes at beginning of chapter.

**TO USE:** Unwrap and turn into a saucepan or casserole, or keep in the foil container. **Either** Thaw for about 15 hr. in the refrigerator, add any remaining ingredients (see variations below) and cook at 375°F

(mark 5) for about 45 min.; season.
**Or** Reheat from frozen. Place in the oven at 375°F (mark 5) for about 1¼ hr., adding any remaining ingredients when the casserole has thawed. Season.

## CRANBERRY LAMB

¼ quantity basic lamb casserole
3 tbsps. cranberry sauce
1 level tsp. dried rosemary

Thaw the casserole. Stir the cranberry sauce and rosemary into the casserole 30 min. before the end of the reheating time.

## MINTED LAMB CASSEROLE

¼ quantity basic lamb casserole
½ level tsp. dried mint
7½-oz. can butter beans, drained
8-oz. can tomatoes, drained

Thaw the casserole. Stir the mint, beans and tomatoes into the casserole 30 min. before the end of the reheating time.

## LAMB AU GRATIN

¼ quantity basic lamb casserole
1 large aubergine (about 6–8 oz.)
salt
½ lb. courgettes
3 tbsps. cooking oil
1 oz. fresh white breadcrumbs
2 oz. mature Cheddar cheese, grated
½ oz. butter

Thaw the casserole for about 15 hr. in the refrigerator. Wipe the aubergine and slice across into ¼-in. slices. Lay on a plate, sprinkle with salt and leave for about 30 min. to extract the bitter juices. Wipe and slice the courgettes. Heat the oil in a frying pan and fry the courgettes on both sides—drain and lay on absorbent paper. Dry the aubergine slices and fry these, adding a little extra oil to the pan if necessary. When brown on both sides remove from the pan and drain on absorbent paper. Place half the aubergines and courgettes in a layer on the base of a casserole, cover with the basic lamb casserole and place the remaining courgettes and aubergines on top. Combine the breadcrumbs and cheese, sprinkle over the casserole and dot with butter. Cook uncovered in the oven at 400°F (mark 6) for about 30 min. until browned on top.

## SCANDINAVIAN LAMB

¼ quantity basic lamb casserole
4 oz. frozen peas
¼ pt. soured cream

Thaw the casserole. Add the frozen peas 30 min. before the end of the reheating time. Stir in the soured cream just before serving and reheat without boiling. Dish up on a bed of buttered noodles.

# Lamb with peppers

2 lb. middle neck of lamb
2 tbsps. cooking oil
3 small onions, skinned and cut into quarters
2 level tbsps. flour
½ level tsp. ground cinnamon
1 level tbsp. Demerara sugar
½ pt. stock
14-oz. can tomatoes
1 green pepper, seeded and cut into large pieces
1 bayleaf
salt and freshly ground black pepper

*Serves 4*

**TO MAKE:** Cut the lamb away from the bone and into 1-in. cubes. Heat the oil and brown the meat all over. Place in a casserole. Brown the onions in the remaining oil. Remove from the heat and sprinkle in the flour, cinnamon and Demerara sugar. Gradually blend in the stock and the tomatoes, with their juice. Add the green pepper, bayleaf and season with salt and freshly ground black pepper. Pour over the meat, cover and place in the oven to cook at 325°F (mark 3) for about 2 hr. Cool quickly.

**TO PACK AND FREEZE:** See notes at beginning of chapter.

**TO USE: Either** Unwrap and thaw in the refrigerator for 12–15 hr. Reheat, covered, in the oven at 375°F (mark 5) for 45 min.

Or Unwrap and reheat from frozen in the oven at 375°F (mark 5) for about 1¾ hr.

# Honeyed lamb

½ oz. butter
1 small onion, skinned and chopped
¼ level tsp. ground ginger
¾ pt. canned tomato juice
½ tsp. Worcestershire sauce
1 bayleaf
pinch of mixed spice
1 level tbsp. clear honey
¾ lb. lean boneless lamb, cubed
1 oz. flour
cooking oil
1 oz. currants

*Serves 2*

TO MAKE: Melt the butter in a frying pan and fry the onion with the ginger until golden brown. Add the tomato juice, Worcestershire sauce, bayleaf, mixed spice and honey. Bring to the boil. Toss the cubed lamb in the flour. Fry in just enough oil to coat the pan, until evenly brown. Combine the meat and sauce in a casserole and add the currants. Cover and cook in the oven at 325°F (mark 3) for 1¼ hr. Cool quickly.

TO PACK AND FREEZE: See notes at the beginning of chapter.

TO USE: Either Unwrap and thaw for 12–15 hr. in the refrigerator. Reheat in the oven at 350°F (mark 4) for about 45 min. Or Unwrap and heat from frozen in the oven at 350°F (mark 4) for about 1¾ hr.

# Hearty lamb soup

1½ lb. scrag of lamb, in pieces
3 pt. stock
bouquet garni
½ lb. small carrots, peeled
2 tbsps. cooking oil
½ lb. leeks, sliced and washed
1 turnip, peeled and diced
2 sticks celery, chopped
3 oz. long grain rice
salt and freshly ground black pepper

*To complete dish from freezer*
chopped parsley

*Serves 4–6*

TO MAKE: Place the meat (still on the bone) in a saucepan, cover with cold water and bring to the boil. Drain and refresh under cold running water; drain well. Replace the drained meat in the pan, cover with the stock and add the bouquet garni and whole carrots. Bring to the boil, reduce the heat, cover and simmer for 1 hr. Cool a little and skim off the excess fat.

Heat the oil in a large saucepan and gently sauté the leeks, turnip and celery for about 10 min. Strain the stock from the lamb over the vegetables and bring to the boil—add the rice, season lightly and continue to boil for about 8–9 min., until the rice is cooked. Strip the lamb from the bones, slice the carrots and return both to the pan. Chill.

TO PACK AND FREEZE: See notes at beginning of chapter.

TO USE: Either Unwrap and turn the frozen soup into a saucepan. Heat very gently at first, stirring occasionally, before bringing to the boil Or Thaw for 8–10 hr. at room temperature before reheating. Adjust seasoning and sprinkle with chopped parsley just before serving.

# Lamb roulades

2 large breasts of lamb, boned
freshly ground black pepper
celery stalks, trimmed
1 tbsp. cooking oil
10½-oz. can condensed mushroom soup
¼ pt. water
½ level tsp. dried rosemary

*Serves 4*

TO MAKE: Cut the boned lamb into 2-in. strips across the width of the breast. Sprinkle with the pepper. Cut the celery stalks to the width of the breasts and place one on each strip of lamb, roll up and secure with string or cocktail sticks. Heat the oil in a frying pan and fry the lamb until brown all over; drain and place in a casserole or foil container. Heat the soup with the water and rosemary and pour it over the lamb. Cover and cook in the oven at 325°F (mark 3) for 1¼ hr. Cool quickly.

**TO PACK AND FREEZE:** See notes at beginning of chapter.

**TO USE:** Unwrap and reheat from frozen in the oven at 350°F (mark 4) for $1\frac{1}{4}$ hr., separating the roulades with a fork halfway through the cooking time. Remove the cocktail sticks or string before serving.

# Stuffed loin of lamb

3–$3\frac{1}{2}$ lb. loin of lamb, boned
1 oz. butter
1 small onion, skinned and finely chopped
4 rashers bacon, rinded and chopped
6 heaped tbsps. fresh white breadcrumbs
grated rind and juice 1 lemon
$\frac{1}{2}$ level tsp. dried rosemary
1 small egg, beaten
salt and freshly ground black pepper

*To complete dish from freezer*
2–3 tbsps. dripping for roasting

*For sauce*
1 onion, skinned and finely sliced
1 level tbsp. plain flour
$\frac{1}{2}$ pt. stock
2 tbsps. redcurrant jelly
a little lemon juice

*Serves 6–8*

**TO MAKE:** Wipe the meat and flatten it by lightly beating with a rolling pin or meat mallet. Melt the butter and gently sauté the onion and bacon. Mix with the breadcrumbs, lemon rind and juice, rosemary and beaten egg. Season lightly with salt and pepper. Spread the stuffing over the meat, roll up and secure in several places with fine string.

**TO PACK AND FREEZE:** See notes at beginning of chapter.

**TO USE:** Unwrap and thaw for 24 hr. in the refrigerator. Melt the dripping in a roasting tin, dredge the meat lightly in flour and roast join side down in the oven at 350°F (mark 4) allowing 40 min. per lb. Keep the joint warm on a serving plate. Pour off most of the fat from the roasting tin, leaving any sediment in the bottom. Add the sliced onion, and cook slowly, until just beginning to colour. Stir in the flour, stock and redcurrant jelly; sharpen the taste with a little lemon juice. Bring to the boil, stirring, cook for a few min. and

strain. Serve with the meat carved and arranged in a warm serving dish.

# Mint stuffed breast of lamb

2 breasts of lamb, boned
2 medium sized cooking apples, peeled and cored
$\frac{1}{2}$ oz. butter
3 tsps. mint sauce
1 oz. fresh white breadcrumbs
salt and freshly ground black pepper

*Serves 4*

**TO MAKE:** Trim the breasts of lamb, removing any excess fat. Coarsely grate the apple and gently sauté it in the butter; mix with the mint sauce and breadcrumbs. Season and spread the stuffing over each breast, roll up tightly and secure with string.

**TO PACK AND FREEZE:** See notes at beginning of chapter.

**TO USE: Either** Thaw overnight in the refrigerator. Unwrap and place the joints in a roasting tin; roast in the oven at 350°F (mark 4) for about $1\frac{1}{2}$ hr. **Or** Roast from frozen in the oven at 375°F (mark 5) for $1$–$1\frac{1}{2}$ hr.; lower the temperature to 350°F (mark 4) and cook for a further 1 hr.

# Stuffed lamb chops

4 thick loin chops
1 lamb's kidney, skinned and chopped
1 oz. mushrooms, chopped
salt and freshly ground black pepper

*Serves 4*

**TO MAKE:** Cut along the edge of the meat of each chop to form a pocket. Mix together the chopped kidney and mushrooms, season with salt and pepper and fill the pockets of the chops.

**TO PACK AND FREEZE:** See notes at beginning of chapter.

**TO USE:** Remove wrappings, place the chops on a lightly greased baking dish and cook from frozen in the oven at 375°F

*Lamb korma*

(mark 5) for about 1 hr. Serve garnished with onion rings.

# Apricot stuffed lamb

3–4 lb. shoulder of lamb or best end of
  neck, boned
salt and freshly ground black pepper

*For the stuffing*
½ oz. butter
½ small onion, skinned and finely chopped
¼ level tsp. curry powder
⅛ level tsp. ground ginger
2 oz. long grain rice, cooked
2 oz. dried apricots, chopped
1 level tbsp. clear honey
salt and freshly ground black pepper

*Serves 6–8*

**TO MAKE:** Wipe the meat and season with salt and freshly ground black pepper. Melt the butter and gently sauté the onion until transparent, add the curry powder and ginger and sauté a little longer. Mix together the rice, spiced onion, chopped apricots and honey and season the stuffing with salt and pepper. Spread this over the meat, roll up and tie securely with string. Weigh the stuffed joint and note the weight on the freezer label.

**TO PACK AND FREEZE:** See notes at beginning of chapter.

**TO USE:** Thaw for 24 hr. in the refrigerator. Roast at 350°F (mark 4) allowing 40 min. per lb. Serve with roast potatoes and glazed whole onions and carrots.

# Lamb korma

(*see picture above*)

1½ lb. lean boneless lamb, from leg or
  shoulder
2 oz. margarine or butter
2 onions, skinned and finely chopped
2 cloves garlic, skinned and crushed with
  1 level tsp. salt
¼ pt. natural yoghurt
2 level tbsps. tomato paste
1 level tsp. ground coriander
½ level tsp. ground ginger
pinch ground cloves
¼ level tsp. ground cinnamon
½ level tsp. ground turmeric
¼ level tsp. ground cardamom
salt
½ pt. water

*Serves 4*

**TO MAKE:** Cut the lamb into even sized pieces of about ½-in. Melt 1 oz. margarine or butter in a pan and sauté the onions and garlic until transparent. Add the yoghurt, tomato paste, spices and salt. Melt the remaining fat in another pan and sauté the meat until brown on all sides. Drain well, and stir into the spice mixture, add the water and simmer in a covered pan for about 45 min. until the meat is tender. Cool rapidly.

**TO PACK AND FREEZE:** See notes at beginning of chapter.

**TO USE: Either** Thaw for 14 hr. in the refrigerator, unwrap and turn into a saucepan and simmer gently for 20–30 min. on top of the stove. **Or** Unwrap and reheat from frozen in the oven at 375°F (mark 5) for 1¼ hr. Alternatively, reheat gently in a saucepan until thawed, then simmer for 20–30 min. Serve with rice and mango chutney.

# Somerset lamb

½ oz. butter
3 oz. button onions, skinned
¼ lb. new carrots, scraped and halved
    lengthwise
2 oz. button mushrooms
¾ lb. lean lamb (boned shoulder or leg),
    cut into cubes
2 level tbsps. flour
½ pt. light stock
¼ pt. dry cider
1 large pinch dried thyme
salt and freshly ground black pepper

*To complete dish from freezer*
4 tbsps. single cream
chopped parsley

*Serves 2*

**TO MAKE:** Melt the butter in a frying pan, and sauté the whole onions and the carrots until beginning to brown. Add the mushrooms to the pan and continue to cook for a few min. Lift the vegetables from the fat with a draining spoon and transfer to a foil lined casserole or a foil pie dish. Add the cubed meat to the fat in the pan and brown it on all sides. Stir the flour into the meat and gradually blend in the stock and cider. Bring to the boil, stirring, and season with thyme, salt and pepper. Cover and cook in oven at 325°F (mark 3) for about 1¼ hr. Cool quickly.

**TO PACK AND FREEZE:** See notes at beginning of chapter.

**TO USE: Either** Thaw for 14 hr. in the refrigerator, turn into a saucepan and reheat on top of the stove until bubbling. Alternatively reheat in the oven at 375°F (mark 5) for about 45 min. **Or** Unwrap and turn the frozen meat into a saucepan and reheat for about 30 min. Alternatively turn into a casserole, cover and cook from frozen in the oven at 375°F (mark 5) for 1¼ hr. Just before serving stir in the cream and sprinkle with chopped parsley.

# Lamb Italian style

2 lb. chump ends of lamb, trimmed
salt and freshly ground black pepper
1 oz. butter or margarine
2 onions, skinned and chopped
2 carrots, peeled and sliced
2 stalks celery, chopped
¼ pt. dry white wine
1 level tbsp. flour
½ pt. stock
¾ lb. tomatoes, skinned
pinch of dried rosemary

*To complete dish from freezer*
grated rind of 1 lemon
2 tbsps. chopped parsley

*Serves 4*

**TO MAKE:** Lightly season the meat with salt and pepper. Melt the butter or margarine in a frying pan and slowly brown the lamb all over. Set it aside on kitchen paper to drain. Fry the vegetables in the reheated pan fat until pale golden brown, then remove them and put them to drain on kitchen paper. Put the meat and vegetables in a saucepan or flameproof casserole, add the wine, cover and simmer for 20 min. Skim off any surplus surface fat.

Blend the flour with a little stock to a smooth cream. Add the remainder of the stock and pour it over the meat. Halve the tomatoes, discard the seeds and add the flesh to the pan with the rosemary. Bring to the boil, cover and simmer for a further 1½ hr. or until meat is tender. Skim off any surplus fat again. Cool quickly.

**TO PACK AND FREEZE:** See notes at beginning of chapter.

**TO USE:** Unwrap and reheat from frozen in a covered casserole in the oven at 375°F (mark 5) for about 2 hr. Just before serving, stir in the lemon rind and sprinkle with chopped parsley.

## PORK

# Basic pork casserole

6 lb. lean boned pork
1½ lb. onions, skinned and sliced
1 pt. stock
salt and freshly ground black pepper

*Makes 4 casseroles, each to serve 4*

**TO MAKE:** Cut the pork into ½-in. cubes. Layer with the rest of the ingredients in 1 or 2 large casseroles. Cook in the oven at 325°F (mark 3) for 1¾ hr. or until tender. Cool quickly.

**TO PACK AND FREEZE:** See notes at beginning of chapter.

**TO USE: Either** Unwrap and thaw for 15 hr. in the refrigerator, add any remaining ingredients (see variations below) and reheat in the oven at 375°F (mark 5) for about 45 min. Adjust seasoning. **Or** Unwrap and reheat from frozen in the oven at 375°F (mark 5) for about 1¼ hr. Add any remaining ingredients when the casserole has thawed. Adjust seasoning.

### CHINESE PORK

¼ quantity basic pork casserole
15-oz. can beanshoots, drained
3 tbsps. soy sauce
a little more stock if necessary

Thaw the casserole and stir in the extra ingredients 30 min. before the end of the reheating time.

### HONEYED PORK WITH APRICOTS

¼ quantity basic pork casserole
15-oz. can apricots, drained
3 level tbsps. clear honey

Thaw the casserole and stir in the extra ingredients 30 min. before the end of cooking.

### DINNER PARTY PORK

¼ quantity basic pork casserole
8-oz. can prunes, drained and stoned
3 tbsps. red wine
3 tbsps. soured cream

Thaw the casserole and add the prunes and wine 30 min. before the end of the reheating time. Stir in the soured cream just before serving.

### CRUSTY PORK CASSEROLE

¼ quantity basic pork casserole
2½ oz. sage and onion stuffing mix

Thaw the casserole. Make up the stuffing according to the directions on the packet and spoon it on to the top of the casserole 30 min. before the end of the reheating time.

# Pork and prune hot pot

4 oz. dried prunes
1 lb. lean boneless pork, leg or blade bone
1 lb. leeks, washed and sliced
1 large onion, skinned and sliced
salt and pepper
1 level tbsp. cornflour
¼ pt. stock (made from pork bones or a
   chicken stock cube)

*To complete dish from freezer*
1 oz. sage and onion stuffing mix
1 medium size cooking apple, peeled,
   cored and sliced

*Serves 4*

**TO MAKE:** Leave prunes to soak overnight in cold water. Next day: cut the meat into small pieces. Cut the soaked prunes in half and remove the stones. Put the pork, leeks, onion and prunes in layers in a casserole, seasoning each layer with salt and pepper.

Blend the cornflour with the stock and add to the casserole. Cover and cook slowly in the oven at 325°F (mark 3) for 1½ hr. or until meat and vegetables are tender. Cool.

**TO PACK AND FREEZE:** See notes at beginning of chapter.

**TO USE:** Unwrap and allow to thaw overnight in the refrigerator. Cover and reheat in the oven at 350°F (mark 4) for 40 min. Meanwhile make up the stuffing as directed on the packet. Remove the casserole from the oven, increase the temperature to 400°F (mark 6). Spread the stuffing over the top of the hot pot and cover with apple slices. Return to the oven and cook until the apples are browned. Serve from the dish with green vegetables or potatoes.

# Braised pork and red cabbage

2 lb. unsalted belly pork
salt and freshly ground black pepper
½ oz. butter
1 tbsp. cooking oil
1 lb. cooking apples, peeled and cored
1 lb. red cabbage, trimmed and finely
  shredded
2 level tbsps. flour
3 tbsps. wine vinegar
1 pt. stock

*Serves 6*

**TO MAKE:** Remove the rind from the pork and cut away the bones, leaving the pork in one piece. Cut the meat lengthwise into strips ½-in. thick. Cut each piece in half again. Season with salt and pepper. In a frying pan, melt the butter with the oil, and heat until bubbling. Fry the pork until well browned on all sides, reduce the heat and cook a further 10–15 min. Cut the apples into rough slices. Line a deep, straight-sided casserole with foil, or use a large foil container. Place a third of the red cabbage in the bottom, add some apple and then some meat. Continue to layer up, finishing with apple. Blend the flour with the vinegar, gradually add the stock; adjust the seasoning and bring to the boil, stirring. Cook for 2–3 min. before pouring over the casserole. Cook, covered, in the oven at 350°F (mark 4) for about 2 hr. or until the pork is tender. Cool rapidly.

**TO PACK AND FREEZE:** See notes at beginning of chapter.

**TO USE:** Unwrap and reheat from frozen in a covered dish in the oven at 350°F (mark 5) for about 1½ hr.

# Cherry topped pork and ham loaf

(*see picture opposite*)

½ lb. boned blade of pork
½ lb. gammon
1 oz. fresh brown breadcrumbs
4 tbsps. milk
1 large onion, skinned and chopped
1 tbsp. chopped parsley
1 level tsp. made mustard
1 large egg, beaten
salt and pepper

*To complete dish from freezer*
8-oz. can stoned red cherries
½ level tbsp. cornflour
large pinch salt
pinch cinnamon
pinch nutmeg

*Serves 4*

**TO MAKE:** Remove the skin and fat from pork and gammon and mince the flesh. Put the breadcrumbs and milk in a bowl. Add the meat with the onion, parsley, mustard, egg, salt and pepper. Mix well. Line a loaf tin 4½ in. by 8½ in. (top measurement) with greaseproof paper. Spoon the meat mixture into the prepared tin.

**TO PACK AND FREEZE:** Cover the tin and freeze rapidly until firm. Remove the loaf from the tin and wrap in foil, overwrap in polythene, seal and label. Return loaf to the freezer.

**TO USE:** Unwrap the loaf, replace in the original tin and leave in the refrigerator to thaw for 24 hr. Bake in the oven at 350°F (mark 4) for 1½ hr. Meanwhile drain the cherries, make the juice up to ¼ pt. with cold water and put it into a pan. Blend the cornflour with a little of the juice, return it to

*Cherry topped pork and ham loaf*

the pan with the salt, cinnamon and nutmeg. Cook over a medium heat, stirring until thick, then add the cherries. Pour the dripping from the loaf, turn the loaf on to a warm plate, spoon some of the sauce over the top and serve the rest separately.

# Boston baked beans

1¼ lb. fat belly of pork
6 oz. butter beans, soaked overnight in
  cold water
2 onions, skinned and sliced
1½ level tsps. dry mustard
1–2 level tsps. salt
good pinch pepper
1 level tbsp. soft brown sugar
2 tbsps. black treacle
1–2 tbsps. cider vinegar
pinch ground cinnamon
pinch ground cloves

*Serves 4*

TO MAKE: Cut the pork into 1-in. cubes. Drain the beans and put with the onions and pork into a large casserole, pour in just enough fresh water to cover the beans, and stir in the remaining ingredients. Cover tightly and cook in the oven at 325°F (mark 3) for 3–4 hr. Stir occasionally, adding more water if the beans dry out whilst cooking. Cool. Remove any fat from the top of the casserole.

TO PACK AND FREEZE: See notes at beginning of chapter.

TO USE: Either Unwrap and thaw for 8–10 hr. at room temperature or 24 hr. in the refrigerator. Reheat in the oven at 375°F (mark 5) for about 45 min. Or Reheat from frozen in the oven at 375°F (mark 5) for 1¾ hr.

# Mustard pork casserole

1½ lb. fresh belly pork
1 level tsp. made mustard
1 oz. lard
2 large onions, skinned and sliced
1 level tbsp. plain flour
1 chicken stock cube
½ pt. boiling water
juice of ½ small lemon
2 level tsps. chopped parsley
1 tsp. Worcestershire sauce
salt and pepper

*Serves 4*

TO MAKE: Wipe the meat, remove the rind, bones and excess fat. Cut the meat into very thin slices and spread with mustard. Melt the lard in a pan and fry the meat until browned. Remove the meat from the pan, reheat the fat and add the onions, fry until golden. Remove from pan. Stir the flour into the remaining fat in the pan. Cook for 2 min. Dissolve the stock cube in boiling water, remove the pan from the heat and stir in the stock, lemon juice, parsley and Worcestershire sauce. Return the pan to the heat, bring to the boil and stir until the sauce thickens. Adjust seasoning. Put the pork and onions in a large casserole and pour the sauce over the top. Cover the casserole, place on a baking sheet and cook in the oven at 325°F (mark 3) for $1\frac{1}{4}$ hr. or until the pork is tender. Cool quickly.

TO PACK AND FREEZE: See notes at beginning of chapter.

TO USE: Unwrap and reheat from frozen in the oven at 350°F (mark 4) for $1\frac{1}{2}$ hr.

# Pork and cranberry curry

$1\frac{1}{4}$ lb. pork fillet (tenderloin)
1 oz. butter
2 tbsps. cooking oil
2 large onions, skinned and chopped
1 clove garlic, skinned and crushed
1 level tbsp. flour
1 level tbsp. curry powder
2 level tbsps. tomato paste
juice of 1 lemon
3 cloves
bayleaf
8-oz. can cranberry sauce
$\frac{1}{2}$ pt. water
salt and pepper

*Serves 4*

TO MAKE: Slice the meat into $\frac{1}{4}$-in. slices. Heat the butter and oil in a pan and sauté the meat, transfer it to a casserole, add the onion and garlic to the pan and fry until pale golden. Remove the pan from the heat and stir in the flour, curry powder, tomato paste, lemon juice and add the cloves, bayleaf and cranberry sauce. Gradually stir in the water, season with salt and pepper. Bring to the boil, stirring, and pour over the meat in the casserole. Cover and cook in the oven at

325°F (mark 3) for 1 hr. Leave to cool.

TO PACK AND FREEZE: See notes at beginning of chapter.

TO USE: Unwrap, turn into a saucepan and thaw gently on top of the stove. Simmer for a further 30 min. and serve with rice.

# Gammon and apricot casserole

1 oz. lard
$1\frac{1}{2}$ lb. lean gammon, cut into $\frac{1}{2}$-in. cubes
1 large onion, skinned and roughly chopped
6 oz. long grain rice
1 pt. chicken stock
3 oz. dried apricots, roughly chopped
1 bayleaf
salt and pepper

*Serves 4*

TO MAKE: Melt the lard in a pan and fry the gammon on all sides to seal. Remove the meat from the pan and place in a casserole. Add the onion to the remaining fat and fry gently for 3 min., stir in the rice and cook for a further 3 min. Add the stock to the pan, bring to the boil, add the apricots, bayleaf, salt and pepper. Pour over the gammon and mix well. Cover the casserole and cook in the oven at 325°F (mark 3) for 1 hr. Cool.

TO PACK AND FREEZE: See notes at beginning of chapter.

TO USE: Unwrap and reheat from frozen in the oven at 325°F (mark 3) for $1\frac{1}{4}$–$1\frac{1}{2}$ hr. Fork the rice through from time to time. *Maximum storage time:* 6 weeks.

# Bacon chops en croûte

(*see picture opposite*)

1 level tbsp. plain flour
1 level tsp. soft brown sugar
1 level tsp. dry mustard
freshly ground black pepper
4 bacon chops
few drops Tabasco sauce
8-oz. can pineapple rings, drained
6 oz. Bel Paese cheese, cut into 4 slices
13-oz. pkt. frozen puff pastry, thawed

*Serves 4*

*Bacon chops en croûte*

**TO MAKE:** Mix together the flour, sugar, mustard and pepper in a small polythene bag. Toss the bacon chops in this mixture, one at a time. Place the bacon on a board, sprinkle each chop with a little Tabasco sauce, place a pineapple ring in the centre and top with a slice of cheese. Roll out the pastry to about $\frac{1}{16}$-in. thick. Cut into 4 squares, large enough to encase the chops. Place a chop in the centre of each piece of pastry. Moisten the edges of the pastry and fold over to cover the meat completely. Press to seal.

**TO PACK AND FREEZE:** See notes at beginning of chapter.

**TO USE:** Unwrap, place on a baking sheet and cook from frozen in the oven at 425°F (mark 7) for $\frac{3}{4}$–1 hr. Serve with buttered courgettes and new potatoes.

*Maximum storage time:* 6 weeks.

# Baconburgers

$\frac{3}{4}$ lb. lean bacon, minced (bacon pieces are ideal)
1 onion, skinned and finely chopped
1 tbsp. tomato ketchup
1 oz. fresh white breadcrumbs
1 small egg, beaten
salt and pepper

*To complete dish from freezer*
4 oz. Cheddar cheese, cut into 4 slices
4 baps

*Serves 4*

**TO MAKE:** Mix the minced bacon, onion, tomato ketchup and breadcrumbs in a bowl with the beaten egg. Season well with salt and pepper and divide the mixture into four and shape into $\frac{1}{2}$-in. thick round cakes.

**TO PACK AND FREEZE:** See notes at beginning of chapter.

**TO USE:** Preheat the grill, cook the frozen baconburgers until brown on both sides, lower the heat of the grill and continue to cook for 15–20 min. Keep hot. Cut the baps in half and toast the insides. Put a bacon-burger on the lower half of each and top with a slice of cheese. Put under the grill and cook for 2 min. or until the cheese has melted. Sandwich with the top of the bap.

*Maximum storage time:* 1 month.

# Bacon pudding

1 lb. (trimmed weight) lean back bacon
$\frac{1}{2}$ oz. butter
1 large onion, skinned and chopped
pepper and salt
large pinch mixed dried herbs
8 oz. self-raising flour
4 oz. shredded suet

*Serves 4–6*

**TO MAKE:** Lightly grease a 3-pt. pudding basin. Remove the rinds from the bacon and fry the rinds gently in a pan to extract fat. Remove rinds from pan, add the butter and fry onion gently for about 5 min. until tender but not golden. Drain. Mince the bacon, add onion, pepper and mixed herbs. Sift the flour with some salt into a bowl, stir in suet and mix to a fairly stiff dough with cold water. Roll out four rounds: one very small to fit the base of the basin, two medium ones and one large one to fit the top of the basin. Put the small round in the base, and add some of the bacon mixture. Place another round of pastry on top. Repeat layers of bacon and pastry using all the bacon and ending with a pastry layer on top. Cover basin with greaseproof paper then foil.

**TO PACK AND FREEZE:** Cover basin with another piece of foil, seal, label and freeze.

**TO USE:** Steam or boil pudding from frozen for 2 hr. Serve with green vegetables and a rich gravy or tomato sauce.

# Basic beef casserole

(*see colour picture facing page 128*)

6 lb. stewing steak
1 oz. seasoned flour
1½ oz. lard
1½ lb. onions, skinned and chopped
1½ pt. stock
salt and freshly ground black pepper
bouquet garni

*Makes 4 casseroles, each to serve 4*

**TO MAKE:** Cut the meat into ½-in. cubes, and toss in the seasoned flour. Heat the lard in a large pan and fry the onions until beginning to brown. Remove from the pan, draining well. Fry the meat a little at a time until brown all over. Place the meat and vegetables in a large casserole, or several smaller ones. Sprinkle any remaining flour into the fat left in the pan, stir in the stock and bring to the boil, season with salt and pepper, add bouquet garni and pour over the meat. Cover and cook in the oven at 325°F (mark 3) for 1½–2 hr.

**TO PACK AND FREEZE:** See notes at beginning of chapter.

**TO USE: Either** Unwrap and thaw for 15 hr. in the refrigerator. Add any remaining ingredients (see variations below). Reheat in the oven at 375°F (mark 5) for about 45 min. **Or** Unwrap and reheat from frozen in the oven at 375°F (mark 5) for about 1¼ hr. Add any remaining ingredients before the end of the reheating time according to recipe.

### STROGANOFF

¼ quantity basic beef casserole
¼ lb. sliced and sauté button mushrooms
2 tbsps. brandy
salt and pepper
¼ pt. soured cream
chopped parsley

Thaw the casserole. 30 min. before the end of reheating time, add the mushrooms and brandy to the basic casserole. Season if necessary. Just before serving, stir in the soured cream, reheat without boiling, adjust seasoning and sprinkle with chopped parsley.

### CURRY

¼ quantity basic beef casserole
8 oz. apple, peeled and chopped
1 oz. sultanas
1 level tbsp. curry powder
1 level tsp. ground turmeric

Thaw the casserole and 30 min. before the end of reheating time, add the remaining ingredients. Serve with rice and curry accompaniments.

### BEEF WITH HERBY DUMPLINGS

¼ quantity basic beef casserole
4 oz. self-raising flour
2 oz. shredded suet
½ small onion, skinned and finely grated
½ level tsp. mixed dried herbs
salt and pepper

Thaw the casserole. Mix the remaining ingredients together, with sufficient cold water to make a soft dough. Divide into 16 portions and roll them into small balls, using a little flour. Add to the basic casserole 15 min. before the end of the reheating time. Cover the pan or casserole with a tightly fitting lid. If using frozen dumplings—add to the casserole and cook them from frozen for the last 30 min. of cooking time.

### BEEF PAPRIKA

¼ quantity basic beef casserole
8-oz. can tomatoes, drained
2 caps canned red pimiento, sliced
½ level tsp. paprika pepper

Thaw the casserole and add the remaining ingredients 20 min. before the end of the reheating time. Serve with rice or noodles, and a green salad.

# Pressed salt beef

3 lb. lean salted brisket or silverside
   of beef
1 onion, skinned
1 large carrot, peeled and sliced
1 bayleaf
sprigs of parsley
8 peppercorns

*Serves 10–12*

TO MAKE: Order the meat in advance so that it can be put into brine.

Put the brisket into cold water; bring to the boil and throw away the water. Replace in fresh cold water to cover. Add the onion, carrot, bayleaf, parsley and peppercorns. Cover and simmer for about 3 hr. or until the beef is tender. Drain the meat, fit it snugly into a casserole or foil lined tin, spoon a few tbsps. of the cooking liquid over and place a board or plate on top with a heavy weight. Cool rapidly.

TO PACK AND FREEZE: See notes at beginning of chapter.

TO USE: Remove polythene, loosen foil wrappings and thaw for 48 hr. in the refrigerator or 24 hr. at room temperature. Serve cold.

# Curried steak pie

1 tbsp. cooking oil
1 onion, skinned and chopped
1 lb. raw minced beef
8-oz. can tomatoes
2 level tsps. curry powder
$\frac{1}{2}$ level tsp. ground ginger
2 level tbsps. chutney
1 level tsp. tomato paste
2 oz. sultanas
salt and freshly ground black pepper
8 oz. shortcrust pastry (i.e. made with
    8 oz. flour etc.)

*Serves 6*

TO MAKE: Heat the oil in a frying pan and sauté the onion for a few min. until transparent. Add the minced beef to the pan and cook until brown all over. Add the remaining ingredients except pastry, bring to the boil, reduce the heat and simmer for a few min. Cool. Roll out two thirds of the pastry and use it to line a 10-in. foil pie plate. Spoon the mince into the pie crust. Roll out the remaining pastry to make a lid. Damp the edges of the pastry and seal the lid in position.

TO PACK AND FREEZE: See notes at beginning of chapter.

TO USE: Unwrap and brush the pie with milk or beaten egg to glaze it and cook from frozen in the oven at 400°F (mark 6) for $\frac{3}{4}$–1 hr.

# Hearty beef casserole

1$\frac{1}{2}$ lb. leg or shin of beef
3 tbsps. wine vinegar
1 pt. tomato juice
bouquet garni
salt
1 level tsp. paprika pepper
$\frac{1}{4}$ Dutch cabbage, sliced (approx. 1 lb.)
3 small onions, skinned and sliced
2 green peppers, seeded and sliced
$\frac{1}{2}$ lb. potatoes, peeled and sliced
2 carrots, peeled and sliced
freshly ground black pepper

*Serves 4*

TO MAKE: Remove any gristle from the meat and cut into small pieces. Mix the vinegar, tomato juice, bouquet garni, salt and paprika and pour half into the base of a large saucepan. Add all the other ingredients and the meat. Pour the remaining juice over the top. Bring to the boil, lower the heat, cover and simmer for about 3 hr. Cool quickly.

TO PACK AND FREEZE: See notes at beginning of chapter.

TO USE: Either Unwrap and thaw for 15 hr. in the refrigerator. Reheat in the oven at 350°F (mark 4) for about 1 hr. Or Unwrap and reheat from frozen in the oven at 375°F (mark 5) for 2 hr.

# Cider beef with mushrooms

1$\frac{1}{2}$ lb. shin of beef, trimmed and cubed
3 onions, skinned and sliced
$\frac{1}{2}$ pt. cider
$\frac{1}{2}$ oz. margarine
1 level tbsp. flour
$\frac{1}{4}$ lb. button mushrooms
salt and pepper

*Serves 4*

TO MAKE: Marinate the meat and onions in the cider overnight. Drain and reserve the marinade. Melt the margarine in a frying pan and sauté the meat and onion until the meat is brown and sealed. Transfer to a 2½-pt. casserole. Remove the pan from the heat and stir the flour into the sediment in the frying pan, then gradually stir in the cider marinade. Return to the heat, bring to the boil, add the mushrooms and season with salt and pepper. Pour over the meat, cover and cook in the oven at 325°F (mark 3) for 3 hr. Cool rapidly.

TO PACK AND FREEZE: See notes at beginning of chapter.

TO USE: Either Unwrap and thaw overnight in the refrigerator. Reheat in the oven at 375°F (mark 5) for about 45 min. Or Unwrap and cook from frozen in the oven at 375°F (mark 5) for 2 hr.

# Beef roulades

1½ lb. lean minced beef
1 large egg
salt and freshly ground black pepper
½ level tsp. dried oregano
6 oz. mature Cheddar cheese, grated
½ oz. lard or dripping
14-oz. can tomatoes
2 oz. onion, skinned and grated
¼ pt. water

*To complete dish from freezer*
chopped parsley

*Serves 4*

TO MAKE: Work the meat with the egg, salt, pepper and oregano in a bowl with a wooden spoon; divide it into 4 portions. On a floured board, pat each portion out to a 6-in. square and cover each with a quarter of the cheese. Roll up firmly and cut each in half crosswise. Fry the meat briskly in the melted fat until golden brown. Transfer to a shallow 3-pt. casserole. Add the tomatoes to the pan, work the residue off the base of the pan with a wooden spoon and add the onion and water. Bring to the boil, season and pour over the meat. Cover with a lid and cook in the oven at 325°F (mark 3) for about 45 min. Cool quickly.

TO PACK AND FREEZE: See notes at beginning of chapter.

TO USE: Either Unwrap and thaw overnight in the refrigerator, reheat in the oven at 350°F (mark 4) for 30 min. or until bubbling well. Or Unwrap and reheat from frozen in the oven at 350°F (mark 4) for 1½ hr. Sprinkle with chopped parsley to serve.

# Tender beef casserole

2 lb. stewing steak
2 tbsps. wine vinegar
2 oz. dripping
1½ oz. flour
¼ lb. button onions, skinned
¼ lb. button mushrooms, stalks removed
¼ lb. streaky bacon, rinded and diced
1 pt. beef stock
bayleaf
1 level tbsp. tomato paste
bouquet garni
salt and pepper

*Serves 6*

TO MAKE: Cut the meat into neat pieces and place in a polythene bag with the vinegar. Toss the meat inside the bag and put the bag in a deep bowl; leave overnight to marinate. Melt the dripping in a frying pan, drain the meat and dry on absorbent paper (reserve juices). Toss in the flour and fry until sealed and brown on all sides. Transfer to a casserole. Add the onions, mushrooms and bacon to the pan and fry for 5 min. Place with the meat and juices in the casserole. Pour the stock into the frying pan, stir to loosen the sediment and add the bayleaf, tomato paste and bouquet garni. Adjust the seasoning. Bring to the boil and pour over the meat. Cover tightly and cook in the oven at 325°F (mark 3) for about 1½ hr. Discard the bayleaf and bouquet garni and cool rapidly.

TO PACK AND FREEZE: See notes at beginning of chapter.

TO USE: Either Unwrap and thaw in the refrigerator for 24 hr. Reheat in the oven at 375°F (mark 5) for ¾–1 hr. Or Unwrap and reheat from frozen in the oven at 350°F (mark 4) for 2 hr.

# Moussaka cauliflower style

1 oz. butter
8 oz. onions, skinned and sliced
1 large pinch ground cinnamon
2 level tbsps. tomato paste
1 lb. raw minced beef
½ level tsp. salt
½ level tsp. pepper
6 oz. mushrooms, sliced
1 large cauliflower, cooked and divided into florets
½ pt. cheese sauce (see page 63)

*To complete dish from freezer*
1 oz. cheese, grated
chopped parsley

*Serves 4*

TO MAKE: Melt ½ oz. of the butter in a pan, add the onions and fry slowly until soft and lightly coloured. Stir in the cinnamon, tomato paste and mince. Cook, stirring frequently, until the beef has separated and coloured. Season with salt and pepper. Sauté the mushrooms in the remaining ½ oz. butter. Layer the mince and mushrooms in a 2½-pt. casserole or foil container, cover with the cauliflower florets, and pour the cheese sauce over. Cool rapidly.

TO PACK AND FREEZE: See notes.

TO USE: Unwrap and reheat from frozen in the oven at 375°F (mark 5) for about 2 hr.; 30 min. before the end of the cooking time, sprinkle with grated cheese. Garnish with the chopped parsley just before serving.

# Bobotie

1 onion, skinned and finely chopped
a little lard or dripping
1 slice white bread, soaked in a little milk
1 oz. stoned raisins, finely chopped
1 lb. lean beef, minced
1 oz. blanched almonds, chopped
2 tsps. vinegar
2 level tsps. Demerara sugar
3 level tsps. curry powder
2 eggs
¼ pt. milk
salt and freshly ground black pepper

*Serves 4*

TO MAKE: Fry the onion slowly in the lard or dripping, until well browned. Squeeze out the soaked bread. Add the chopped raisins to the beef, with the onions, bread, almonds, vinegar, sugar, curry powder, and 1 egg, beaten. Stir well. Turn the mixture into a 2½-pt. pie dish or foil container, ensuring that the dish is well filled. Beat the other egg with the milk, season with salt and pepper and pour over the meat mixture. Stand the pie dish in another container with hot water half-way up and cook in the oven at 375°F (mark 5) for about 1 hr., until lightly set. Cool quickly.

TO PACK AND FREEZE: See notes at beginning of chapter.

THE USE: Unwrap and reheat from frozen in the oven at 350°F (mark 4) for 1 hr. Serve hot or cold with fluffy rice, mango chutney and a green salad.

# Beef naranja

2 lb. blade bone/chuck steak
1½ oz. margarine
4 oz. streaky bacon rashers, rinded and chopped
1 oz. flour
¼ pt. red wine
½ pt. stock
pinch thyme
½ bayleaf
salt and pepper
2 sticks celery, chopped
grated rind of 1 lemon

*To complete dish from freezer*
6 slices fresh orange

*Serves 6*

TO MAKE: Cut the meat into large cubes. Melt the margarine in a frying pan and fry the meat on all sides until brown. Transfer the meat to a large casserole. Fry the bacon, and when golden add the flour and cook until it begins to colour. Gradually stir in the wine and stock and bring to the boil. Add thyme, bayleaf, salt and pepper. Pour the sauce over the meat, cover and cook in the oven at 325°F (mark 3) for 2 hr. Remove from the oven, add the chopped celery and lemon rind. Return to the oven, reduce temperature to 300°F (mark 2) and cook for a further 30 min. Cool rapidly.

**TO PACK AND FREEZE:** See notes at beginning of chapter.

**TO USE: Either** Unwrap and thaw for about 15 hr. in the refrigerator. Reheat in the oven at 375°F (mark 5) for 45 min. **Or** Unwrap and reheat from frozen in the oven at 350°F (mark 4) for 1½ hr. Garnish with orange slices.

# Beef Milano

½ oz. butter
6 oz. onion, skinned and sliced
1 small clove garlic, skinned and finely chopped or crushed
2 level tbsps. tomato paste
½ level tsp. salt
½ level tsp. pepper
1 lb. raw minced beef
4 oz. short macaroni, cooked
½ lb. tomatoes, skinned and sliced
½ pt. cheese sauce (see page 63)

*To complete dish from freezer*
2 large tomatoes, sliced
small bunch parsley or watercress

*Serves 4*

**TO MAKE:** Heat the butter in a pan, add the onion and fry slowly until soft and lightly coloured. Stir in the garlic and tomato paste. Add the salt and pepper and the beef and cook for about 15 min., stirring frequently, until the mince has separated and coloured. Lightly grease a 1-lb. loaf tin. Divide the beef and macaroni in half. Place half the pasta on the base of the tin and cover with half the mince. Place the sliced tomatoes in a layer on top, covering with the remainder of the mince. Add the remaining pasta to the cheese sauce and pour over the top of the meat.

**TO PACK AND FREEZE:** See notes at beginning of chapter.

**TO USE:** Unwrap but leave in the tin and cook from frozen in the oven at 375°F (mark 5) for 2 hr. Turn out on to a serving dish and garnish with parsley or watercress and a few extra tomato slices.

# Beef and onion igloo

*For filling*
1 oz. lard
1 onion, skinned and chopped
1½ lb. raw lean minced beef
½ oz. cornflour
2 tbsps. Worcestershire sauce
2 tbsps. water
1 level tbsp. tomato paste
salt and pepper

*For pie case*
4½-oz. pkt. instant potato
¾ pt. boiling water
2 eggs, beaten
1 oz. flour

*To complete dish from freezer*
parsley sprigs

*Serves 4–6*

**TO MAKE:** Heat the lard in a pan and fry the onion for 2 min. Add the minced beef and fry gently for 5 min. Cover the pan and cook slowly for 30 min. Blend the cornflour with the Worcestershire sauce and water and stir into the mince, with the tomato paste. Bring to the boil, cook for 1 min., season and cool.

Mix the instant potato with water and beat in the eggs and flour. Line a 7-in. diameter, loose bottomed cake tin with foil, grease well and spread one third of the potato mixture on the base. Use one third of the mixture to line the sides of the tin to a height of approximately 2½ in. Spoon in the mince and cover with the remaining one third of potato mixture. Make sure the meat is completely covered with potato. Bake at 375°F (mark 5) for 45 min. until the top of the potato is well browned. Cool rapidly.

**TO PACK AND FREEZE:** Keep in tin and wrap in foil and overwrap in polythene bag, seal and label—freeze.

**TO USE:** Unwrap and reheat in the tin from frozen in the oven at 375°F (mark 5) for 1¾ hr. Turn out. Garnish with parsley and serve with vegetables.

# Curried beef balls

2 cloves garlic, skinned
1 level tsp. chopped stem ginger
1 level tsp. ground fennel
1 level tsp. ground cinnamon
$\frac{1}{4}$ level tsp. ground cloves
freshly ground black pepper
1 lb. lean raw minced beef
4 oz. bread soaked in $\frac{1}{4}$ pt. milk
1 onion, skinned and chopped
1 level tsp. salt
1 egg, beaten
beaten egg and breadcrumbs for coating
cooking oil

*For curry sauce*
1 onion, skinned and chopped
1 tbsp. cooking oil
1 clove garlic, skinned and crushed
1 level tbsp. curry powder
1 level tsp. ground ginger
1 level tsp. turmeric
1 oz. desiccated coconut
1 pt. milk
salt and freshly ground black pepper
lemon juice

*Serves 4*

**TO MAKE:** Pound together the garlic, ginger, fennel, cinnamon, cloves and pepper with a pestle or the butt of a rolling pin. Add the beef. Squeeze out the bread and work into the beef with the onion, salt and egg. Divide the mixture into 14 balls, coat with the egg and breadcrumbs and refrigerate.

For the sauce, fry the onion in oil until soft, stir in the garlic, curry powder, ginger and turmeric. Cook gently for 5 min. Add coconut and milk and simmer until creamy. Season with salt and pepper. Fry the meat balls in enough oil to cover the base of the pan, until brown all over. Add to the sauce, cover and cook gently for about 20 min. Sharpen with lemon juice. Cool quickly.

**TO PACK AND FREEZE:** See notes at beginning of chapter.

**TO USE: Either** Unwrap and thaw overnight in the refrigerator, turn into a large saucepan and heat on top of the cooker until bubbling. **Or** Unwrap and reheat from frozen in the oven at 350°F (mark 4) for $1\frac{1}{2}$–$1\frac{3}{4}$ hr.

# Beef and chestnut casserole

3–$3\frac{1}{2}$ lb. chuck steak
2 oz. flour
2 oz. lard
1 onion, skinned and sliced
15-oz. can tomatoes
4-oz. can pimientos, drained and sliced
$\frac{1}{4}$ pt. hot water
1 beef stock cube
$\frac{1}{4}$ pt. red wine
salt and pepper
3 oz. sliced garlic sausage

*To complete dish from freezer*
10-oz. can whole chestnuts, drained
butter
chopped parsley

*Serves 6*

**TO MAKE:** Trim any fat from the steak and cut the meat into 2-in. pieces. Toss in flour. Melt the lard in a frying pan and fry the meat, browning a few pieces at a time. Transfer the meat to a large casserole. Add the onion to the pan and fry gently until beginning to brown. Stir in any excess flour, then gradually add the tomatoes and pimientos. Blend the water and stock cube and add with the wine to the frying pan. Bring to the boil and adjust seasoning. Cut the sausage into strips and add to the beef in the casserole; pour the sauce over. Cover the casserole and cook in the oven at 325°F (mark 3) for about $2\frac{1}{2}$ hr. Cool.

**TO PACK AND FREEZE:** See notes at beginning of chapter.

**TO USE: Either** Unwrap and thaw for 12 hr. at room temperature. Reheat on top of the cooker in a flameproof casserole for about 30 min. until bubbling. **Or** Unwrap and reheat from frozen in the oven at 350°F (mark 4) for about $1\frac{1}{2}$–$1\frac{3}{4}$ hr. As the food thaws, fork it over gently to separate the pieces.

Sauté the drained chestnuts in the melted butter until browned. Add to the casserole and garnish with parsley.

# Bavarian beef

1 oz. margarine
1½ lb. shin of beef, cubed
1 lb. onions, skinned and chopped
1 level tbsp. flour
8-oz. can tomatoes
¾ pt. stock
½ lb. smoked sausage, sliced
salt and pepper

*Serves 4*

**TO MAKE:** Melt the margarine in a large frying pan, fry the meat quickly until brown on all sides. Drain and place in a 2½-pt. casserole. In the frying pan, sauté the onions until beginning to brown, drain and add these to the meat. Off the heat, stir in the flour, add the can of tomatoes and gradually stir in the stock. Bring to the boil, add the sausage slices, season with salt and pepper and cook for a few min., stirring. Pour over the meat. Cover and cook in the oven at 325°F (mark 3) for 3 hr. Cool quickly.

**TO PACK AND FREEZE:** See notes at beginning of chapter.

**TO USE: Either** Unwrap and thaw for 12–15 hr. in the refrigerator. Reheat in the oven at 375°F (mark 5) for 45 min. **Or** Unwrap and reheat from frozen in the oven at 375°F (mark 5) for 2 hr.

# PÂTÉS

## Rilettes de porc

2 lb. frozen belly or neck of pork, rinded
   and boned, thawed
salt
1 lb. back pork fat
1 clove garlic, bruised
bouquet garni
freshly ground black pepper
water

*Serves 4–6*

**TO MAKE:** Rub the meat well with salt and leave it to stand for 4–6 hr. Cut the meat into thin strips along the bone cavities. Put these into a casserole with the pork fat also cut into small strips. Bury the garlic and bouquet garni in the centre, season with a little pepper and add 3 fl. oz. water. Cover with a lid and cook in the oven at 300°F (mark 2) for about 4 hr. Discard the bouquet garni and garlic. Season well, strain the fat from the meat. Pound the meat with the butt of a rolling pin then pull into fine shreds with 2 forks. Pile into a foil lined casserole or rigid foil container. Pour the fat over the top of the rilettes. Cover with foil, cool rapidly.

**TO PACK AND FREEZE:** See notes at beginning of chapter.

**TO USE:** Remove polythene and loosen the foil. Thaw overnight in the refrigerator, and then return to room temperature.

*Maximum storage time:* 2 months.

## Thrifty pâté

1¼ lb. lean belly pork
½ lb. pigs' or ox liver
¼ lb. lean streaky bacon, rinded
4 oz. onion, skinned and chopped
1 small clove garlic, skinned and crushed
1 level tsp. salt
freshly ground black pepper
pinch of ground coriander

*Serves 6*

**TO MAKE:** Remove the rind and any bones from the belly pork and dice the meat. Rinse the liver under cold running water. Dry on absorbent paper. Cut into fairly large pieces. Mince the pork, liver, bacon, onion and garlic together three times. Work in the salt, pepper and coriander. Turn the mixture into a 2-pt. terrine or small casserole and place in a small roasting tin half full of water. Cover and bake in the oven at 300°F (mark 2) for about 1½ hr. Remove the lid, lay a double sheet of foil over the top, add weights and weight down until quite cold, preferably in a refrigerator. Remove weights and covering.

TO PACK AND FREEZE: Turn the pâté out when quite cold. Wrap in foil and a heavy gauge polythene bag, seal and label. Freeze.

TO USE: Remove polythene, loosen foil and thaw out overnight in a cool place. Slice and serve cold.

*Maximum storage time:* 1 month.

# Country meat loaf

2 lb. lean raw minced beef
2 oz. rolled oats
1 small clove garlic, skinned and crushed
½ level tsp. mixed herbs
1 tbsp. Worcestershire sauce
3 level tsps. Dijon mustard
¼ pt. water
1 beef stock cube
freshly ground black pepper
¼ level tsp. salt
1 egg

*To complete loaf from freezer*
2 level tsps. cornflour
14-oz. can tomato juice
salt and freshly ground black pepper
chopped parsley to garnish

*Serves 4–6*

TO MAKE: In a bowl, combine the beef, oats, garlic, herbs, Worcestershire sauce and mustard. Heat the water and dissolve the beef stock cube, add to the other ingredients stirring until fully blended. Adjust the seasoning. Break the egg into the mixture and bind together.

Line an 8½-in. by 4½-in. (top measurement) loaf tin with non-stick paper and lightly press the mixture in. Cover with foil and bake in the oven at 400°F (mark 6) for about 40 min. Cool and invert on to a piece of foil.

TO PACK AND FREEZE: Wrap in foil and overwrap in polythene, seal and label.

TO USE: Remove wrappings and replace loaf in the tin. Reheat from frozen in the oven at 350°F (mark 4) for 2 hr.

Blend the cornflour with a little juice from the can. Add the remainder, bring to the boil and cook, stirring until clear. Adjust seasoning. Turn out the hot meat loaf on to a serving dish, pour tomato sauce over, sprinkle with parsley and serve.

# Rich liver pâté

½ pt. milk
1 bayleaf
sprig dried thyme
3 peppercorns
6 oz. thin rashers streaky bacon, rinded
1 lb. pig's liver
4 oz. fat streaky bacon, rinded
1 clove garlic, skinned
1 small onion, skinned
1 oz. butter
1 oz. flour
salt and pepper

*Serves 4–6*

TO MAKE: Put the milk in a pan, add the bayleaf, thyme and peppercorns. Bring almost to boil, remove from the heat, cover and leave to stand.

Line the base and sides of a 2-lb. loaf tin with the thin rashers of bacon. Wash the liver and drain it. Mince the liver and fat bacon twice with the garlic and onion. Strain the milk, discarding the bayleaf and thyme. Melt butter in a pan, stir in the flour and cook gently for 2 min. Remove from the heat, stir in the flavoured milk and return to the heat and bring to the boil, stirring. Cook for 2 min. Beat the minced liver mixture into the sauce, season with salt and pepper and spoon the mixture into prepared loaf tin and smooth the top. Stand the loaf tin in a roasting tin containing 1 in. of water. Cover and cook in the oven at 325°F (mark 3) for about 2 hr. until firm. Cool in the tin.

TO PACK AND FREEZE: Turn out the pâté, wrap and freeze—see notes at beginning of chapter.

TO USE: Remove polythene, loosen the foil wrapping and thaw for about 48 hr. in the refrigerator or about 24 hr. at room temperature. Serve sliced with salad or freshly made toast.

## Herb pâté

1 lb. pig's liver
¾ lb. streaky bacon rashers or pieces
1 small onion, skinned and finely chopped
pinch of ground black pepper
¼ level tsp. ground mace
pinch of grated nutmeg
¼ level tsp. mixed dried herbs
1 small egg, beaten

*Serves 6*

**TO MAKE:** Wash the liver well; cut the rind from the bacon and set it aside, remove any excess fat and put on one side. Fry the bacon rind and fat trimmings gently until the fat runs. Discard the rinds and pieces of fat and add the onion to the pan. Fry the onion until soft and transparent, for about 4 min. Mince the liver and bacon three times to give a fine smooth texture. Put in a large bowl with the onion, pepper, mace, nutmeg and mixed herbs. Add the egg and mix the ingredients together until well blended. Spoon the mixture into an 8½-in. by 4½-in. loaf tin (top measurement) lined with greaseproof paper. Cover the top with aluminium foil and place in a roasting tin containing 2 in. water. Cook in the oven at 350°F (mark 4) for 1¾ hr. or until firm. Leave in the tin with a weight on top to become cold.

**TO PACK AND FREEZE:** Turn out, remove the paper, wrap and freeze according to notes at beginning of chapter.

**TO USE:** Remove polythene, loosen foil wrapping and thaw for up to 24 hr. at room temperature or 48 hr. in the refrigerator. Serve with hot toast.

## Beef and bacon mould

2 lb. chuck steak
¾ lb. back bacon, rinded
4 oz. fresh white breadcrumbs
2 oz. onion, skinned and chopped
3 tbsps. chopped parsley
8 tbsps. water
1 beef stock cube
3 level tsps. powdered gelatine
salt and pepper

*Serves 10*

**TO MAKE:** Trim the beef and remove any excess fat. Mince the beef and bacon together. In a bowl, combine the meats, breadcrumbs, chopped onion and parsley. Place the water, crumbled stock cube and gelatine in a basin over hot water and leave until the gelatine has dissolved. Stir the liquid into the meat mixture. Season well with the salt and pepper and press into a greased 3½-pt. pudding basin. Make a small depression in the centre. Cover with greased foil, securing it with string. Place in a pan with boiling water to come halfway up the basin and boil gently for about 1½ hr. Top up with more boiling water as necessary. When the mould is cooked, allow to cool until warm, then loosen the foil and add a 2-lb. weight to press the mould into the cooking liquid. Chill.

**TO PACK AND FREEZE:** Remove any surplus fat, ease round the edge and turn out. Wrap and freeze—see notes at beginning of chapter.

**TO USE:** Remove polythene, loosen the foil and thaw for at least 24 hr. in the refrigerator.

# Basic oxtail casserole

3 oz. margarine or butter
4 oxtails, cut up
6 onions, skinned and sliced
3 oz. plain flour
2¼ pt. stock
1 level tsp. dried mixed herbs
2 bayleaves
6 tsps. lemon juice
8 carrots, peeled and sliced
salt and freshly ground black pepper

*Makes 4 casseroles, each to serve 4*

**TO MAKE:** Melt the margarine or butter and fry the oxtail until golden brown on all sides, frying a few pieces at a time and transferring to a very large casserole. Fry the onions and add to the meat. Remove from the heat, sprinkle the flour into the fat remaining in the pan, gradually add the stock, herbs and lemon juice, bring to the boil and pour over the meat. Add the carrots and season with salt and pepper; cover and cook in the oven at 375°F (mark 5) for 30 min. then reduce the temperature to 300°F (mark 1–2) and simmer for 2½ hr. Chill and skim off the surplus fat.

**TO PACK AND FREEZE:** See notes at beginning of chapter.

**TO USE: Either** Unwrap and thaw for about 15–18 hr. in the refrigerator. Reheat in the oven at 350°F (mark 4) for 1 hr., adding any remaining ingredients (see variations below) in the last 30 min. **Or** Unwrap, cover lightly with foil or a lid and reheat from frozen in the oven at 400°F (mark 6) for about 1½ hr. Add any remaining ingredients to the dish when it has thawed and continue to cook for a further 30 min.

### OXTAIL PAYSANNE

¼ quantity basic oxtail casserole
8-oz. can tomatoes
1 level tsp. dried basil
8 stuffed olives

Thaw the casserole and add the remaining ingredients 30 min. before the end of re-heating time.

### OXTAIL AU VIN

¼ quantity basic oxtail casserole
3 oz. streaky bacon rashers, rinded and chopped
4 oz. button mushrooms
4 tbsps. red wine

Thaw the casserole. Gently sauté the bacon in a dry frying pan, add the mushrooms and continue to cook for a few min. Add to the casserole with the red wine, 30 min. before the end of the reheating time.

### DEVILLED OXTAIL CASSEROLE

¼ quantity basic oxtail casserole
1 level tbsp. tomato paste
½ level tsp. paprika pepper
2 tbsps. Worcestershire sauce
few drops Tabasco sauce
1 level tsp. made mustard
1 level tbsp. horseradish sauce

Thaw the casserole and add the chilli powder, tomatoes and beans 30 min. before the end of the reheating time.

### CHILLI OXTAIL CASSEROLE

¼ quantity basic oxtail casserole
1 level tsp. chilli powder
8-oz. can tomatoes
15¼-oz. can red kidney beans

Thaw the casserole. Mix all the remaining ingredients together and add to the casserole 30 min. before the end of the reheating time.

# Pressed tongue

1 salted ox tongue, approx. 3 lb.
8 peppercorns
1 carrot, peeled and sliced
1 onion, skinned and studded with 3 cloves
1 bayleaf

*Serves 10–12*

TO MAKE: Wash the tongue and allow to soak for 24 hr. if highly salted. Put the tongue into a saucepan, just cover with cold water, cover with a lid and bring to the boil. Remove any scum with a spoon. Add the peppercorns, carrot, onion and bayleaf. Cover the pan, bring to the boil again, reduce the heat and simmer the tongue until tender, allowing approximately 1 hr. per lb. Plunge the tongue briefly into cold water.

Ease off the skin while the tongue is still hot and remove the small bones from the back of the tongue. Return the meat to the cooking liquid and leave to cool. Curl the tongue into a round, deep, foil lined cake tin. The container used should be of such a size as to take the tongue leaving few gaps. Just cover the meat with cool, strained, cooking liquid. Press either with a plate, heavily weighted, or a tongue press. Chill until the juices have set. Ease from the tin.

TO PACK AND FREEZE: Wrap and freeze— see notes at beginning of chapter.

TO USE: Thaw for about 1 day in the refrigerator, unwrap and slice.

*Maximum storage time:* 2 months.

# Savoury liver jalousie

(*see picture below*)

1 level tsp. dried thyme
8 oz. plain flour
a pinch of salt
2 oz. margarine
2 oz. lard
cold water to mix

*For filling*
1 oz. dripping
1 onion, skinned and chopped
6 oz. lambs' liver, sliced
6 oz. bacon rashers, rinded and chopped
2 oz. mushrooms, chopped
1 small cooking apple, peeled, cored and
  chopped
1 tbsp. chopped parsley
salt and freshly ground black pepper

*To complete dish from freezer*
beaten egg or milk to glaze
$\frac{1}{2}$ pt. tomato sauce (see page 67)

*Serves 6*

TO MAKE: Sift the thyme with the flour and salt. Rub in the fat, and add water to mix to a soft dough. In a pan, melt the dripping, lightly fry the onion, liver and bacon. Drain, turn on to a board and chop

*Savoury liver jalousie*

finely. Add the mushrooms, apple, parsley, salt and pepper. Roll out half the pastry into a rectangle 12 in. by 8 in. and place it on a baking sheet. Turn the filling on to the pastry base, spreading it evenly to within $\frac{1}{4}$ in. of the edges. Roll out the remaining pastry to the same size as the base. Damp the edges of the base with water and cover with the pastry lid, pressing the sides together to seal it.

**TO PACK AND FREEZE:** Open freeze—see notes at beginning of chapter.

**TO USE:** Unwrap, place on a baking sheet and brush with beaten egg or milk and carefully slash the top of the jalousie with a sharp knife at 1-in. intervals. Cook in the oven at 400°F (mark 6) for 45 min. Reduce the temperature to 350°F (mark 4) for the last 15 min. Serve with a well flavoured tomato sauce.

# Rich casserole of hearts

4 lambs' hearts
1 oz. butter or margarine
4 oz. streaky bacon rashers, rinded and
    roughly chopped
4 oz. shallots, skinned
1 oz. flour
1 pt. rich stock
salt and pepper
bouquet garni

*Stuffing balls*
4 oz. fresh white breadcrumbs
2 oz. shredded suet
grated rind of 1 orange
2 tbsps. chopped parsley
1 egg beaten
salt and pepper

*Serves 4*

**TO MAKE:** Wash the hearts, cut open and remove any tubes and gristle; wash again and dry. Cut into $\frac{1}{2}$-in. slices. Melt the butter and fry the slices of heart a little at a time until lightly browned; put into a casserole. Fry the bacon and shallots until golden; add to the casserole. Add the flour to the residue in the pan and blend together; gradually pour in the stock, bring to the boil, stirring, and cook for 2–3 min. Adjust seasoning. Pour into the casserole, add the bouquet garni, cover and cook in the oven at 325°F

(mark 3) for about $2\frac{1}{2}$ hr. Skim any excess fat from the surface and remove the bouquet garni. Cool rapidly. In a bowl, combine the ingredients for the stuffing. Season to taste. Shape into 10 medium sized balls.

**TO PACK AND FREEZE:** Open freeze the stuffing balls until solid, pack in a heavy duty polythene bag, seal and label. Freeze the casserole separately—see notes at beginning of chapter.

**TO USE:** Unwrap and reheat from frozen in the oven at 375°F (mark 5) for 2 hr.; 30 min. before the end of cooking time, add the frozen stuffing balls. Before serving, skim the surface again to remove any traces of fat.

# Kidney casserole

1 lb. ox kidney
1 oz. seasoned flour
$\frac{1}{2}$ lb. pork sausages
1 oz. butter
1 onion, skinned and sliced
$\frac{1}{2}$ pt. unseasoned stock
1 level tbsp. tomato paste
salt and freshly ground black pepper
1 bayleaf
5 tbsps. dry sherry

*To complete dish from freezer*
1 tbsp. chopped parsley
freshly boiled rice

*Serves 4*

**TO MAKE:** Remove the core from the kidneys, and cut into 1-in. pieces; wash and dry thoroughly on absorbent paper, toss in flour. Cut the sausages into chunks. Melt the butter in a frying pan and sauté the onion until transparent. Drain well and transfer to a casserole. Fry the kidney and sausage in the reheated fat until beginning to brown— fry the kidney fairly slowly so as not to toughen it. Drain and place in the casserole with the onion. Remove from the heat and stir any excess flour into the pan juices. Blend in the stock and tomato paste and bring to the boil, stirring. Add salt, pepper, bayleaf and sherry. Pour into the casserole and mix with the other ingredients. Cover and cook in the oven at 350°F (mark 4) for 1 hr. until tender. Cool rapidly.

**TO PACK AND FREEZE:** See notes at beginning of chapter.

**TO USE:** Remove wrappings and reheat from frozen in a covered container in the oven at 400°F (mark 6) for about 1½ hr. Garnish with chopped parsley and serve with rice.

## RABBIT AND SAUSAGE

# Sweet-sour rabbit with prunes

(*see picture opposite*)

2 lb. rabbit, jointed
6 oz. onions, skinned and sliced
½ pt. dry white wine
½ pt. chicken stock
1 bayleaf
2 tbsps. redcurrant jelly
a few peppercorns
8 whole prunes, stoned
2 oz. seedless raisins

*To complete dish from freezer*
2 level tsps. cornflour
1 tbsp. malt vinegar
salt and freshly ground black pepper
chopped parsley
fried almonds

*Serves 4*

**TO MAKE:** Marinate the rabbit overnight with the onion and wine. Discard the onion and place the wine and rabbit in a flameproof casserole; add the chicken stock, bayleaf, redcurrant jelly and a few peppercorns. Bring to the boil. Submerge the prunes and raisins in the liquid. Cover the casserole tightly and cook in the oven at 325°F (mark 3) for about 1½ hr., or until the rabbit is tender and the prunes plump. Remove meat and discard bones. Place the meat in a foil lined casserole or rigid foil container. Add the prunes and raisins and strain the cooking liquid over. Cool.

**TO PACK AND FREEZE:** See notes at beginning of chapter.

**TO USE: Either** Thaw overnight in the refrigerator, unwrap and turn into a saucepan, reheat until bubbling and heated through. Arrange the rabbit, prunes and raisins in a warm serving dish, add the cornflour blended with the vinegar to the sauce, adjust seasoning and boil 1–2 min. Pour over the rabbit and garnish with parsley and fried almonds. **Or** Unwrap and reheat from frozen in the oven at 350°F (mark 4) for 1½ hr. Finish as above.

# Rabbit and bacon pie

¾ lb. streaky bacon, rinded and diced
1 oz. lard or dripping
2 lb. rabbit, jointed
3 level tbsps. flour
1 beef stock cube
¾ pt. water
1 oz. margarine or butter
6 oz. flat mushrooms
6 oz. rich shortcrust pastry (i.e. made with 6 oz. flour, etc.)

*Serves 6–8*

**TO MAKE:** Fry the bacon until crisp. Drain and place in a casserole. Add the lard or dripping to the pan and heat. Coat the rabbit in 2 tbsps. flour and fry on both sides until well-browned. Add the rabbit to the casserole, stir the excess flour into the pan and slowly stir in the stock made from the cube and water. Bring to the boil and pour over the rabbit. Cover tightly and cook in the oven at 300°F (mark 2) for about 1½ hr. or until the rabbit is tender.

Strip the flesh from the rabbit and place in a 9-in. (2-pt.) fluted flan dish, similar size pie dish or foil dish. Melt the margarine or butter and sauté the whole mushrooms; stir in 1 level tsp. flour and the meat juices. Bring to the boil and pour over rabbit. Cool. Roll out pastry and use to cover the rabbit.

**TO PACK AND FREEZE:** See notes at beginning of chapter.

*Sweet-sour rabbit with prunes*

**TO USE:** Unwrap the pie, brush with milk and bake from frozen in the oven at 375°F (mark 5) for 40–60 min. until golden and heated through.

*Maximum storage time:* 6 weeks.

# Sausage and tomato pie

8 oz. plain flour
large pinch of salt
2 oz. lard
2 oz. margarine
water
1 oz. butter
1 large onion, skinned and roughly
 chopped
1 lb. sausage meat
2 large tomatoes, skinned and roughly
 chopped
pinch mixed dried herbs
salt and freshly ground black pepper

*Serves 4–6*

**TO MAKE:** Sift the flour and salt into a bowl, rub in the lard and margarine until mixture resembles fine breadcrumbs. Add enough cold water to make a stiff dough. Roll out two-thirds of the pastry into a circle and use to line an 8-in. pie dish. Melt the butter in a frying pan and fry the onion until tender. Allow to cool. Add the onion to the sausage meat, with the tomatoes and mixed herbs; adjust seasoning. Put the meat mixture into the pastry lined dish. Roll out the remaining pastry, moisten the edges and use to cover the pie. Trim the pastry and decorate.

**TO PACK AND FREEZE:** Open freeze—see notes at beginning of chapter.

**TO USE:** Remove wrappings and cook from frozen in the oven at 350°F (mark 4) for 1hr. Turn the oven to 375°F (mark 5) and cook for a further 30 min.

# DISHES USING FROZEN MEAT

## Crown roast of pork with sausage stuffing

frozen crown of pork, thawed (7-ribbed
  loin of pork joints, shaped into a crown,
  see below)

*For stuffing*
$\frac{3}{4}$ oz. butter
1 large onion, skinned and finely chopped
1 stick celery, finely chopped
1 large cooking apple, peeled, cored and
  chopped
$\frac{1}{2}$ oz. fresh white breadcrumbs
8 oz. minced lean pork
4 oz. sausage meat
1 tbsp. chopped parsley
pinch dried sage
$\frac{3}{4}$ level tsp. salt
freshly ground black pepper

*Serves 12–14*

First make the stuffing. Melt the butter in a
pan, stir in the onion and cook, stirring
frequently for 5 min. Add the celery and
apple and cook without browning for 5 min.
Turn the onion and celery mixture into a
bowl and add the breadcrumbs, minced
pork, sausage meat, parsley, sage, salt and
pepper. Mix ingredients well together.

Place the crown on a piece of foil and fill
the centre with the stuffing. Weigh the
stuffed joint, and place it in a large roasting
tin. Cover the top of the stuffing with a piece
of foil and twist small pieces of foil around
the ends of the bones to prevent them
burning. Cook in the oven at 375°F (mark 5)
for 35 min. per lb. plus 25 min., basting
frequently. When the meat is cooked,
remove the foil from the bones and replace
with cutlet frills.

*To prepare the meat for a crown roast:*
Chop through the main bone between each
rib. Score across both joints $1\frac{1}{2}$ in. down from
the narrow end. Remove the fatty ends and
carefully scrape the bone tips free of flesh.
Position the joints back to back in a crown
and sew the joint ends together, using a
trussing needle and fine string. Tie the
string. This joint will need 36 hr. to thaw in
a refrigerator.

## Indian style pork chops

4 large frozen spare rib pork chops,
  thawed
1 oz. lard
2 oz. butter
1 onion, skinned and finely chopped
1 oz. flour
3 level tsps. curry powder
2 level tsps. salt
2 small apples, chopped, but not peeled
2 level tbsps. orange marmalade
1 tbsp. raisins
1 level tsp. granulated sugar
2 tbsps. lemon juice
$\frac{3}{4}$ pt. water

*Serves 4*

Wipe the chops and remove any excess fat.
Melt the lard in a large pan and brown the
chops well on both sides. Remove from pan,
pour off the fat. Add the butter to pan, melt
and stir in the onion; sauté for about 5 min.
until golden. Stir in the flour, curry powder
and salt until well blended. Add the apples,
marmalade, raisins, sugar and lemon juice.
Stir in $\frac{3}{4}$ pt. water and bring to the boil,
stirring all the time. Return the chops to the
pan, cover and simmer gently for 45 min.
until tender. Serve with boiled rice or
noodles.

## Pork with paprika sauce

8 frozen spare rib pork chops about $\frac{3}{4}$ in.
  thick, thawed
1 oz. seasoned flour
$1\frac{1}{2}$ oz. lard
2 onions, skinned and chopped
1 small clove garlic, skinned and chopped
2 level tbsps. paprika pepper
$\frac{1}{2}$ pt. chicken stock
2 level tbsps. cornflour
4 tbsps. double cream
4 tbsps. soured cream
fresh dill or chopped parsley

*Serves 8*

Toss the chops in seasoned flour, heat the
lard in a pan and fry the chops for about
3 min. on each side. Remove the chops from

the pan, add the onions and cook gently until lightly coloured; add the garlic. Remove the pan from the heat, stir in the paprika and stock; return the pan to the heat and bring to the boil, stirring all the time. Return the chops to the pan, cover and simmer very gently for 1 hr. Remove chops and keep warm. Blend the cornflour with the double and soured cream, add a little juice from the pan and pour it all into the pan. Simmer for 2–3 min., stirring constantly until sauce is thick. Add chopped dill or parsley and pour over the chops.

## Barbecued pork

8 oz. frozen pork fillet or tenderloin, thawed
1 level tbsp. flour
1 oz. butter
7½-oz. can apricot halves
1 tbsp. Worcestershire sauce
1 level tbsp. Demerara sugar
2 tbsps. cider vinegar
4 tsps. water
salt and freshly ground black pepper

*Serves 2*

Slice the pork and toss in 1 level tbsp. flour. Melt the butter in a frying pan and fry the pork until golden on both sides and cooked through—about 5 min. Drain and slice the apricot halves. Mix 4 tbsps. of the juice with the Worcestershire sauce, sugar, cider vinegar and water. Stir any excess flour into the pork and slowly stir in barbecue sauce, seasonings and fruit. Bring to the boil, stirring. Adjust seasoning and simmer for about 10 min.

## Brawn

1 frozen pig's head, salted and halved, thawed
2 frozen pig's trotters, thawed
1½ lb. frozen shin of beef, thawed
1 large bouquet garni
1 large onion, skinned
1 level tsp. ground mixed spice
thinly pared rind of 1 lemon
salt and pepper

*Serves 6–8*

Rinse the pig's head and trotters in cold water and put in a large pan with the beef, bouquet garni, onion, mixed spice and lemon rind. Barely cover with water, cover with a tightly fitting lid, bring to the boil, reduce heat and simmer for 2½–3 hr. or until head and skin are very tender. Cool a little and lift the meat on to a large dish.

Remove the flesh from the head, discarding any excess fat. Return the bones to the liquor and reduce by fast boiling to about 1 pt. Skim off the excess fat. Dice all the meat finely and place it in a 3½-pt. pudding basin. Adjust the seasoning of the stock, strain it over the meat to cover. Leave overnight in a cold place. To serve, leave plain or coat with brown breadcrumbs.

## Pork with tomato and garlic sauce

2 tbsps. cooking oil
4 frozen large loin pork chops, thawed
1 small clove garlic, skinned and finely chopped
½ level tsp. dried oregano
¼ level tsp. dried thyme
1 small bayleaf
1 level tsp. salt
4 tbsps. red wine
15-oz. can tomatoes, drained
1 level tbsp. tomato paste
1 large green pepper, seeded and cut into strips
4 oz. button mushrooms, quartered

*Serves 4*

Heat 1 tbsp. oil in a pan and brown the chops for about 3 min. on either side. Remove from the pan, and pour off most of the fat. Return the pan to the heat, add the garlic, oregano, thyme, bayleaf and salt and cook gently for 1 min., stirring continuously. Add the wine, tomatoes and tomato paste and return chops to the pan, baste them with the sauce, cover the pan and simmer gently for about 40 min. Meanwhile heat the remaining oil in another pan and fry the pepper for 5 min. Add the mushrooms and cook for a further 2 min. Add to the pan with the chops and simmer for a further 10–15 min. or until the pork is cooked and the sauce thick.

# Plum sauce spare ribs

4 frozen spare rib pork chops, thawed
2 level tsps. salt
1 oz. butter
1 small onion, skinned and chopped
14-oz. can dark red plums, stoned
1 tbsp. lemon juice
2 tbsps. soy sauce
1 tsp. Worcestershire sauce
1 level tsp. made mustard
½ level tsp. ground ginger
4 drops Tabasco sauce

*Serves 4*

Put the spare ribs into a large, shallow, ovenproof dish, sprinkle with salt and roast in the oven at 400°F (mark 6) for 30 min. Heat butter in a pan and sauté the onion. Purée the plums and juice and add to the onion with the lemon juice, soy sauce, Worcestershire sauce, mustard, ginger and Tabasco. Simmer, uncovered, for about 5 min. Drain the fat from the spare ribs and pour the plum sauce over them. Continue to cook for a further 1 hr. at 400°F (mark 6), turning half way through cooking time and basting frequently. Cover with foil if overbrowning. Serve with boiled rice.

# Pork boulangère

3½–4 lb. frozen hand or leg of pork, thawed
cooking oil
salt
2½ lb. potatoes, peeled
¾ lb. onions, skinned
½ pt. stock
1 clove garlic, skinned and crushed
½ level tsp. dried thyme
pepper
1 lb. frozen sliced courgettes
½ oz. butter, melted

*Serves 8*

Score pork rind, rub oil and salt into the skin. Roast in the oven at 425°F (mark 7) for 1 hr. Meanwhile slice the potatoes and onions about ¼ in. thick. Arrange the onions in the base of a large casserole, arrange potatoes on the top. Lay the pork on the potatoes. Pour the stock, with the garlic and thyme, over the potatoes and onions. Season well with pepper; cover with a lid. Cook in the oven at 375°F (mark 5) for about 45 min. Add the frozen courgettes, fork into the potatoes, brush with butter and continue cooking, uncovered, for about 10 min.

# Citrus pork chops

*(see picture opposite)*

4 frozen loin pork chops, thawed
salt and pepper
1 tbsp. cooking oil
1 clove garlic, skinned and crushed
2 tbsps. granulated sugar
2 level tsps. cornflour
¾ pt. stock
½ level tsp. dried rosemary
3 tbsps. lemon juice
3 tbsps. orange juice
1 small orange, cut into 4 thick slices

*Serves 4*

Sprinkle the chops with salt and pepper. Heat the oil in a large pan and add the chops. Brown the chops on both sides and remove them from the pan. To the fat remaining, add the garlic, sugar and corn-flour. Gradually stir in the stock and bring to the boil, stirring continually and cook for a few min. Stir in the rosemary, lemon and orange juice. Place an orange slice on top of each chop and arrange in the sauce. Cover the pan and cook over a low heat for 45 min. or until the pork is tender. Skim off any excess fat and serve with a green vegetable and new potatoes.

# Pork chops in cider

4 large frozen spare rib chops, thawed
1 oz. seasoned flour
2 tbsps. cooking oil
1 oz. butter
1 large onion, skinned and finely chopped
3 large cooking apples, peeled and cored
large pinch mixed dried herbs
salt and pepper
½ pt. cider

*Serves 4*

Trim any excess fat from the chops and toss in the seasoned flour. Heat the oil and

*Citrus pork
chops*

butter in a pan and fry the chops on both sides until brown; transfer them to a casserole, add the onion to the fat remaining in the pan and fry gently for 5 min. Strain off most of the fat. Add any remaining seasoned flour to the pan, chop the apples and add with the herbs, salt, pepper and cider; bring to the boil, stirring, and pour the sauce over the chops. Cover the casserole with a lid or foil. Cook in the oven at 375°F (mark 5) for 1 hr. 15 min. or until chops are cooked through. Skim off any excess fat. Serve with purée potatoes and a green vegetable.

# Pork fillet in white wine

2 lb. frozen fillet of pork, thawed
1 large clove of garlic, skinned and
  chopped
$\frac{1}{4}$ pt. dry white wine
1 bayleaf, crushed
4 whole cloves
1 level tsp. dried marjoram
salt and freshly ground black pepper
1 level tbsp. flour
$\frac{1}{2}$ level tsp. paprika pepper
1 oz. butter

*Serves 6*

Cut the pork into slices $\frac{1}{2}$ in. thick and cut these into strips approx. $1\frac{1}{2}$ in. long and $\frac{1}{4}$ in. wide. Put the garlic, wine, bayleaf, cloves, marjoram, salt and pepper in a bowl. Drop in the pork strips and turn them over until well moistened. Cover and leave to marinate for at least 4 hr. at room temperature, or 8 hr. in the refrigerator.

Remove the meat from the marinade, dry on absorbent paper, reserving the marinade; toss the meat in the flour and paprika. Melt the butter in a frying pan and brown the pork well all over. Pour in the strained marinade and bring to the boil, stirring well. Reduce the heat, cover and simmer for 20 min., until the meat is tender. Serve with rice and a green salad.

# Spiced pressed pork

3 lb. frozen lean belly of pork, thawed and
  boned
4 tbsps. vinegar
2 tbsps. corn oil
$\frac{1}{2}$ level tsp. salt
$\frac{1}{2}$ level tsp. ground cinnamon
$\frac{1}{2}$ level tsp. ground ginger
$\frac{1}{4}$ level tsp. freshly ground black pepper

*Serves 4–6*

Score the pork rind in diamond shapes with a sharp knife. Combine together the remaining ingredients. Put the pork into a shallow dish, pour the marinade over and allow to stand for at least 3 hr. at room temperature. Turn the meat frequently. Drain the joint and dry on absorbent paper. Wrap it in a loose parcel of foil and place in a roasting tin. Cook in the oven at 350°F (mark 4) for 2 hr. Remove the foil and place the meat between two flat plates; put a heavy weight on top and leave until cold. Slice the meat fairly thickly and serve with jacket potatoes and a green salad.

# Lamb paprika

$1\frac{1}{2}$ oz. butter
8 frozen middle or best end of neck chops
8 oz. onions, skinned and chopped
1 lb. tomatoes, skinned and sliced
1 tbsp. chopped parsley
1–2 level tsps. paprika pepper
salt
$\frac{1}{4}$ pt. soured cream or natural yoghurt

Heat the butter, brown the frozen chops on both sides and remove from the pan. Fry the onions in the fat for about 5 min. until golden brown; add the tomatoes and parsley, with paprika and salt to taste, replace the chops, cover and simmer gently for 1 hr. or cook in a covered casserole in the oven at 325°F (mark 3) for about $1\frac{1}{2}$ hr. Stir in the soured cream or yoghurt, adjust seasoning and reheat carefully, without boiling.

# Lamb with pineapple

4 large frozen lamb chump chops
8-oz. can pineapple rings
2 oz. button mushrooms, thickly sliced
$\frac{1}{4}$ lb. gammon, rinded and thinly sliced
salt and pepper
$\frac{1}{4}$ pt. light stock
chopped parsley

*Serves 4*

Trim the chops and grill slowly until golden brown on both sides. Place the chops in a single layer in a large ovenproof dish, lay a slice of pineapple on each chop and scatter the thickly sliced mushrooms over the top. Add the gammon. Sprinkle with salt and pepper and pour in the stock and $\frac{1}{4}$ pt. juice from the pineapple can. Cover and cook in the oven at 350°F (mark 4) for 1 hr. Serve sprinkled with parsley.

# Daube d'agneau

3 lb. frozen leg of lamb
3 fl. oz. olive oil
1 pt. red wine
4 oz. onions, skinned and chopped
3 carrots, peeled and diced
1 level tsp. salt
$\frac{1}{4}$ level tsp. freshly ground black pepper
5 sprigs parsley
2 sprigs thyme
1 bayleaf
$\frac{1}{2}$ lb. salt pork
1 pig's foot, split
2 cloves garlic, skinned and crushed
4 firm red tomatoes, skinned, seeded and quartered

*Serves 6–8*

Place the frozen lamb in a heavy gauge polythene bag, or place one thin bag inside another. Place in a large bowl, cover the meat with the oil, wine, onions, carrots, seasoning and herbs. Loosely close the polythene bag. Marinate for 18–24 hr. in the refrigerator, turning the joint in the marinade from time to time. Scrub the rind of the salt pork, place it in a pan with the pig's foot, cover with water and bring to the boil. Pour off the water and stand in fresh cold water overnight.

Next day place the lamb and the marinade in a deep casserole. Cut the rind off the pork and scissor snip into small squares; cut the flesh into narrow strips. Add with the pig's foot, garlic, and 2 tomatoes to the lamb. Cover tightly, cook in the oven at 300°F (mark 2) for about $3\frac{1}{2}$ hr. Baste the joint occasionally. Remove the lamb and keep warm. Skim the surface fat from the juices. Strain the diced vegetables and spoon them round the joint. Boil the juices rapidly to reduce them by a quarter and pour them over the meat. Garnish with the remaining tomato quarters.

*Guard of
honour*

# Guard of honour

(*see picture above*)

2 frozen best ends of neck of lamb,
    each of 7 cutlets, thawed and chined
$1\frac{1}{2}$ oz. butter
4 oz. onions, skinned and chopped
8 oz. long grain rice
1 level tsp. turmeric
1 bayleaf
1 pt. lamb stock, made from trimmings
salt and freshly ground black pepper
1–2 oz. flaked almonds
parsley sprigs

*Serves 6*

Prepare the lamb by trimming off most of the
fat, and cleaning the bones down to the nut
of meat. Then tie these together with the fat
side inside so that the bones cross. Tie
together firmly with string. Place the lamb
in a roasting tin and roast in the oven at
375°F (mark 5) for 25 min. per lb.

    Melt 1 oz. butter in a pan, sauté the
onions until tender, add the rice and cook
for 5 min. more until beginning to brown.
Stir in the turmeric, add the bayleaf and
pour over the seasoned stock; bring to the
boil, turn into a casserole and cook, covered,
in the oven alongside the lamb for the last
30 min. of cooking time. Arrange the rice
mixture on a serving dish and place the

lamb on top. Garnish the bone tips with
cutlet frills, sprinkle the nuts over the rice
and add parsley sprigs.

# Barbecued topside

3 oz. fat or dripping
$3\frac{1}{2}$–4 lb. frozen topside
1 pt. frozen tomato sauce (see page 67)
2 level tbsps. cornflour
2 tbsps. water
1 tbsp. vinegar
1 oz. soft brown sugar
a dash of Worcestershire sauce
$\frac{1}{2}$ lb. frozen green beans
1-lb. 3-oz. can whole new potatoes,
    drained

*Serves 8*

Melt the fat or dripping in a roasting tin in
the oven at 475°F (mark 8). When the fat is
hot and sizzling, place the frozen joint in it
and cook uncovered in the oven for about
20 min., turning once to seal it. Whilst the
meat is being sealed, make the barbecue
sauce. Thaw the tomato sauce in a saucepan,
stirring occasionally. Blend the cornflour
with the water to a smooth paste, add the
vinegar, then stir in the warm tomato sauce.
Add the sugar and Worcestershire sauce to
taste, bring to the boil and simmer for a few

*Cooking meat from frozen using a thermometer*

min. stirring. Pour off the fat from the meat and pour the barbecue sauce over it. Reduce the oven temperature to 350°F (mark 4), cover the meat and cook for 50 min. per lb., basting occasionally. About 20 min. before the end of the cooking time skim the surface fat from the sauce, add the beans and potatoes and complete the cooking.

# Braised sweetbreads

1 lb. frozen lambs' sweetbreads, thawed
1 large carrot, peeled and diced
1 onion, skinned and diced
2 stalks celery, diced
1 tbsp. corn oil
salt and freshly ground black pepper
approx. ½ pt. white stock
2–3 oz. green streaky bacon rashers, rinded
2 level tsps. cornflour
parsley

*Serves 4*

Soak the sweetbreads for at least 4 hr., changing the water several times. Put into fresh cold water, bring slowly to the boil, then lift out the sweetbreads, using a draining spoon. Rinse them under running cold water and remove the black veins and skin. Wrap the sweetbreads lightly in a cloth or muslin and cool, pressed between two weighted plates.

Sauté the prepared vegetables in the oil until half cooked. Place them in the base of a casserole just large enough to take the sweetbreads. Add enough seasoned stock to just cover the vegetables and arrange the sliced sweetbreads on top. Place the rashers of streaky bacon overlapping on the sweetbreads. Cover and cook in the oven at 375°F (mark 5) for 30–45 min.; baste occasionally with the stock. Increase the oven temperature to 425°F (mark 7) and remove the lid for the last 10 min. of cooking time. Strain the cooking liquid from the casserole. Mix the cornflour with a little cold water, stir into the cooking liquid and heat, stirring, until thickened. Pour the sauce over the sweetbreads and garnish with parsley.

# Liver en papillote

1 lb. frozen lambs' liver, thawed and sliced
½ oz. butter
4 oz. lean bacon rashers, rinded and chopped
1 onion, skinned and chopped
1 tbsp. finely chopped parsley
½ level tbsp. grated orange rind
salt and freshly ground black pepper
orange slices to garnish

*Serves 4*

Lightly butter four squares of foil and divide the slices of liver between the foil. Melt the ½ oz. butter and gently sauté the bacon and onion until transparent. Remove from the heat, add the parsley and grated orange rind and season lightly with salt and pepper. Pile equally on top of the liver slices. Wrap up loosely in parcels, place on a baking sheet and cook in oven at 350°F (mark 4) for 30 min.

To serve, fold back the foil and garnish with the slices of orange.

# Pot roast Polynesian

2 onions, skinned and sliced into rings
3 lb. frozen boned and rolled rib of beef
$\frac{1}{2}$ pt. canned sweetened pineapple juice
3 tbsps. soy sauce
1$\frac{1}{2}$ level tsps. ground ginger
$\frac{1}{2}$ level tsp. salt
4 stalks celery, sliced diagonally
4 carrots, peeled and sliced thinly
    lengthwise
1 level tbsp. cornflour

*Serves 6–8*

Start preparation the evening before the dish is required. Place the onions in the bottom of a large casserole and place the frozen meat on top. Combine the pineapple juice, soy sauce, ginger and salt; pour over the meat, cover, and leave to thaw in the marinade overnight in the refrigerator. Cover the casserole and cook in the oven at 325°F (mark 3) for about 2$\frac{1}{2}$ hr. or until tender. Add the celery and carrots to the pan about 10 min. before the end of the cooking time. Remove the meat and vegetables to a warm serving dish and keep warm. Blend the cornflour with 2 tsps. cold water, stir it into the gravy and cook, stirring, for a few min. until thickened. Carve the joint and serve the gravy separately.

This meat is also excellent when served cold.

## GAME

**Freezing game**

Game birds are considered to improve with freezing. Bleed the bird as soon as possible after shooting. Hang undrawn and unplucked prior to freezing. The hanging time depends on the weather and individual taste, varying from a week in 'muggy' weather to 2–3 weeks in frosty weather. If necessary a bird can be frozen for a *short* time with its feathers on, and plucked after thawing. Pluck, removing the feathers in the opposite direction to that in which they grow. Draw, wash, drain and thoroughly dry.

It is very important that only really fresh birds are hung and frozen. They should not be stuffed before freezing. Cover any protruding bones with greaseproof paper or foil, wrap in foil, and overwrap in a polythene bag, seal, label and freeze. Thaw in the wrapping, preferably in the refrigerator. Thaw a small bird overnight, birds up to 4 lb. up to 12 hr.

Water birds should be plucked, drawn and frozen quickly after killing—having removed the oil sac from the base of the tail. The storage time is 6–8 months.

Venison requires some skill in the art of butchery so it is best to ask the butcher to prepare it for you. It should be chilled, hung up for 6 days, jointed, packed, sealed, labelled and frozen. Thaw as for other meats. The storage life is up to 12 months.

A hare may be hung for 7–10 days prior to freezing if a gamey flavour is preferred. It is probably best to pack the hare in portions—discarding the more boney part (which can be used for casseroles, paté and stock). Pack, freeze and thaw as for meat, and cook as usual. Recommended storage life is 6 months.

# Braised pigeons

2 small oranges
4 frozen plump pigeons, thawed
4 rashers streaky bacon, rinded
butter
$\frac{1}{4}$ lb. mushrooms, sliced
1 small onion, skinned and sliced
2 level tsps. flour
$\frac{1}{4}$ pt. Marsala
$\frac{1}{2}$ pt. stock
salt and freshly ground black pepper
bouquet garni
chopped parsley

*Serves 4*

Peel the oranges free of all white pith and cut each in half. Insert a piece of orange in each bird. Truss the pigeons and wrap each in a rasher of bacon. Fry the birds in a little butter until beginning to brown, drain and place in a 4-pt. casserole. Fry the mushrooms and onion in the same fat. Stir in the flour, cook for 3 min. then gradually add the wine and stock. Bring to the boil, season and pour over the birds. Add the bouquet garni, cover and cook in the oven at 350°F (mark 4) for about 1½ hr. until really tender.

To serve, drain the birds from the gravy, arrange on a serving dish and strain the gravy over. Garnish with chopped parsley.

# Casserole of game

2 frozen partridges or pheasants, thawed
1 oz. butter
2 tbsps. cooking oil
1 small onion, skinned and chopped
4 oz. cooked lean veal, minced
4 oz. lean ham, minced
1 small cooking apple, peeled, cored and
    sliced
1 clove garlic, skinned and crushed
salt and pepper
1 bayleaf
4 tbsps. stock
¼ pt. single cream
2 tbsps. brandy
lemon juice (optional)
¼ lb. button mushrooms
butter
watercress

*Serves 4*

Wipe and truss the birds and fry evenly in the butter and oil until well sealed and browned. Take out of the pan and set aside. To the pan add the onion, veal, ham, apple and garlic. Cook gently for 5 min. Turn the mixture into a casserole just large enough to take the birds. Season with salt and pepper and add the bayleaf. Place the birds on top, breast sides down, and pour over the stock. Cover and cook in the oven at 325°F (mark 3) for about 1 hr., until the birds begin to be tender. Remove from the oven, discard the bayleaf and turn the birds over. Pour in the cream; warm the brandy, ignite it and add to

casserole. Cover tightly and return to the oven for 15 min.

To serve, carve the birds, arrange on a serving dish and keep hot. Adjust the seasoning of the gravy, skim off any surplus fat and sharpen if necessary with lemon juice. Sauté the mushrooms in a little butter, drain well and add to the gravy; reheat but do not boil. Spoon the gravy over the birds. Garnish with watercress and serve accompanied by game chips and celery.

# Roast grouse with gaufrette potatoes

6 oz. streaky bacon rashers, rinded
2 frozen grouse, approx. 1 lb. each, thawed
1½ lb. large potatoes, peeled
2 oz. butter
4 oz. fresh white breadcrumbs
¼ pt. chicken stock
1 level tbsp. redcurrant jelly
salt and freshly ground black pepper
watercress for garnish

*Serves 4*

Lay bacon rashers along the length of each bird and tie in position with string. Wrap the feet in foil to prevent them burning and place the birds in a roasting tin. Roast at 400°F (mark 6) for about 40 min.

To make gaufrette potatoes, slice the potatoes very thinly using a mandoline slicer and quarter-turning the potato between each slice to form a lattice. Keep the fingers tucked in over the potato, discard it when too thin. Keep the slices in water during preparation; drain and dry on absorbent paper before using. Melt the butter in a large frying pan, stir in the breadcrumbs and cook over a moderate heat until golden. Deep fat fry the lattice potatoes; drop a few at a time into the fat heated to 360°F and fry until golden. Drain on absorbent paper and keep warm until served.

Remove the birds from the roasting tin and discard the foil from the feet. Snip the string and remove the bacon. Halve the birds through the breastbone, allowing one half per person. Keep warm with the breadcrumbs and potatoes. Pour off any fat and stir the stock into the sediment in the pan. Add the redcurrant jelly, season well. Garnish the birds with watercress and serve the gravy separately.

# Pheasant with chestnuts

1 frozen pheasant, thawed and jointed
1 tbsp. olive oil
1 oz. butter
½ lb. fresh chestnuts, peeled
2 onions, skinned and sliced
3 level tbsps. flour
¾ pt. stock
1 wineglass Burgundy
salt and pepper
grated rind and juice of ½ orange
2 tsps. redcurrant jelly
bouquet garni
chopped parsley

*Serves 4*

Fry the pheasant in the oil and butter for 5–6 min., until golden brown. Remove from the pan with a slotted spoon and put in a casserole. Fry the chestnuts and onions in the reheated pan fat for about 5 min., until golden brown, and add to the pheasant. Stir the flour into the remaining fat and cook for 2–3 min. Remove the pan from the heat and add the stock and wine gradually; bring to the boil and continue to stir until it thickens. Season with salt and pepper and pour over the pheasants. Add the orange rind and juice, redcurrant jelly and bouquet garni; cover and cook in the oven at 350°F (mark 4) for about 1 hr. until the pheasant is tender. Remove the bouquet garni before serving and adjust the seasoning to taste. Sprinkle with chopped parsley.

# Quail à la crème

4 frozen quail, thawed
seasoned flour
2 oz. butter
4 oz. button mushrooms
4 tbsps. dry sherry
salt and pepper
¼ pt. soured cream
chopped parsley

*Serves 4*

Coat the quail in the seasoned flour. Melt the butter in a flameproof casserole and brown the birds evenly. Add the mushrooms and sauté them, then add the sherry, salt and pepper. Cover and cook in the oven at 375°F (mark 5) for 40 min. Stir in the soured cream, adjust the seasoning and serve sprinkled with chopped parsley.

# Partridge with cabbage

2 frozen partridges, thawed
butter or bacon fat
1 firm cabbage
4–6 oz. streaky bacon rashers, rinded
salt and pepper
1 carrot, peeled and roughly chopped
1 onion, skinned
2–3 cloves
bouquet garni
stock
1–2 small smoked sausages (optional)

*Serves 4*

This is one of the best ways of serving old partridges; the red-legged type can be cooked very well this way.

Fry the partridges in butter or bacon fat until golden brown. Cut the cabbage in quarters, removing the outside leaves and any hard pieces of stalk, wash it well and cook for 5 min. in boiling salted water, then drain. Line a casserole with the bacon and lay half the cabbage over it, with salt and pepper to taste. Put the partridges on the top, with the carrot, the onion stuck with the cloves and the bouquet garni; add the rest of the cabbage and more salt and pepper. Cover with stock, put on a lid and cook in the oven at 350°F (mark 4) for 1–1½ hr., or until the birds are tender.

As a variation, 1–2 lightly fried smoked sausages may be added to the casserole before it is put into the oven.

To serve, remove the partridges, bacon and sausages (if used) from the casserole, cut the birds into neat joints and the sausages into pieces, remove the carrot, onion and bouquet garni and cut the cabbage in shreds with a sharp knife. Serve the cabbage with the pieces of partridge on top of it and the bacon and sausage around. Serve sauce separately.

# 6 Ways With Chicken

This chapter includes recipes for complete cooked dishes ready for reheating and recipes to make the best use of frozen raw chickens and chicken portions. Plainly roasted chicken is best not frozen as it tends to dry out, but of course leftovers can be kept to use up in salads or such dishes as Chicken à la King.

### Freezing and packing raw chicken
Commercially frozen chickens are readily available and most people will buy these. However, if you prefer to freeze your own, make sure that the chickens bought are fresh, not frozen, thawed ones; check this with your supplier. The chicken should be plucked and drawn, wrapped tightly in heavy duty polythene and frozen quickly. The giblets are best wrapped and frozen separately.

### Thawing
It is important that frozen poultry be completely thawed before cooking, and then thoroughly cooked (this is in contrast to other meats, which may be cooked from frozen). Don't forget that it is perfectly acceptable to thaw a frozen chicken, cook it and then freeze the cooked dish. Frozen chickens should be left in their wrappings and allowed to thaw in the refrigerator, although the thawing may begin at room temperature. Once thawed, cook the bird at once before deterioration starts.

Provided they are thoroughly cooked, small joints of chicken may be cooked from frozen but this should not be done with whole birds.

### Thawing times
Joints require about 6 hr., chickens up to 4 lb. approximately 16 hr. and chickens over 4 lb. may take up to 24 hr.

(Turkeys under 18 lb. need 2–3 days; those over 18 lb. need 3–4 days.)

If you need to speed up the defrosting process, hold the bird under cold (not hot) running water, but this should be used only as an emergency measure.

### Packing and freezing cooked chicken dishes
This can be done in one of two ways, depending which method suits you best. Before packing, the dish should be cooled rapidly and if necessary skimmed to remove any excess fat. Leave $\frac{1}{2}$ in. headspace between the food and lid to allow for expansion during freezing.

128

*Basic beef casserole* (page 104)

**Either**

Spoon the chicken mixture into a rigid foil container. Make sure that solids are below the surface of the liquid. Place the lid lightly on top and freeze until solid. Then secure the lid, label the package and return it to the freezer. When required, unseal the lid but leave the container lightly covered with foil during thawing and reheating.

**Note:** If the lid is not airtight use freezer tape to seal it securely.

**Or**

Line with foil or a heavy polythene bag a casserole suitable for serving the dish; pour or spoon in the cooked food and freeze it until solid. Remove the pre-formed package from the casserole, cover the top and, if foil was used, over-wrap with a polythene bag. Seal and label the package and return it to the freezer. When required, unwrap the chicken and return it frozen to the original casserole. Reheat, covered, according to the recipe instructions.

**Note:** Cooking times will vary a little from those given in recipes, depending on the depth of the frozen mixture. Do not freeze more than a 2-pt. quantity in any one container.

# COOKED DISHES FOR THE FREEZER

## Basic chicken casserole

2 fresh oven-ready chickens, 4–4½ lb. each
6 level tbsps. plain flour
salt and freshly ground black pepper
1 tbsp. cooking oil
2 oz. butter
3 onions, skinned and sliced
1½ pt. chicken stock

*Makes 4 casseroles, each serving 4*

**TO MAKE:** Cut each chicken into 8 joints. Place the joints with the flour and salt and pepper in a polythene bag and toss until the joints are well coated. Heat the oil with 1 oz. of the butter and fry the onions until golden. Remove from the pan, draining well. Fry half the chicken joints until golden on all sides.

Melt the remaining 1 oz. butter and fry the rest of the chicken joints. Transfer all the chicken and onions to one or two casseroles, pour in the boiling stock, cover and cook in the oven at 350°F (mark 4) for 1–1¼ hr., until just tender. Cool rapidly.

**TO PACK AND FREEZE:** See notes at beginning of chapter.

**TO USE: Either** Unwrap and thaw for about 15–18 hr. in the refrigerator. Add any remaining ingredients (see variations below) and simmer on top of the cooker for about 30 min. Season and serve. **Or** Unwrap, cover lightly with foil and reheat from frozen in the oven at 375°F (mark 5) for about 1½ hr. Add any remaining ingredients to the dish when it has thawed, i.e. after about 1 hr.

*Chicken and pâté pie (page 132); Chicken and peach salad (page 146);*
*Barbecued roast chicken (page 142)*

## LEMON CHICKEN PILAFF

¼ quantity basic chicken casserole
½ pt. stock
grated rind of ½ lemon
¼ level tsp. turmeric
3 oz. long grain rice
2 oz. cashew nuts
½ oz. butter

Add the stock, lemon rind and turmeric to the thawed casserole. Bring to the boil and stir in the rice. Sauté the nuts in the butter and add half of them to the casserole. Continue to cook the casserole until the rice is cooked and the chicken heated through. Turn the chicken on to a heated serving dish and scatter the remainder of the nuts over. Serve with a green salad.

## CHICKEN WITH PRUNES

¼ quantity basic chicken casserole
16 prunes, boiled until tender and plump
8 oz. streaky bacon rashers, rinded
¼ pt. chicken stock
1 level tbsp. cornflour
2 tbsps. red wine

First thaw the basic chicken casserole. Remove the stones from the prunes. Stretch each rasher of bacon with a round bladed knife, cut in half and wrap around a prune, securing with a cocktail stick. Fry the prunes until beginning to brown, adding a little fat to the pan only if necessary. Remove the cocktail sticks and add the prunes and bacon to the thawed casserole with the stock. Bring to the boil and cook until heated through. Blend the cornflour with the wine, add a little stock from the pan and stir into the chicken casserole. Bring back to the boil and serve.

## PINEAPPLE CHICKEN

¼ quantity basic chicken casserole
1 green pepper, seeded and sliced
butter
8-oz. can pineapple chunks
1 level tbsp. cornflour

Gently sauté the pepper in a little butter and add to the thawed casserole with the drained pineapple chunks. Blend the cornflour with the juice from the pineapple can, stir into the casserole and cook for a few minutes, stirring.

# Peppered pollo cacciatore

3 tbsps. cooking oil
1 oz. butter
4 chicken joints, skinned
8 oz. onions, skinned and sliced
14-oz. can tomatoes
2 tbsps. chopped parsley
½ level tsp. salt
½ level tsp. dried basil
freshly ground black pepper
¼ pt. red wine
1 red pepper, seeded and sliced

*To complete dish from freezer*
chopped parsley

*Serves 4*

TO MAKE: Heat the oil and butter in a large pan. Fry the chicken pieces until brown on all sides, then remove them from the pan. Add the onion to the fat and fry until golden. Add the tomatoes with their juice, parsley, salt, basil and pepper and bring to the boil. Return the joints to the pan, add the wine and bring back to the boil. Cover the pan, reduce the heat and simmer for 25 min. Add the sliced pepper and simmer for a further 20 min. Cool rapidly and skim off any surplus fat.

TO PACK AND FREEZE: See notes at beginning of chapter.

TO USE: Unwrap and reheat, covered, from frozen in the oven at 375°F (mark 5) for about 1½ hr. Sprinkle with chopped parsley before serving.

# Chicken with cranberries and red wine

2 tbsps. cooking oil
1 onion, skinned and finely chopped
4 chicken joints
8-oz. can cranberry sauce
10½-oz. can condensed tomato soup
¼ pt. red wine
juice of 1 lemon
4 oz. button mushrooms
salt and freshly ground black pepper

*Serves 4*

TO MAKE: Heat the oil in a large frying pan and sauté the onion until golden. Add the chicken joints and fry until brown on all sides, then remove the joints from the pan. Add the cranberry sauce, soup and red wine gradually, then add the lemon juice. Bring to the boil, stirring, and boil for 2 min. Add the mushrooms, replace the chicken, cover the pan and simmer for 45 min. Cool rapidly.

TO PACK AND FREEZE: See notes at beginning of chapter.

TO USE: Unwrap and reheat from frozen in the oven at 375°F (mark 5) for about 1½ hr. Adjust seasoning.


# Cassoulet

(*see picture overleaf*)

1 lb. haricot beans
2½–3-lb. oven-ready chicken
4–6 whole cloves
1 lb. onions, skinned
bayleaf
½ lb. streaky bacon, in a piece
parsley stalks
cooking oil
freshly ground black pepper
2 oz. butter
2 cloves garlic, skinned and crushed
1 level tsp. dried thyme
½ pt. milk

*To complete dish from freezer*
½ lb. garlic sausage, skinned and sliced
salt and pepper
chopped parsley

*Serves 8*

TO MAKE: If you buy loose beans, spread them over a flat surface to remove any discoloured beans or grit. Sprinkle the beans into water and rub them through your hands to get rid of the dust. Change the water at least twice until it remains clear. Soak the beans overnight in enough fresh water to cover, then drain. Use the chicken giblets to make stock.

Push the cloves into the surface of an onion. Put the beans in a large pan and add the whole onion, bayleaf, rind cut from the bacon and parsley stalks. Cover with water. Put a lid on the pan, bring to the boil, reduce the heat and simmer for about 2 hr., when

the water should be almost absorbed. Drain well. Remove the onion, bayleaf, parsley and rind. Wipe the chicken with a damp cloth. Place it in a roasting tin and brush the surface liberally with oil. Season with pepper, either sprinkled over or rubbed into the flesh slightly. Place the chicken in the centre of the oven and cook at 400°F (mark 6) for about 1 hr. Remove from the oven and allow to cool until it can be handled. Divide the chicken into 8 portions or ease the flesh away from the bones for smaller, bite-size pieces. Remove the skin if preferred.

Still using a large pan, melt the butter. Slice the remaining onions thinly and fry them in the butter until golden brown. Remove from the heat and stir in the garlic. Cut the bacon into small cubes. Lightly stir the bacon and beans into the onion. Sprinkle in the thyme. Turn this mixture into a large (8-pt.) casserole. Using a wooden spoon, make a small depression in the beans and embed the chicken pieces. Heat the milk and pour it over while hot. Cover with a tightly fitting lid. Cook in the oven at 325°F (mark 3) for about 1 hr. Fork through the beans once or twice during cooking. Cool rapidly.

TO PACK AND FREEZE: See notes at beginning of chapter.

TO USE: Unwrap and cook, covered, from frozen in the oven at 375°F (mark 5) for 2 hr. Add the skinned, sliced sausage, pepper and salt during cooking. The cassoulet needs to be forked through occasionally and may need a little extra liquid added. Sprinkle with chopped parsley and serve with a tossed green salad.


# Chicken with prunes and pepper

1 oz. butter
1 tbsp. cooking oil
4 chicken joints
15½-oz. can prunes
¼–½ pt. white wine
1 level tbsp. flour
salt and pepper
2 green peppers, seeded and sliced

*Serves 4*

131

*Cassoulet*

TO MAKE: Heat the butter and oil in a pan and fry the chicken joints until brown on all sides. Remove the joints and place them in a casserole. Drain and stone the prunes and make up the juice to $\frac{3}{4}$ pt. with the wine. Stir the flour into the fat and cook for 1 min. Pour in the prune juice and wine, bring to the boil, stirring, and boil for 2 min. Add the prunes. Adjust the seasoning. Pour the sauce over the chicken in the casserole, cover and cook in the oven at 350°F (mark 4) for 30 min. Add the sliced peppers and cook for a further 30 min. Cool rapidly and skim well to remove surplus fat.

TO PACK AND FREEZE: See notes at beginning of chapter.

TO USE: Unwrap and reheat from frozen in the oven at 375°F (mark 5) for $1\frac{1}{2}$ hr.

# Chicken and pâté pie

(*see colour picture facing page 129*)

2$\frac{1}{2}$-lb. oven-ready chicken
1 onion, skinned
2 carrots, peeled
bouquet garni
salt and pepper
3 oz. butter
3 oz. flour
1 lb. shortcrust pastry (i.e. made with 1 lb. flour etc.)
$\frac{3}{4}$ lb. liver sausage
7-oz. can pimientos, drained and sliced
beaten egg

*Serves 12*

TO MAKE: Place the chicken in a pan with the onion, carrots, bouquet garni, salt, pepper and about $1\frac{1}{2}$ in. of water. Bring to the boil, cover and simmer for about 1 hr.

Strain off the stock and make up to 1 pt. with water if necessary. Discard the vegetables and herbs, skin the chicken and carve the flesh into thick slices. Melt the butter in a pan, remove from the heat and stir in the flour. Return to a low heat and cook for 2 min. Gradually add the chicken stock, stirring. Bring to the boil and cook for 2 min. Season with salt and pepper. Combine with the carved chicken, cover and cool.

Roll out two-thirds of the pastry and use it to line the base and sides of an $8\frac{1}{2}$-in. loose-bottomed spring-release cake tin (fitted with a plain base). Divide the liver sausage into walnut-sized pieces and roll them into balls; place these in the base of the tin and cover with pimiento. Pour the sauce and chicken over. Roll out the remaining pastry and use to make a lid for the pie; roll out any trimmings thinly and cut into strips to arrange in a lattice or similar decoration. Brush the pastry with beaten egg and cook in the oven at 400°F (mark 6) for about $1\frac{1}{4}$ hr. Cool rapidly.

**TO PACK AND FREEZE:** Remove the cold pie from the tin and freeze it uncovered until firm. Wrap in foil and overwrap with a polythene bag. Seal, label and return to the freezer.

**TO USE:** Unwrap and thaw in the refrigerator for 24 hr. Bring back to room temperature for a short time before serving.

# Chicken and orange casserole

(*see picture overleaf*)

4 chicken thighs, 6 oz. each
3 level tbsps. flour
2 oz. butter
2 oz. onion, skinned and very finely chopped
4 oz. bacon rashers, rinded and chopped
$\frac{1}{2}$ pt. chicken stock
$\frac{1}{4}$ pt. milk
finely grated rind and juice of 1 orange
1 level tbsp. finely chopped capers
salt and pepper

*To complete dish from freezer*
fresh orange segments
parsley

*Serves 4*

**TO MAKE:** Place the chicken thighs in a polythene bag with the flour and toss together until the chicken is well coated. Heat the butter in a thick frying pan. Add the chicken and fry until evenly browned on all sides. Remove from the pan and place in a casserole.

Sauté the onion and bacon in the pan for 5 min. Stir in any excess flour and cook for 2–3 min. Add the stock and milk all at once and bring to the boil, stirring. Add the orange rind, juice, capers, salt and pepper and pour the sauce over the chicken. Cover and cook in the oven at 325°F (mark 3) for about $1\frac{1}{2}$ hr. until tender. Cool rapidly.

**TO PACK AND FREEZE:** See notes at beginning of chapter.

**TO USE:** Unwrap and reheat, covered, from frozen in the oven at 375°F (mark 5) for about $1\frac{1}{2}$ hr. To serve, garnish with the fresh orange segments and parsley.

# Chicken pilaf with rich tomato sauce

8 oz. chicken livers
2 oz. butter
1 onion, skinned and finely chopped
12 oz. long grain rice
pinch grated nutmeg
$\frac{1}{4}$ level tsp. dried marjoram
1 bayleaf
$\frac{1}{4}$ pt. dry white wine
1 pt. chicken stock
salt and pepper
4 oz. freshly cooked chicken meat, cut in bite-sized pieces

*For the sauce*
1 oz. butter
1 small onion, skinned and grated
1 small carrot, peeled and grated
$\frac{1}{2}$ oz. flour
2 level tbsps. tomato paste
1 lb. tomatoes, skinned and roughly chopped, or a 15-oz. can, drained
$\frac{1}{2}$ pt. chicken stock
bayleaf
1 clove garlic, skinned and crushed
salt and pepper

*Serves 4*

**TO MAKE:** Wash the chicken livers and

*Chicken
and orange
casserole*

drain them. Melt 1 oz. butter, add the livers and cook until they are firm. Remove livers and cut into small pieces. Melt the rest of the butter in the pan, sauté the onion for a few min. Add the rice, and cook until evenly browned, stirring frequently. Add the nutmeg, marjoram and bayleaf, pour in the wine and stock. Adjust seasoning and add the livers and chicken meat. Transfer to a casserole, cover and cook in the oven at 350°F (mark 4) for about 40 min.

Meanwhile make the tomato sauce. Melt the butter and fry the onion and carrot for 3 min. Stir in the flour, tomato paste, tomatoes, stock, bayleaf and garlic. Bring to the boil, turn into a small ovenproof dish, cover and cook in the oven alongside the pilaf. Sieve, and check seasoning.

Cool the pilaf and sauce rapidly. There should be no liquid left in the pilaf.

**TO PACK AND FREEZE:** Pour the sauce into a preformed polythene bag or a small waxed or plastic carton and freeze. Pack the pilaf according to the notes at the beginning of the chapter.

**TO USE:** Unwrap the pilaf and reheat, covered, from frozen in the oven at 350°F (mark 4) for 1¾ hr. (There is no need to add water.) Place the frozen block of sauce unwrapped in a saucepan and heat through

for about 30 min. Serve the sauce separately from the pilaf but spoon 2 tbsps. sauce over the pilaf to add colour.

# Sherry chicken

**6 chicken portions**
**¼ pt. dry sherry**
**2 level tsps. caster sugar**
**3 tbsps. cooking oil**
**3 oz. mushrooms, sliced**
**1 level tsp. dried oregano**
**1 pt. chicken stock**
**salt and pepper**

*To complete dish from freezer*
**2 level tbsps. cornflour**

*Serves 6*

**TO MAKE:** Marinade the chicken pieces in the sherry and sugar for 30 min. Heat the oil in a frying pan and fry the chicken until brown on all sides. Add the mushrooms, oregano, marinade and stock. Bring to the boil, season with salt and pepper, cover and simmer for 1 hr. Cool rapidly and skim off any excess fat.

**TO PACK AND FREEZE:** See notes at beginning of chapter.

**TO USE:** Unwrap and cook, covered, from

frozen in the oven at 375°F (mark 5) for about 1½ hr. Place the chicken pieces on a warm serving dish and keep hot. Blend the cornflour with a little of the cooking liquid, add to the remaining liquid. Bring to the boil and cook until thickened. Pour over the chicken.

## Summer casserole

3¾–4-lb. oven-ready chicken
3 level tbsps. flour
3 tbsps. cooking oil
1 oz. butter
4 oz. streaky bacon, rinded and diced
¼ lb. young carrots, scraped and sliced
4 oz. onions, skinned and chopped
½ pt. chicken stock
good pinch of dried thyme
salt and freshly ground black pepper

*To complete dish from freezer*
4 oz. frozen peas
¾ lb. tomatoes, peeled and quartered, or a
  15-oz. can tomatoes, drained

*Serves 6*

TO MAKE: Joint the chicken into 12 pieces. Cut each leg in half and use the wings as separate joints; cut the breast in half along the breast bone and divide each half into 3. Serve one piece each of dark and light meat per person.

Put the joints in a polythene bag with the flour and toss together until the chicken is well coated. Heat the oil in a large frying pan, add the butter and, when on the point of browning, add the chicken skin side down and fry until well browned. Remove from the pan and place in a shallow casserole. Add the bacon, carrots and onion to the pan and sauté for 10 min. Stir in any excess flour and the stock, add thyme and adjust the seasoning. Bring to the boil, stirring, and spoon the sauce over the chicken. Cover and cook in the oven at 300°F (mark 2) for 1¼ hr. Cool rapidly and skim off any surplus fat.

TO PACK AND FREEZE: See notes at beginning of chapter.

TO USE: Unwrap, cover and reheat in the oven at 375°F (mark 5) for about 1½ hr.; add peas and tomatoes and heat for a further 1 hr.

## Chicken and chilli pie

4-lb. oven-ready fresh chicken
1 onion, skinned and quartered
1 carrot, peeled and halved
1 stick celery, chopped
salt
6 peppercorns
1 bayleaf
parsley sprigs
13-oz. pkt. frozen puff pastry, thawed

*For sauce*
2 oz. butter
6 oz. onions, skinned and chopped
½ lb. sweet red peppers, seeded and finely
  sliced
2 oz. green chillies, halved and seeded
4 level tbsps. flour
1 pt. chicken stock
4 oz. mature Cheddar cheese, grated
salt and freshly ground black pepper

*To complete dish from freezer*
beaten egg

*Serves 6*

TO MAKE: Simmer the chicken in water to come half way up the bird, with the flavouring vegetables and herbs, for about 2 hr. Remove the chicken from the pan and reduce the cooking liquid to 1 pt. by rapid boiling; strain. Melt the butter in a saucepan, sauté the onion, peppers and chillies for 10 min. If preferred, remove the chillies at this stage. Carve the chicken, cutting it into bite-sized pieces and discarding the skin. Place the meat in a 2-pt. pie dish. Stir the flour into the sauté vegetables, slowly add the strained stock, stirring, and bring to the boil. When thickened, add the cheese. Adjust seasoning and spoon over the chicken. Cool rapidly. Roll out the pastry and make a lid to cover the pie.

TO PACK AND FREEZE: See notes at beginning of chapter.

TO USE: Place on a baking sheet. Allow to stand for 10 min. Unwrap the pie and make a slit in the pastry. Brush the pastry with beaten egg and place the frozen pie in the oven at 425°F (mark 7) for 30 min. Reduce the temperature to 375°F (mark 5) and cook for a further 1½ hr. Cover the pastry with foil during the cooking to prevent over-browning.

# Arroz con pollo

(*see picture above*)

3-lb. oven-ready chicken
2 level tbsps. seasoned flour
4 tbsps. cooking oil
1 onion, skinned and chopped
14-oz. can tomatoes
6-oz. can pimientos, drained and sliced
2 chicken bouillon cubes
8 stuffed olives
6 oz. long grain rice
salt and pepper
½ lb. pork chipolata sausages, cut into
  ½-in. slices

*To complete dish from freezer*
4 oz. frozen peas
2 large tomatoes, skinned
watercress

*Serves 4*

**TO MAKE:** Cut the chicken into 8 pieces.
Put in a polythene bag with the seasoned
flour and toss together until the chicken is
well coated. Heat the oil in a large saucepan,
brown the chicken well on all sides, drain
and set aside. Add the onion to the pan and
brown. Drain the tomatoes and make up the
juice to ¾ pt. with water. Add this to the
saucepan with the tomatoes, pimientos,
crumbled bouillon cubes, olives, rice, seasoning and sausages. Mix well. Arrange the
chicken joints on top. Cover and simmer for
30 min., occasionally lifting the rice with a
fork to prevent sticking. Cool rapidly.

**TO PACK AND FREEZE:** See notes at
beginning of chapter.

**TO USE:** Unwrap and reheat, covered, from
frozen in the oven at 375°F (mark 5) for about
2 hr. Fork through occasionally and add the
peas after 1½ hr. and garnish with tomatoes
and watercress before serving.

# Spiced drumsticks

(*see picture opposite*)

6 chicken drumsticks (4 oz. each)
1 oz. seasoned flour
1 tbsp. cooking oil
1 oz. butter
2 oz. onion, skinned and finely chopped
1 level tsp. curry powder
15-oz. can apricot halves, drained and
  sliced
5 whole cloves
1½ tbsps. lemon juice
¼ pt. chicken stock

*To complete dish from freezer*
salt and freshly ground black pepper
chopped parsley

*Serves 3*

**TO MAKE:** Put the drumsticks in a polythene bag with the seasoned flour and toss together until the chicken is well coated. Heat the oil in a heavy based pan, add the butter and when on the point of browning add the drumsticks. Fry them until evenly browned all over then transfer to a casserole.

Sauté the onion in the fat until beginning to brown, stir in the excess flour with the curry powder, sliced apricots and juice, cloves, lemon juice and stock. Bring to the boil, stirring, and pour over the drumsticks. Cover and cook in the oven at 325°F (mark 3) for about 2 hr. Cool rapidly and skim off any surplus fat.

**TO PACK AND FREEZE:** See notes at the beginning of the chapter.

**TO USE:** Unwrap and reheat, covered, from frozen in the oven at 375°F (mark 5) for about 2 hr. It may be necessary to reduce the sauce by boiling fast for a few min. Adjust seasoning. Arrange the drumsticks on a serving plate and pour the sauce over. Garnish with chopped parsley. This may be served with a piped potato border, or with plainly boiled rice.

# Cherkes tavagu (from Turkey)

3–3½-lb. oven-ready fresh chicken
water
1 large carrot, peeled
1 onion, skinned
salt and pepper
1 bayleaf

*For walnut sauce*
2 oz. butter
6 oz. walnut halves
½ lb. onions, skinned and sliced
¼ level tsp. paprika pepper
¼ pt. chicken stock

*To complete dish from freezer*
½ pt. natural yoghurt

*Serves 4*

**TO MAKE:** Place the chicken in a large saucepan and cover with cold water. Cut the carrot and onion into thick slices, add to the pan together with the salt, pepper and bayleaf. Bring to the boil, then cover, reduce the heat and simmer for about 1½ hr. Leave the bird to cool in the stock for an hour.

*Spiced drumsticks*

Remove the skin from the chicken and discard it. Cut the chicken meat into chunks.

In a frying pan, heat the butter and add 2 oz. walnut halves; fry gently until light golden brown then remove from the pan. Add the onions and sauté until soft but not brown, then add the paprika and chicken stock. Grind the remaining 4 oz. walnuts, using an electric blender or a Mouli nut mill, and add these to the sauce. Adjust seasoning. Add the walnut halves and chicken and cool rapidly.

**TO PACK AND FREEZE:** See notes at beginning of chapter.

**TO USE:** Unwrap and reheat from frozen in a covered dish in the oven at 375°F (mark 5), for about 1½ hr. Stir in the yoghurt, heat through and serve.

# Chicken Maryland

2½–3-lb. oven-ready fresh chicken
3 level tbsps. seasoned flour
1 egg, beaten
dry breadcrumbs
2 oz. butter
1–2 tbsps. cooking oil

*To complete dish from freezer*
*Corn fritters*
4 oz. flour
pinch of salt
1 egg
¼ pt. milk
6-oz. pkt. frozen sweetcorn kernels
a little fat

*Fried bananas*
4 bananas
butter

*Bacon rolls*
4 rashers of streaky bacon, rinded

*Serves 4*

**TO MAKE:** Joint the chicken into 8. Coat the joints with seasoned flour, dip in beaten egg and coat with breadcrumbs. Fry the chicken in the butter and oil in a large frying pan until lightly browned. Reduce the heat and continue frying gently, turning the pieces once, for about 20 min., or until tender. Alternatively fry them in deep fat for 5–10 min. (The fat should be hot enough to brown a 1-in. cube of bread in 60–70 sec.) Drain well and cool rapidly.

**TO PACK AND FREEZE:** Wrap the chicken pieces in foil, with non-stick paper between each. Overwrap with a polythene bag, seal, label and freeze.

**TO USE:** Unwrap the chicken pieces and place them on a baking sheet. Cover with foil. While still frozen, place them in the oven at 375°F (mark 5) for about 1½ hr. Meanwhile make up a fritter batter. Sift the flour and salt together into a bowl, make a well in the centre and break in the egg. Add half the milk and beat the mixture until smooth. Gradually add the rest of the milk and beat until well mixed. Just before the chicken is ready fold the frozen corn kernels into the batter. Drop a spoonful at a time into hot fat and fry until crisp and golden, turning once. Drain well on crumpled kitchen paper and keep hot.

Peel and slice the bananas lengthways and then in half again. Fry gently for about 3 min. in a little butter until tinged brown. Keep hot. Stretch the bacon rashers with the back of a knife, cut in half, roll up and grill or fry for a few min. Keep hot.

To serve, arrange the chicken on a platter with the fritters, bananas and bacon rolls.

# RECIPES USING
# CHICKEN FROM THE FREEZER

## Oriental chicken

3½-lb. frozen chicken, thawed
3 tbsps. soy sauce
1 tbsp. cooking oil
1 oz. butter
3 oz. onion, skinned and finely chopped
1 level tbsp. flour
¼ pt. giblet stock
2 tbsps. lemon juice
3 tbsps. stem ginger syrup
1 bayleaf
6 pieces stem ginger, chopped
chopped parsley

*Serves 4–5*

Joint the chicken into 6 and remove the skin. Place the joints in a polythene bag resting in a bowl, add the soy sauce and marinade for 2 hr., turning occasionally. Remove the chicken and dry it on absorbent paper. Heat the oil and butter in a flameproof casserole, fry the chicken briskly until brown on all sides. Remove the joints from the pan and fry the onion in the remaining fat. Stir the flour into the fat, add the stock, lemon juice, ginger syrup and bayleaf. Return the chicken joints, cover and cook in the oven at 300°F (mark 2) for about 1 hr., until tender. 15 min. before serving, add the roughly chopped ginger. Return to the oven for the remainder of the cooking time. To serve, sprinkle with chopped parsley.

## Chicken roulades

4 frozen chicken breasts, thawed and boned
4 slices of cooked ham
3 oz. Gruyère cheese, grated
a few spikes of fresh rosemary
2 eggs, beaten
4 oz. fresh white breadcrumbs
2 oz. Gruyère cheese, grated
3 oz. butter
1 tbsp. corn oil

*Serves 4*

Place the chicken breasts between 2 sheets of waxed paper and beat them with a cutlet bat or rolling pin until flattened. Cut the slices of ham in half and place 2 halves on each breast, so that the ham does not overlap the chicken. Sprinkle each with a quarter of the first measure of cheese, add the rosemary and roll up tightly and neatly; secure with cocktail sticks. Brush each roulade with beaten egg then coat in a mixture of breadcrumbs and the second measure of cheese. Pat the coating on firmly and coat a second time if wished.

Melt butter with the oil in a frying pan, fry the roulades quickly until evenly browned, then reduce the heat and cook slowly for about 15 min. Drain well and serve plain or with a fresh tomato or hollandaise sauce.

## Buttery garlic chicken

3 oz. butter
2 tbsps. malt vinegar
2 tbsps. clear honey
2 cloves garlic, skinned and crushed
2 level tsps. salt
½ level tsp. dried marjoram
½ level tsp. dry mustard
freshly ground black pepper
4 frozen chicken joints, thawed
chopped chives

*Serves 4*

Warm the butter in a small saucepan until just melted; add the vinegar, honey, garlic, salt, marjoram, dry mustard and a few turns of freshly ground black pepper. Line the grill pan with foil and arrange the joints in a single layer, fleshy side uppermost. Spoon over the butter and leave for 30 min. for the flavour to penetrate.

Grill under a fierce heat for 5 min., reduce the heat and cook for a further 10 min. Turn over and cook for 10 min. Cover the joints with foil and grill for a further 15 min. Brush with the pan juices at intervals. To ensure that the chicken is thoroughly cooked, pierce the thickest part of the flesh with a skewer. If the juice runs clear, the chicken is cooked; if the juice runs pink, cook for a little longer.

*Ingredients for Chicken ratatouille*

Sprinkle with chives before serving.

# Chicken ratatouille

(*see picture above*)

2 large aubergines
salt
3½-lb. frozen chicken, thawed
½ oz. flour
8 tbsps. cooking oil
1 lb. onions, skinned and sliced
1 green pepper, seeded and sliced
1 red pepper, seeded and sliced
½ lb. courgettes, thickly sliced
½ lb. tomatoes, skinned
4 tbsps. white wine
pepper
chopped parsley

*Serves 4–5*

Wipe the aubergines and slice them thinly. Leave the slices on a flat plate, sprinkled with salt, for about an hour to extract the bitter juices. Joint the chicken into 8 and sprinkle each joint with flour.

Heat 3 tbsps. oil in a frying pan and fry the chicken joints on all sides until evenly browned. Put to one side. Heat the rest of the oil in the pan, add the drained aubergines and fry for a few min. Put into a casserole. Fry the onion, peppers and courgettes. Using a slotted spoon, transfer the vegetables to the casserole and spoon the chicken joints between the vegetables. Cover with thickly sliced tomatoes and pour the wine over. Season well with pepper, cover and cook in the oven at 325°F (mark 3) for about 45 min. Garnish with chopped parsley.

# Chicken with ginger

4 frozen chicken joints, thawed
1 oz. seasoned flour
2 tbsps. cooking oil
2 oz. butter
salt and freshly ground black pepper
1 large onion, skinned and sliced
1 level tsp. ground ginger
2 level tsps. French mustard
¾ pt. chicken stock
2 tbsps. medium sherry
¼ lb. button mushrooms, stalks removed
4 tbsps. single cream
chopped parsley

*Serves 4*

Put the chicken and seasoned flour in a polythene bag and toss together until the joints are well coated. Heat the oil and 1 oz. butter in a frying pan. Fry the chicken until evenly browned on all sides. Drain well and place in a casserole. Season with salt and pepper. Fry the onions in the pan. Stir in any excess flour with the ginger and mustard; cook for a few min. Remove from the heat, stir in the stock and sherry and bring to the boil, stirring. Adjust seasoning and pour over the chicken.

In a clean pan, melt the remaining butter and quickly sauté the mushrooms; add them to the casserole. Cover and cook in the oven at 325°F (mark 3) for about 1½ hr. Blend the cream with a little of the sauce, stir it into the casserole and reheat in the oven for 15 min. Garnish with chopped parsley.

# Danish chicken

3-lb. frozen chicken, thawed
1 clove garlic, skinned and halved
salt and freshly ground black pepper
pinch of ground ginger
3 oz. butter
½ pt. water
1 green pepper, seeded and cut into thin
    rings
2 tomatoes, peeled and thickly sliced
¼ pt. double cream
2 oz. Samsoe cheese, grated

*Serves 4*

Joint the chicken into 8 and remove the skin. Rub the joints with garlic, then season with

salt, pepper and ginger. Melt the butter in a large pan, add the chicken pieces and fry, turning until browned all over. Add the water and bring to the boil then reduce the heat, cover the pan and simmer for 30–40 min. until tender. Drain the joints, place them in a serving dish and keep warm.

Add the pepper rings to the stock in the pan and cook until soft (about 5 min.) then add the tomatoes and cook for a further few min. Drain the vegetables and spoon them over the chicken. Stir the cream into the juices in the pan and pour over the chicken. Sprinkle with cheese and place under a hot grill until golden brown. Serve with French bread or plain boiled rice.

# Chicken vol-au-vent

3½-lb. frozen chicken, thawed
1 onion, skinned and sliced
1 carrot, peeled and sliced
1 bayleaf
water
1 lb. frozen puff pastry, thawed
beaten egg
2 oz. butter
3 level tbsps. flour
¼ pt. milk
½ pt. chicken stock
3–4 level tbsps. smooth liver pâté
1 tbsp. dry sherry
salt and pepper
¼ lb. mushrooms, stalked and sliced

*Serves 6*

Place the chicken in a pan with the onion, carrot and bayleaf and just enough water to cover and simmer for about 1½ hr. Cool in the cooking liquid. Drain the chicken and reduce the stock to ½ pt. by fast boiling; strain, chill and remove any fat. Meanwhile, roll out the pastry into a 10-in. round and trim it using a pan lid. Using an 8-in. lid, cut out the centre and roll this out to another 10 in. round. Place the round on a damp baking sheet, brush the edge with egg and place the pastry band in position on top. Prick the base and brush the band only with egg. Bake at 425°F (mark 7) for about 30 min.

Strip the meat from the cold chicken and cut it into small pieces. Melt 1½ oz. butter, stir in the flour and cook for a few min.

Remove from the heat and add milk and stock. Bring to the boil, stirring. Add the pâté and sherry, salt and pepper. Sauté the mushrooms in the remaining ½ oz. butter and add to the hot sauce with the chicken. Reheat the mixture and use it to fill the hot vol-au-vent case.

# Barbecued roast chicken

(*see colour picture facing page 129*)

*For the barbecue sauce*
5 tbsps. red wine
4 tbsps. chicken stock
2 tbsps. red wine vinegar
1 tsp. soy sauce
1 tsp. Worcestershire sauce
1 level tsp. tomato paste
1 level tsp. French mustard
1 level tbsp. sugar

*For the forcemeat*
4 oz. onions, skinned and chopped
4 oz. bacon, rinded and chopped
2 oz. butter
4 oz. fresh white breadcrumbs
1 lb. pie veal, minced
pinch of fresh thyme
salt and freshly ground black pepper
1 large egg, beaten

3½–4-lb. frozen boned chicken, thawed
6 oz. sliced tongue

*Serves 8*

Combine all the sauce ingredients together in a saucepan and boil until reduced by half For the forcemeat, fry the onion and bacon in 1 oz. of the butter and add this to the breadcrumbs with the minced veal and thyme. Season with salt and pepper and bind together with the beaten egg.

Lay out the chicken on a board, skin side down. Cover with half the forcemeat. Cut the tongue into strips and place over the forcemeat; cover with the remaining forcemeat. Draw the sides of the chicken together and sew up with a trussing needle and fine string. Place in a roasting tin, brush with 1 oz. melted butter and cook in the oven at 400°F (mark 6) for 1 hr. Brush the chicken with barbecue sauce, return to the oven and baste every 15 min. for a further 1 hr.—cover loosely with foil if it starts to overbrown. Serve hot.

# Poulet en cocotte

*For the stuffing*
4 oz. sausage meat
2 level tbsps. fresh white breadcrumbs
1 chicken liver, chopped
2 level tbsps. chopped parsley

3–3½-lb. frozen chicken, thawed
salt and freshly ground black pepper
2½ oz. butter
8 oz. lean back bacon, in one slice
1 lb. potatoes, peeled
6 oz. shallots, skinned
1 lb. small new carrots, scraped
chopped parsley

*Serves 4*

In a bowl, mix together all the stuffing ingredients until well blended. Stuff the chicken at the neck end. Truss the bird as for roasting. Sprinkle with salt and pepper. Melt the butter in a large frying pan, add the chicken and fry, turning it until browned all over. Transfer the chicken and butter to a large casserole. Rind the bacon and cut it into ¾-in. cubes. Add to the casserole, cover and cook in the oven at 350°F (mark 4) for 15 min. Meanwhile, cut the potatoes into 1-in. dice.

Remove the casserole from the oven and baste the chicken. Surround it with the potatoes, shallots and carrots, turning them in the fat. Season with salt and pepper. Cover, return the casserole to the oven and cook for a further 1½ hr. Garnish with chopped parsley.

To serve, have a plate to hand for carving the bird and serve the vegetables and juices straight from the casserole.

# Chicken in a pot

1 oz. butter
½ lb. carrots, peeled and sliced
½ lb. button onions, skinned
¼ lb. streaky bacon, rinded and diced
thinly pared rind of 1 lemon
good pinch of thyme
1 clove garlic, skinned
4-lb. frozen chicken, thawed
salt and freshly ground black pepper
1 lb. Jerusalem artichokes
chopped parsley

*Serves 6*

Melt the butter in a flameproof casserole a little larger than the bird. Add the carrots, onions and bacon and sauté for 10 min. Stir in the lemon rind, thyme and garlic. Push the vegetables to the side of the casserole. Place the chicken in the casserole, surround with vegetables, season lightly with salt and pepper. Cover and cook in the oven at 300°F (mark 2) for about 2 hr. Just before the end of the cooking time, peel the artichokes, placing them in salted water to prevent discoloration. Drain, chop roughly and tuck them in amongst the other vegetables in the casserole. Return the chicken to the oven, without the lid, for a further 30 min.

Place the chicken on a serving dish, spoon the vegetables round and discard the garlic and lemon rind. Skim the fat from the juices, bring to the boil and spoon over the chicken. Garnish with parsley.

# Barbecued chicken

6 frozen chicken joints, thawed
1 oz. butter

*To coat joints*
½ level tsp. salt
½ level tsp. caster sugar
½ level tsp. ground ginger
freshly ground black pepper

*For the sauce*
1 level tbsp. cornflour
2 tbsps. malt vinegar
2 tbsps. soy sauce
1 tbsp. Worcestershire sauce
2 level tsps. caster sugar
8 tbsps. tomato ketchup
2 bayleaves
2 cloves garlic, skinned and crushed
few drops of Tabasco sauce

*To serve*
12–14 oz. noodles
chopped parsley

*Serves 6*

Place the chicken and coating ingredients in a polythene bag and toss the joints to coat them evenly. Melt the butter in a large pan and fry the joints for 10 min. to brown. Transfer them to a casserole Blend the cornflour with the other sauce ingredients in a basin and pour them over the joints. Cover

and cook in the oven at 350°F (mark 4) for 1 hr.

Cook the noodles in boiling salted water for about 12 min., drain and arrange over the base of a serving dish. Skim any surplus fat from the casserole and remove the bayleaves. Spoon the joints and sauce over the noodles and sprinkle with chopped parsley.

# Hsing jen chi ting (from China)

1 lb. frozen chicken breasts, thawed
3 tbsps. cooking oil
1 level tsp. salt
2 tbsps. soy sauce
9½-oz. can bean sprouts, drained
2–3 sticks celery, finely sliced
2 oz. button mushrooms
3½-oz. can pineapple pieces, drained
¼ pt. chicken stock
1 level tbsp. cornflour
salt and pepper
2 oz. flaked almonds, toasted

*Serves 3–4*

Cut the chicken meat into 1-in. pieces. Heat the oil in a large frying pan, add the chicken and salt. Sauté for 3-5 min. Add the soy sauce and blend well. Add the bean sprouts, celery, mushrooms, pineapple and stock. Cover and simmer for 10 min. Blend the cornflour with a little water and stir into the chicken. Bring slowly to the boil, stirring; season to taste and sprinkle with almonds. Serve hot with plain boiled rice.

# Brazilian chicken Rio

3 oranges
water
4 frozen chicken portions, thawed
salt
½ level tsp. paprika pepper
2 oz. butter
2 level tbsps. flour
½ level tbsp. soft brown sugar
¼ level tsp. ground ginger

*Serves 4*

Pare the rind from 1 orange, free of all white pith, and cut it into fine shreds. Squeeze the juice from the rinded orange and 1 other and add water to make up to $\frac{1}{2}$ pt. Divide each chicken joint into 2 sprinkle with salt and paprika.

Melt the butter in a large deep frying pan, add the chicken and brown well. Remove the joints. Stir in the flour, together with the sugar and ginger. Blend well, stir in the orange juice and rind, bring to the boil and boil for 2–3 min., stirring. Adjust seasoning. Return the chicken to the sauce, cover and simmer for about 45 min., until the chicken is tender.

Peel the third orange and divide into segments free of membrane. Add to the chicken 5 min. before the end of the cooking time. Serve with plain boiled rice.

# Chicken with peaches (from Holland)

3-lb. frozen chicken, thawed
4 oz. butter
2 tbsps. port or sherry
3 tbsps. water
1 level tbsp. cornflour
$\frac{1}{4}$ pt. water
$\frac{1}{4}$ pt. double cream
salt and freshly ground black pepper
2 bananas
3 fresh peaches, skinned and sliced

*Serves 4*

Quarter the chicken. Heat 2 oz. butter in a large pan, add the chicken and fry, turning until browned all over. Add the port or sherry and 3 tbsps. water and bring to the boil. Reduce the heat, cover the pan and simmer for about 50 min. until tender. Remove the chicken pieces from the pan and keep warm. Blend together the cornflour and $\frac{1}{4}$ pt. water and add with the cream to the pan juices. Cook over a very low heat until the sauce thickens. Season to taste with salt and pepper. Return the chicken to the sauce and keep hot over a low heat.

In a frying pan, heat the remaining 2 oz. butter and add the bananas, skinned and halved lengthwise, and the peaches. Cook over a low heat, turning, for 5 min. Arrange the chicken portions in a serving dish with the bananas and peaches and pour the sauce over. Serve with plain boiled rice or potatoes.

# Frango piri-piri

*For piri-piri*
2 fresh or dried red chillies, seeded and sliced
$\frac{1}{4}$ level tsp. dried tarragon
$\frac{1}{4}$ level tsp. dried marjoram
$\frac{1}{4}$ level tsp. dried basil
$\frac{1}{4}$ level tsp. dried thyme
1 small bayleaf, crushed
2 tbsps. raisins
salt and freshly ground black pepper
$\frac{1}{4}$ pt. olive oil
2 tbsps. lemon juice

$2\frac{1}{2}$-lb. frozen chicken, thawed
lemon wedges and parsley sprigs, to garnish

*Serves 2*

In a large bowl, mix together all the ingredients for piri-piri. Allow to stand for 2 hr.

Split the chicken down the back but do not separate the 2 halves entirely; press them out flat, using a weight if necessary. Fix with crossed skewers to maintain a good shape during grilling. Brush the chicken inside and out with some of the piri-piri.

Cook, cage side uppermost, under a moderate grill for 15 min. Turn and baste with the piri-piri at intervals until the chicken is cooked through (about a further 20 min.). Serve on a bed of plain boiled rice with a garnish of lemon wedges and parsley sprigs.

# Cider chicken with apple

2 tbsps. cooking oil
1 oz. butter
4 frozen chicken joints, thawed
salt and freshly ground black pepper
1 onion, skinned and sliced
2 level tbsps. flour
$\frac{1}{4}$ pt. cider
$\frac{1}{2}$ pt. chicken stock
4 fresh sage leaves
2 large cooking apples, peeled, cored and sliced thickly into rings

*Serves 4*

Heat together the oil and butter. Season the chicken well with salt and pepper and fry the joints gently in the fat until well browned on all sides. Drain and place them in a casserole. Fry the onion in the remaining fat, add the flour, stir well and cook for 2–3 min. Remove from the heat and stir in the cider and stock. Return to the heat and stir well until boiling. Add the fresh sage and adjust seasoning. Pour the sauce over the chicken, cover and cook in the oven at 325°F (mark 3) for about 1½ hr. Add the apple rings and cook for a further 30 min. Skim off any excess fat before serving.

# Chicken en croûte

3 oz. long grain rice
1 oz. butter
3 oz. onion, skinned and chopped
3 oz. lean streaky bacon rashers, rinded
   and diced
1 level tsp. curry powder
3 oz. dried apricots, finely chopped
1 banana, skinned and chopped
salt and pepper
3½-lb. frozen chicken, thawed
1 lb. shortcrust pastry (i.e. made with
   1 lb. flour etc.)
1 egg, beaten
¼ pt. single cream

*Serves 6*

Cook the rice in boiling salted water for about 12 min., until tender; drain. Melt the butter and sauté the onion and bacon. Stir in the curry powder, apricots and banana, cook for a few min., add the rice and season with salt and pepper. Leave until cold.

Skin the chicken and stuff it with the cold rice mixture. Secure the wings with cocktail sticks. Roll out the pastry to a large oblong about ¼ in. thick. Place the chicken in the centre of the pastry; brush each long side of the pastry with egg, bring the sides together over the breast and seal; parcel the ends and seal with egg. Roll out any pastry trimmings and use these to decorate the croûte. Glaze with beaten egg. Place on a baking sheet and bake in the oven at 450°F (mark 8) for 20 min. Reduce the heat to 375°F (mark 5) and cook for about a further 1 hr. Transfer to an ovenproof plate. Make a small hole in the centre of the croûte, pour in the cream and return the chicken to the oven for 10 min. to heat through.

# Chicken à la King

1½ oz. butter
1 onion, skinned and finely sliced
1 red pepper, seeded and diced
2 oz. button mushrooms
12 oz. frozen cooked chicken meat, thawed
10½-oz. can condensed mushroom soup
salt and pepper
6 oz. noodles
chopped parsley

*Serves 4*

Melt 1 oz. butter in a frying pan and fry the onion until lightly browned. Add the diced pepper and mushrooms and cook for 2–3 min. Cut the chicken meat into chunks and add to the pan with the soup. Adjust seasoning, cover and simmer for about 20 min. Meanwhile cook the noodles in boiling salted water for about 7 min., drain well, add the ½ oz. butter in flakes and the parsley.

Arrange the noodles down either side of a warm serving dish and place the chicken mixture in the centre.

# Poulet à l'estragon

3-lb. frozen chicken, thawed

*To cook the chicken*
½ lemon, sliced
½ level tsp. salt
freshly ground black pepper
¼ level tsp. dried tarragon
3 carrots, peeled and quartered
1 small onion, skinned and quartered
1 bayleaf
2 sprigs parsley
¾ pt. chicken stock
6 tbsps. dry white wine

*For tarragon sauce*
1 oz. butter
2 level tbsps. flour
½ level tsp. dried tarragon
2 egg yolks
5 tbsps. single cream
salt and freshly ground black pepper

*Serves 4*

Cut the chicken into 8 joints. Place them in a large pan together with the giblets, lemon, seasoning, herbs, vegetables, stock and wine. Cover and simmer on top of the cooker for about 30 min., or until the chicken is cooked. Remove the joints from the pan, discard the giblets and remove the skin from the joints; transfer the joints to a serving dish and keep hot.

Strain the cooking liquid into a measuring jug and make up to $\frac{3}{4}$ pt. with water if necessary. Heat the butter in a pan, blend in the flour and cook for 1 min. Gradually add the cooking liquid and bring to the boil, stirring. Add the tarragon and simmer for 3 min. In a bowl, blend together the egg yolks and cream. Add 4 tbsps. sauce and return it to the pan; adjust the seasoning. Pour the tarragon sauce over the chicken pieces and serve at once.

# CHICKEN SALADS

*Recipes in this section use the following dressings, which are not suitable for freezing.*

## MAYONNAISE

1 egg yolk
$\frac{1}{2}$ level tsp. dry mustard
$\frac{1}{2}$ level tsp. salt
$\frac{1}{4}$ level tsp. pepper
$\frac{1}{4}$ level tsp. sugar
approx. $\frac{1}{4}$ pt. salad oil
1 tbsp. white vinegar

Put the egg yolk into a basin with the mustard, salt, pepper and sugar. Mix thoroughly, then add the oil drop by drop, stirring briskly with a wooden spoon the whole time or using a whisk, until the sauce is thick and smooth. If it becomes too thick, add a little of the vinegar. When all the oil has been added, add the vinegar gradually and mix thoroughly.

Should the mayonnaise curdle during making, put another egg yolk into a clean basin and add the curdled sauce **very** slowly to it, in the same way as the oil was added to the first yolk.

## LEMON MAYONNAISE

Follow the recipe above, using lemon juice instead of vinegar.

## FRENCH DRESSING

$\frac{1}{4}$ level tsp. salt
$\frac{1}{8}$ level tsp. pepper
$\frac{1}{4}$ level tsp. dry mustard
$\frac{1}{4}$ level tsp. sugar
1 tbsp. vinegar
2 tbsps. salad oil

Put the salt, pepper, mustard and sugar in a bowl, add the vinegar and stir until well blended. Beat in the oil gradually with a fork. The dressing separates if left to stand and will require whipping again immediately before use. Alternatively, store the dressing in a container with a lid, so that it can be shaken vigorously just before serving.

*Note:* The proportion of oil to vinegar varies with individual taste, but use vinegar sparingly. Malt, wine, cider or herbed vinegar may be used.

# Chicken and peach salad

(*see colour picture facing page 129*)

8 oz. long grain rice
a pinch of saffron or turmeric
salt and freshly ground black pepper
3–4 tbsps. French dressing
2$\frac{1}{2}$-lb. whole cooked chicken, frozen and
  thawed
15$\frac{1}{2}$-oz. can peach halves
3 oz. cream cheese
1 oz. shelled walnuts, chopped
bunch of watercress

*Serves 4*

Cook the rice with the saffron and salt in boiling water for about 12 min. Drain well, leave until cold and toss with well seasoned French dressing. Carve the breast meat from the chicken into slices and cut into strips. Carve the dark meat and chop; add all the chicken meat to the rice. Drain the peaches, retaining 4 halves, dice the remainder and add to the rice. Beat the cream cheese until it is soft and smooth, stir in the walnuts and season. Divide the cheese mixture between the reserved peach halves, filling the hollows. Check the rice mixture for seasoning and spoon it on to a platter. Top with the peaches and garnish with watercress.

# Chicken, rice and corn salad

¾–1 lb. frozen cooked chicken meat, thawed
3 tbsps. French dressing
¼ pt. lemon mayonnaise
1 tbsp. lemon juice
1 tbsp. single cream
salt and pepper
6 oz. cooked long grain rice
12 oz. frozen sweetcorn kernels, cooked
    and cooled
1 green pepper, seeded and diced
1 Cos lettuce, separated and washed

*Serves 4*

Cut the chicken into 1-in. pieces and toss in
the French dressing. Put the mayonnaise,
lemon juice, cream and seasoning in a bowl.
Gently fold in the chicken, rice, corn and
diced pepper. Arrange the salad on a platter,
or in a bowl, lined with lettuce leaves.

# Chicken mousse

*For béchamel sauce*
¼ pt. milk
a small piece of onion, skinned
a small piece of carrot, peeled
a 2-in. piece of celery
½ a bayleaf
2 peppercorns
½ oz. butter
½ oz. flour
salt and pepper

¼ pt. aspic jelly, made from aspic jelly
    powder
8 oz. smoked streaky bacon rashers,
    rinded and chopped
12 oz. cooked chicken meat, minced
1 tbsp. lemon juice
½ level tsp. dried sage
2–3 tbsps. thick mayonnaise
½ pt. double cream, whipped

*To complete dish from freezer*
1 lettuce, separated and washed
½ cucumber, sliced
2 tomatoes, sliced
1 lemon, sliced

*Serves 4–6*

**TO MAKE:** Put the milk, vegetables and
flavourings for the sauce in a saucepan and
bring slowly to the boil. Remove from the
heat, cover and leave to infuse for 15 min.
Using the strained, infused milk, make the
sauce. Melt the butter, add the flour, cook
for 2–3 min. then stir in the milk gradually.
Bring to the boil, season and cook for 2 min.
Cool before using. Meanwhile, spoon a little
of the cooled but still liquid aspic into a
1½-pt. mould; hold it over a bowl of ice
cubes and tilt and turn the mould so that
the aspic coats the sides and sets as a lining.

Fry the bacon in its own fat then mix the
chicken, bacon, sauce, lemon juice, sage and
mayonnaise and add the remainder of the
aspic jelly. When the mixture is on the point
of setting, fold in the whipped cream. Turn
the mixture into the mould and leave to set.

**TO PACK AND FREEZE:** When the mousse
is set, leave it in the mould and wrap with
foil. Overwrap with a polythene bag, seal,
label and freeze. As the mousse contains
bacon it is wise to limit storage time to 6
weeks.

**TO USE:** Unwrap and thaw for 24 hr. in the
refrigerator. Turn out and serve on a bed of
lettuce and garnish with slices of cucumber,
tomato and lemon.

# Chicken and banana salad

6 oz. frozen cooked chicken meat,
    thawed
6 oz. cooked ham
6 oz. cooked long grain rice
2 bananas, skinned and sliced
1 small onion, skinned and finely chopped
3 tomatoes, skinned, seeded and chopped
2 tbsps. thick mayonnaise
2 tbsps. soured or double cream
salt and pepper
1 lettuce, separated and washed

*Serves 4*

Dice the chicken and ham and place in a
large bowl. Add the rice, bananas, onion and
tomatoes. Blend together the mayonnaise
and cream and adjust the seasoning. Fold
the mayonnaise through the chicken mixture
and serve the salad on a platter lined with
lettuce leaves.

# 7 Ways With Vegetables

Most vegetables freeze remarkably well, either fresh or in cooked dishes. When freezing uncooked vegetables, do make sure they are absolutely fresh either from your own garden or from a nearby grower. Shop vegetables are inevitably 2–3 days old by the time they reach the consumer and this is too late to attempt freezing. This fastidiousness is less important when preparing cooked dishes for the freezer; normally good quality shop-bought vegetables are quite acceptable.

## Packing vegetables for the freezer

When preparing fresh vegetables for the freezer it is essential to blanch them first. For details of this process in relation to individual vegetables, refer to your manufacturer's handbook or to the Good Housekeeping Home Freezer Cook Book. Once prepared and cooled, most vegetables can be packed in heavy duty polythene bags or heavy duty foil, sealed, labelled and frozen. Pack in small quantities suitable for 1 meal, as it may be difficult to break off the portions required if you pack in bulk. Vegetables cooked in bundles, like asparagus, can be frozen several bundles together, separated by non-stick paper.

If the shape of the vegetable is awkward and you do not want it broken (e.g. artichokes or broccoli), pack it in rigid boxes, either waxed, foil or plastic; you may also find boxes useful for vegetables containing a large amount of water, such as marrow or spinach—in this case leave $\frac{1}{2}$ in. headspace for expansion during freezing. Do the same for vegetable purées and juices.

**Note** If you are freezing onions or other strong smelling vegetables, pack in foil or polythene and overwrap with polythene to prevent the smell filtering out.

## Cooking frozen vegetables

When cooking commercially frozen vegetables, follow the instructions on the packet. When cooking home frozen vegetables, put the frozen vegetable in boiling salted water, bring it back to the boil and time the cooking according to the chart below, taking the cooking time from the moment the salted water returns to the boil. Alternative methods of cooking are given where suitable. Take care not to overcook frozen vegetables. The variations in time shown on the chart reflect differences in size, type and quality of the vegetable; the vegetable is cooked when it is just tender but still crisp.

**Cooking times**

| | | |
|---|---|---|
| Asparagus .. .. .. | 5–10 min. | |
| Beans, broad .. .. | 7–10 min. | |
| runner .. .. | 5–8 min. | |
| French .. .. | 5–8 min. | |
| Broccoli .. .. .. | 5–8 min. | |
| Brussels sprouts .. | 5–9 min. | |
| Carrots .. .. .. | 5–10 min. | |
| Cauliflower .. .. | 5–8 min. | |
| Corn on the cob .. | 5–8 min. | |
| Corn kernels .. .. | 3–5 min. | |
| Courgettes .. .. .. | 3 min. | |
| Marrow .. .. .. | about 5 min. | |
| Mushrooms .. .. | about 3 min. Add frozen cooked mushrooms to a prepared dish. | |
| Onions .. .. .. | Add frozen to soups and stews. | |
| Peas .. .. .. .. | 4–7 min. | |

| | |
|---|---|
| Mange-tout .. .. | 5–7 min. |
| Peppers | 5–8 min. |
| Potatoes, raw new | 7–10 min. |
| cooked new | Thaw and reheat in melted butter. |
| duchesse .. | Thaw, glaze with beaten egg and brown in the oven at 400°F (mark 6) for 10–15 min. |
| croquettes .. | Thaw and fry in deep fat for about 3–5 min., until browned; cooking time depends on the thickness of the croquettes. |
| chip .. .. | 3–5 min. |
| Spinach (leaf) .. .. | 5–7 min. |

## Packing cooked vegetable dishes for the freezer

Cool the food thoroughly before packing in one of the following ways:

### Either

Spoon the food into a casserole or other container lined with foil or a polythene bag. Freeze quickly, then remove the pre-formed package, wrap the foil or polythene over the top and overwrap with heavy duty polythene; seal, label and return to the freezer. When required, unwrap and return the frozen food to the original casserole or put it in a saucepan and follow the recipe instructions.

### Or

Spoon the food into a rigid foil container, allowing $\frac{1}{2}$ in. headspace between the food and the lid. Seal the lid with freezer tape if necessary, label and freeze. When required, unseal the lid and either complete cooking in the container or dip it in water to release the contents and turn the frozen food into a saucepan; then follow the recipe instructions.

### Or

Open freeze on a baking sheet. This method is suitable for quiches, pies, pancakes etc. Freeze the food uncovered until solid, then wrap it in foil, overwrap in polythene, seal, label and return it to the freezer. When required, unwrap and follow the recipe instructions.

# HOT VEGETABLES
## TO SERVE WITH MEAT OR FISH

## Sunshine carrots

(*see colour picture facing page 160*)

1 lb. small frozen carrots
1 level tbsp. soft brown sugar
1 level tsp. cornflour
$\frac{1}{4}$ level tsp. salt
$\frac{1}{4}$ level tsp. ground ginger
$\frac{1}{4}$ pt. orange juice from a small can of frozen concentrated juice
1 oz. butter
chopped parsley, to garnish

*Serves 4*

Cook the carrots in boiling salted water until just tender. Drain well. Meanwhile combine the sugar, cornflour, salt and ginger in a small saucepan. Blend in the orange juice and bring to the boil, stirring constantly until the mixture thickens and bubbles. Stir in the butter and when it has completely melted, pour the sauce over the carrots, tossing to coat them evenly. Garnish with chopped parsley. Serve with pork, gammon chops or boiled bacon.

## Broccoli bake

1 lb. frozen broccoli spears
3 oz. butter
2 oz. fresh white breadcrumbs
$7\frac{1}{2}$ fl. oz. soured cream
salt and freshly ground black pepper
2 eggs, hard-boiled and sliced
paprika pepper

*Serves 4*

Cook the broccoli in boiling salted water until tender. Drain and keep warm. Melt 2 oz. butter in a frying pan and fry the crumbs until evenly golden, stirring frequently. Mix in the soured cream, salt and pepper. Carefully lay the broccoli spears, each with a good coating of the crumb mixture, in an ovenproof dish, with egg slices in between. Dot the remaining butter over the top. Bake in the oven for 20 min. at 350°F (mark 4). Sprinkle with paprika just before serving. Serve with plain smoked haddock or grilled white fish.

## Broccoli with almonds

1 lb. frozen broccoli
2 oz. butter
1 onion, skinned and finely chopped
1 clove garlic, skinned and crushed with a little salt
salt and freshly ground black pepper
$\frac{1}{2}$ tsp. lemon juice
$1\frac{1}{2}$ oz. almonds, blanched, shredded and toasted

*Serves 4*

Cook the broccoli in boiling salted water until just tender, drain it well and arrange in a warm serving dish. Keep it warm. Melt the butter and gently sauté the onion and garlic until transparent and add salt, pepper and lemon juice. Spoon the onion and butter over the broccoli and scatter with the almonds, Serve with any roast meat, grilled mackerel or trout.

## Curried broad beans

1 lb. frozen shelled broad beans
2 oz. butter
4 oz. onion, skinned and finely chopped
1 level tbsp. curry powder
2 level tsps. curry paste
1 tbsp. lemon juice

*Serves 4*

Cook the beans in boiling salted water until tender, drain and keep warm. Melt the butter in a pan, sauté the onion until soft but not coloured, add the curry powder and paste and lemon juice. Cook a little longer then add the beans. Toss together until the beans are well coated with the curry mixture. Turn the beans into a warm serving dish and serve immediately with cold meats.

## Beans lyonnaise

$\frac{3}{4}$ lb. frozen shelled broad beans
6 oz. onion, skinned and thinly sliced
1 oz. butter
salt and freshly ground black pepper
poppy seeds

*Serves 2–3*

Cook the beans in boiling salted water until tender. Sauté the onion in the butter until tender but not coloured, drain the beans thoroughly and fork them through the onion. Add salt and pepper to taste. Transfer to a warmed serving dish and dust lightly with poppy seeds. Serve with grilled meats.

# Broad beans with cream

1 lb. frozen shelled broad beans
1 oz. butter
1 small onion, skinned and finely chopped
2–3 rashers streaky bacon, rinded and
  chopped
¼ pt. single cream
salt and freshly ground black pepper
chopped parsley

*Serves 4*

Cook the beans in boiling salted water until tender. Drain well. Melt the butter in a saucepan, add the finely chopped onion and cook until soft but not brown. Add the chopped bacon, cook until crisp then add the cream and broad beans. Toss over a low heat for a few min. and add salt and pepper. Serve sprinkled with chopped parsley, to accompany any bacon or pork dish.

# Sweetcorn fritters

8 oz. frozen sweetcorn kernels
2 oz. flour
pinch salt
pinch paprika pepper
1 egg
4 tbsps. milk
fat for deep frying

*Serves 3–4*

Cook the sweetcorn in boiling salted water; drain thoroughly. Sift the flour, salt and paprika into a bowl, make a well in the centre, drop in the egg and gradually add the milk, drawing in the flour from all sides and mixing and beating until smooth. Add milk until the batter coats the back of a wooden spoon. Add the corn to the batter, heat the deep fat to 360°F, drop in the batter from a spoon and cook the fritters until golden brown—about 5 min. Drain well on absorbent paper. Serve with chicken Maryland.

# French beans with cheese

1 lb. frozen French beans
½ clove garlic, skinned

*For cheese dressing*
2 oz. butter
2 level tbsps. freshly grated Parmesan
  cheese
freshly ground black pepper

*Serves 4*

Cook the French beans in boiling salted water until tender. Drain well and place them in an ovenproof dish, rubbed with the half clove of garlic. Add the butter, in shavings, sprinkle the cheese over the top and add a good turn of pepper. Place in the oven at 400°F (mark 6) for 10–15 min. to melt the cheese. Serve with grilled, baked or poached fish.

# Beans with pimiento

1 lb. frozen haricot verts
butter
lemon juice
1 cap canned pimiento, sliced

*Serves 4*

Cook the beans in boiling salted water until just tender. Drain thoroughly. Melt a little butter in the pan and add a squeeze of lemon juice. Return the beans and toss lightly. Turn into a hot serving dish and garnish with thin strips of pimiento. Serve with fish.

# Whole bean provençale

(*see colour picture facing page 160*)

6 oz. onions, skinned and thinly sliced
3 tbsps. cooking oil
1 clove garlic, skinned and crushed
14-oz. can tomatoes, drained
1 lb. frozen whole green beans
1 level tsp. dried basil
salt and freshly ground black pepper

*Serves 4*

Sauté the onions in the oil until soft but not coloured. Add the garlic and tomatoes, then the frozen beans, basil, salt and pepper. Cover and simmer gently until the beans are cooked. Alternatively cook in the oven at

350°F (mark 4) for about 30 min. Serve with roast beef or lamb.

# Mange-tout

1 lb. frozen mange-tout (sugar peas)
1 oz. butter
2 oz. mushrooms, thickly sliced
3 tomatoes, skinned and quartered
salt and freshly ground black pepper

*Serves 4*

Cook the mange-tout in boiling salted water and drain well. Melt the butter in a frying pan and when hot sauté the mushrooms for a few min. Add the mange-tout and tomatoes to the pan, mix well and season with salt and pepper. Serve with any meat or fish, as an alternative to peas.

# Sweet and sour Brussels sprouts

(*see colour picture facing page 160*)

1 lb. frozen button Brussels sprouts
2 oz. seedless raisins
2 oz. Demerara sugar
3 tbsps. wine vinegar
pinch dry mustard
salt and freshly ground black pepper
½ level tsp. cornflour

*Serves 4*

Cook the sprouts in boiling salted water until just tender. Drain them and keep warm in a serving dish. Place all the remaining ingredients in a saucepan and heat until thickened, stirring. Pour the sauce over the sprouts and serve immediately, with roast chicken or turkey.

# Minted vegetable fritters

*For batter*
6 level tbsps. self-raising flour
2 eggs
2 tbsps. milk
salt and freshly ground black pepper
pinch paprika pepper
½ level tsp. dried mint
2 tsps. Worcestershire sauce

8 oz. frozen mixed vegetables
1 tbsp. cooking oil
1½ oz. butter

*Makes 8 fritters*

Beat together the batter ingredients. Add the frozen vegetables. Heat the oil in a frying pan, add the butter and, when melted, fry the batter a spoonful at a time, for about 5 min. on either side until golden. Serve at once, with grilled lamb cutlets or cold roast lamb.

# COLD VEGETABLE DISHES TO SERVE WITH MEAT OR FISH

# Sprout salad

(*see picture opposite*)

1 lb. frozen Brussels sprouts
1½ oz. shelled walnuts
1 red dessert apple
1 celery heart
½ small onion, skinned and finely chopped
1 tsp. lemon juice
salt and pepper
4 tbsps. thick mayonnaise

*Serves 4*

Cook the sprouts in boiling salted water until tender, drain and refresh under cold running water. Chop the walnuts and apple roughly (do not peel the apple) and chop the celery heart. Mix the walnuts, apple, celery, onion and lemon juice, add salt and pepper and just before serving mix in the mayonnaise and sprouts. Serve at once with cold roast chicken or turkey.

*Sprout salad*

# Broccoli in soured cream

1 lb. frozen broccoli
2 tbsps. cream cheese
1 tbsp. tomato paste
¼ pt. soured cream
salt
3 tbsps. chopped green olives

*Serves 4*

Cook the broccoli in boiling salted water until it is tender. Drain it well and leave to get cold. Beat the cheese until it is smooth and soft, stir in the tomato paste, soured cream and salt to taste. Fold in the olives and spoon the dressing over the broccoli. Serve with grilled lamb cutlets or pork chops.

# Broccoli vinaigrette

1 lb. frozen broccoli
lettuce or watercress
4 tomatoes, skinned, seeded and chopped

*For dressing*
3 tbsps. salad oil
1 tbsp. vinegar
salt and freshly ground black pepper
pinch dry mustard

*Serves 4*

Cook the broccoli in boiling salted water until just tender. Whisk together the dressing ingredients. Drain the broccoli thoroughly and pour the dressing over while it is still hot. Chill and serve on a bed of lettuce or watercress, with the chopped tomatoes scattered over the top. Serve with cold meats.

## Red cabbage and sweetcorn salad

1½ lb. red cabbage
4 tbsps. French dressing (see page 147)
11 oz. frozen sweetcorn kernels
½ cucumber, peeled and diced
salt and freshly ground black pepper
¼ level tsp. finely grated lemon rind
2 tsps. clear honey

*Serves 8*

Prepare the cabbage, discard any coarse stems and shred it finely. Place it in a large bowl with the dressing, toss and leave for 2 hr. Cook the corn in boiling salted water, drain and leave to cool. Place the diced cucumber, seasoned with salt and pepper, in a bowl and leave for 2 hr. Drain off any excess moisture from the cucumber and add to the cabbage, with the corn, lemon rind and honey. Toss thoroughly. Serve with fried chicken.

# VEGETABLE STARTERS

## Corn on the cob maître d'hôtel

2 frozen corn cobs
1 oz. butter
1 tbsp. chopped fresh chervil or tarragon,
    or ½ level tsp. dried chervil or tarragon
salt and freshly ground black pepper

*Serves 2*

Cook the frozen corn in boiling salted water for 5 min. Drain and place each cob in the centre of a piece of buttered foil. Melt the 1 oz. butter, add the herbs and use to brush the corn. Wrap the foil round the corn, making a double fold along the top and at each end to seal the butter in; put on a baking sheet and cook in the oven at 325°F (mark 3) for 10 min. Serve in the foil, seasoned with salt and pepper.

## Asparagus and corn salad

8 oz. frozen asparagus
6 oz. frozen sweetcorn kernels
1 tbsp. chopped capers
¼ pt. thick mayonnaise (see page 147)
salt and freshly ground black pepper
4 eggs, hard-boiled and sliced
paprika pepper

*Serves 4*

Cook the asparagus in boiling salted water until tender, drain and allow to cool. Cook the corn the same way, and cool. Blend the corn, capers and mayonnaise together and add salt and pepper. Cut off and discard the woody ends of the asparagus; arrange the tips on individual plates, coating the stems with a few spoonsful of the corn dressing. Garnish with sliced eggs and a sprinkling of paprika.

## Spicy cauliflower

(*see colour picture facing page 160*)

1 large egg
salt and pepper
1 lb. frozen cauliflower florets
browned breadcrumbs
deep oil for frying

*Vinegar sauce*
6 tbsps. oil
6 tbsps. vinegar
½ level tsp. paprika pepper
2 small cloves garlic, peeled and crushed
chopped parsley ro garnish

*Serves 4*

Beat the egg lightly with the seasoning. Dip the florets into the egg, then into bread-crumbs, and coat thoroughly. Heat the oil to 375°F and deep fry the coated florets for 5–7 min. until golden brown. Drain on absorbent paper. Keep warm. Meanwhile, have ready the sauce ingredients in a small pan. Bring to the boil and pour over the crisp florets. Garnish with chopped parsley.

# Spinach soufflé

4 oz. onion, skinned and thinly sliced
3½ oz. butter
6-oz. pkt. frozen leaf spinach, thawed
salt and freshly ground black pepper
¼ level tsp. grated nutmeg
1 oz. flour
½ pt. milk
3 large eggs, separated
1½ oz. Parmesan cheese, grated

*Serves 4 (or 2 as a supper dish)*

Gently sauté the onion in 1½ oz. butter until tender. Add the spinach and sauté a little longer. Season with salt, pepper and nutmeg. Melt the remainder of the butter in another pan, stir in the flour and cook for 1 min. Add the milk all at once, beat well and bring to the boil. Add salt and pepper. Pour half the sauce on to the spinach and onion. Turn this into the base of a buttered 6-in. (2-pt.) soufflé dish. Beat the yolks and most of the cheese into the remaining sauce, reserving a little cheese to sprinkle over the soufflé. Stiffly whisk the whites and fold them gently through the sauce using a metal spoon. Pour this on to the spinach base. Sprinkle with the reserved cheese. Place the soufflé dish on a baking sheet and cook in the oven at 375°F (mark 5) for about 40 min. Serve at once.

# Asparagus maltaise

8 oz. frozen asparagus spears
1 oz. butter
2 level tbsps. flour
grated rind and juice of 1 small orange
3 tbsps. single cream
1 egg yolk
salt and freshly ground black pepper

*Serves 2*

Cook the asparagus spears in boiling salted water until tender. Drain and reserve ¼ pt. cooking liquid. Melt the butter, stir in the flour and cook for 1 min. Add the asparagus liquid and bring to the boil. Remove from the heat, stir in the orange rind and juice, cream and egg yolk. Season to taste and chop the asparagus into the sauce. Serve on toast.

# COOKED VEGETABLE DISHES
# FOR THE FREEZER

# Stuffed onions

4 large onions, skinned
½ lb. fresh minced lean beef
9 stuffed olives, halved
½ level tsp. dried thyme
salt and freshly ground black pepper
1 heaped tbsp. fresh white breadcrumbs
1 egg, beaten

*To complete dish from freezer*
2 tbsps. cooking oil

*Serves 4*

**TO MAKE:** Cook the onions in boiling salted water for 30–45 min., according to size, until just tender but not soft. Drain them and carefully ease out the centres. Sauté the minced beef for 8–10 min. without using any extra fat and mix it with the olives, thyme, salt, pepper, breadcrumbs and beaten egg. Fill the onions with the meat mixture. Cool.

**TO PACK AND FREEZE:** Pack each onion individually in foil, freeze and wrap them in a large polythene bag; seal, label and return to the freezer.

**TO USE:** Remove the polythene, open the foil and brush the onions with oil. Re-cover with foil and bake from frozen at 400°F (mark 6) for 1 hr. Serve in the foil cases with a piquant tomato sauce.

# Onion quiche

8 oz. shortcrust pastry (i.e. made with
    8 oz. flour etc.)
3 oz. butter
1¼ lb. onions, skinned and thinly sliced
3 small eggs
1½ oz. flour
½ pt. milk
freshly ground black pepper
2-oz. can anchovy fillets, drained and
    halved lengthwise
stoned black olives

*Serves 4–6*

TO MAKE: Roll out the pastry and use it to
line two 7-in. flan rings placed on baking
sheets. Bake them blind at 400°F (mark 6) for
about 20 min.; leave the flan rings in
position.

   Melt the butter in a frying pan and sauté
the onions until soft, being careful not to
brown them. Divide the onions between the
flan cases. Beat together the eggs, flour,
milk and pepper and pour the mixture over
the onions. Arrange the anchovy fillets on
top in a criss-cross pattern and place a black
olive in each space. Bake at 350°F (mark 4)
for 20–30 min. until set. Remove the flan
rings and cool rapidly on a wire rack.

TO PACK AND FREEZE: Open freeze—see
notes at beginning of chapter.

TO USE: Remove the wrappings and place
the quiche on a baking sheet. Reheat from
frozen in the oven at 400°F (mark 6) for
about 1 hr.

# Bubble and squeak bake

2 lb. old potatoes, peeled
butter
milk
salt and pepper
1 lb. firm green cabbage
½ lb. onions, skinned and finely chopped
¼ lb. carrots, peeled and grated
6 oz. mature Cheddar cheese, grated

*Serves 4–6*

TO MAKE: Cook the potatoes in boiling
salted water until tender. Drain them and
mash well; cream with a little butter and

milk and season with salt and pepper. Cut
the cabbage into wedges, cook in boiling
salted water until just tender, then drain well
and chop. Cook the onion and carrot in
boiling salted water for 5 min., and drain.

   Blend the cabbage and potato together
and turn half into a buttered shallow baking
dish, 9 in. in diameter (1¾-pt.). Cover with
onion and carrot and top with 4 oz. grated
cheese. Spread evenly and top with the
remainder of the potato mixture. Smooth the
surface with a knife, mark with a fork and
sprinkle the remainder of the cheese over.

TO PACK AND FREEZE: See notes at
beginning of chapter.

TO USE: Unwrap the bubble and squeak
and reheat from frozen in the oven at 375°F
(mark 5) for 1 hr. Flash under a pre-heated
grill for a few minutes to brown the top.

# Chicken stuffed peppers

(*see picture opposite*)

1 oz. butter
2 oz. lean streaky bacon, rinded and
    chopped
2 oz. onion, skinned and roughly chopped
2 oz. button mushrooms, trimmed and
    quartered
large pinch dried thyme
½ level tsp. paprika pepper
4 tbsps. chicken stock
½ level tbsp. cornflour
1 tbsp. water
4 tbsps. single cream
¼ lb. cooked white chicken
    meat, diced
2 large red peppers
salt and pepper

*Serves 2*

TO MAKE: Melt ½ oz. of the butter and
sauté the bacon and onion for 5 min. Add
the remaining butter and the mushrooms
and sauté for 5 min. longer. Stir in the
thyme, paprika and stock and bring to the
boil. Blend the cornflour with water, pour
it into the bubbling juices, stirring. Remove
the pan from the heat, add the cream and
gently reheat without boiling. Add the

*Chicken stuffed peppers*

chicken, taste for seasoning and leave to cool.

Blanch the whole peppers in boiling water for 5 min., then drain. Cut off the tops and scoop out the seeds. Spoon the chicken mixture into the peppers and cool quickly.

**TO PACK AND FREEZE:** Open freeze—see notes at beginning of chapter.

**TO USE:** Unwrap the peppers and place them in a baking dish. Cook from frozen in the oven at 350°F (mark 4) for 30 min.; cover them with lightly buttered foil and cook for a further 30 min.

# Braised red cabbage

1 small red cabbage
1 large Spanish onion, skinned and very
   thinly sliced
½ pt. red wine
salt and freshly ground black pepper

*Serves 4–6*

**TO MAKE:** Shred the cabbage, mix it with the onion and place in a large casserole. Moisten with the wine, season with salt and freshly ground black pepper, put on the lid and cook in the oven at 300°F (mark 2) for about 3 hr. Cool rapidly.

**TO PACK AND FREEZE:** See notes at beginning of chapter.

**TO USE:** Unwrap and reheat from frozen, covered, at 350°F (mark 4) for about $1\frac{1}{4}$ hr. Serve with roast pork or game.

## Top-crust vegetable pie

(*see picture below*)

$\frac{1}{2}$ lb. onions, skinned
$\frac{1}{2}$ lb. leeks, trimmed
$\frac{1}{2}$ lb. carrots, peeled
$\frac{1}{2}$ lb. turnips, peeled
3 oz. butter or margarine
$\frac{1}{4}$ pt. water
$\frac{3}{4}$ pt. milk (approx.)
$\frac{1}{2}$ lb. tomatoes, skinned and quartered
$15\frac{1}{2}$-oz. can butter beans, drained
salt and pepper
2 heaped tbsps. chopped parsley
$1\frac{1}{2}$ oz. flour
6 oz. mature Cheddar cheese, grated
8 oz. shortcrust pastry (i.e. made with 8 oz. flour etc.)

*Serves 4–5*

**TO MAKE:** Roughly chop the onions. Leave some green on the leeks, cut into $\frac{1}{4}$-in. slices and wash thoroughly. Cut the carrots into matchsticks and dice the turnips. Melt $1\frac{1}{2}$ oz. butter or margarine in a pan, add the vegetables; cover the pan and cook very gently for 10 min. Add $\frac{1}{4}$ pt. water and simmer for a further 10 min. Drain the vegetables and make the juices up to $\frac{3}{4}$ pt. with milk.

Layer the cooked vegetables with the tomatoes and butter beans in a $3\frac{1}{2}$-pt. pie dish, adding a little salt and pepper between the layers. Sprinkle the parsley over the top. Rinse out the pan and heat the remaining fat. When it is melted, stir in the flour. Remove from the heat and gradually beat in the cooking liquid and milk. Bring to the boil and boil for 1–2 min. Stir in the cheese, adjust the seasoning and pour the sauce over the vegetables. Leave to cool.

Roll out the pastry to make a lid for the pie and place it over the dish; scallop the edges.

**TO PACK AND FREEZE:** See notes at beginning of chapter.

**TO USE:** Remove the wrappings and thaw the pie overnight in the refrigerator. Place on a baking sheet and bake in the oven at 400°F (mark 6) for about $\frac{3}{4}$-1 hr.

*Top-crust vegetable pie*

# Onion and apple flan

½ lb. cooking apples, peeled and cored
1 onion, skinned and chopped
6 oz. shortcrust pastry (i.e. made with
    6 oz. flour etc.)
3 oz. Cheddar cheese, grated
2 eggs
¼ pt. milk
salt and freshly ground black pepper

*To complete dish from freezer*
3 tomatoes, sliced

*Serves 4*

**TO MAKE:** Slice the apples and put with
the onion in a pan with very little water.
Simmer gently until tender. Roll out the
pastry and use it to line a 7½-in. flan ring
placed on a baking sheet. Cool the apple and
onion mixture and pour it into the pastry
case. Sprinkle with cheese. Beat the eggs,
add the milk, season and pour over the flan
filling. Bake in the oven at 400°F (mark 6) for
10 min. to set the pastry, turn the oven
temperature down to 350°F (mark 4) and
cook for a further 20–30 min. until the filling
is set. Remove the flan ring and cool rapidly
on a wire rack.

**TO PACK AND FREEZE:** See notes at
beginning of chapter.

**TO USE:** Unwrap the flan, place it on a
baking sheet and reheat from frozen in the
oven at 400°F (mark 6) for about 1 hr.
Arrange the slices of tomato on top halfway
through the cooking. Serve hot as a starter
or a supper dish.

# Asparagus quiche

*For pastry*
5 oz. plain flour
1½ oz. butter or margarine
1 oz. lard
1 oz. Cheddar cheese, grated
pinch salt

*For filling*
8-oz. pkt. frozen asparagus spears, thawed
4 tbsps. double cream
¼ pt. single cream
salt and pepper
4 eggs, beaten
1 oz. Parmesan cheese, freshly grated

*Serves 4–6*

**TO MAKE:** Make up the cheese pastry as if
making shortcrust, adding the cheese after
the fat. Roll it out and use to line an 8-in.
plain flan ring placed on a baking sheet. Trim
the asparagus to fit the pastry case; cut up
the extra pieces and place them in the base
of the flan. Arrange the trimmed spears as
the spokes of a wheel on top. Mix together
the creams, salt, pepper, eggs and Parmesan.
Pour the mixture over the asparagus. Bake
at 400°F (mark 6) for about 40 min. until the
filling is set and the pastry golden. Cool
quickly on a wire rack and remove the flan
ring.

**TO PACK AND FREEZE:** See notes at
beginning of chapter.

**TO USE:** Unwrap and thaw at room temperature for 2–3 hr., then place the flan on a
baking sheet and refresh in the oven at
350°F (mark 4) for about 30 min. Serve the
quiche warm.

# Risotto

½ lb. lean streaky bacon rashers, rinded
½ lb. onions, skinned and thinly sliced
10 oz. long grain rice
1½ pt. well flavoured stock
salt and pepper

*Serves 5–6*

**TO MAKE:** Cut the bacon into small pieces
with scissors. Put it in a frying pan without
fat, heat gently until the fat runs out of the
bacon and cook until crisp; remove the
bacon from the pan with a slotted spoon. Fry
the onion and rice in the bacon fat until well
coloured. Pour in the stock, bring to the boil
and adjust the seasoning.

Turn this mixture into a casserole with a
close-fitting lid. Place it in the oven at
375°F (mark 5) for about 40 min. then stir in
the bacon pieces. Allow to cool.

**TO PACK AND FREEZE:** See notes at
beginning of chapter.

**TO USE:** Remove polythene but leave the
risotto covered with foil. Place it in the oven
at 350°F (mark 4) for about 50 min.–1 hr.
Turn lightly with a fork from time to time.

# Spinach quiche

6 oz. shortcrust pastry (i.e. made with
  6 oz. flour etc.)
$\frac{3}{4}$ lb. frozen chopped spinach
8 oz. cottage cheese
3 eggs, beaten
$1\frac{1}{2}$ oz. Parmesan cheese, grated
4 tbsps. double cream
freshly grated nutmeg
salt and freshly ground black pepper

*Serves 6–8*

**TO MAKE:** Roll out the pastry and use it to
line an 8-in. flan ring placed on a baking
sheet; bake it blind at 400°F (mark 6) for
20 min. and cool. Cook the spinach in
boiling salted water until just tender, drain
it thoroughly and add the cottage cheese,
beaten eggs, grated Parmesan, cream, nutmeg, salt and pepper. Turn the mixture into
the pastry shell and bake at 375°F (mark 5)
for about 35 min. until the filling has set and
is beginning to brown. Remove the flan ring
and cool on a wire rack.

**TO PACK AND FREEZE:** See notes at
beginning of chapter.

**TO USE:** Unwrap, cover lightly with foil
and reheat from frozen in the oven at 375°F
(mark 5) for about 1 hr.

# Leek, celery and cheese pie

1 lb. leeks, trimmed
3 sticks celery, trimmed
$\frac{3}{4}$ pt. milk
salt and pepper
1 oz. butter
1 oz. flour
4 oz. Cheddar cheese, grated
4 oz. cut macaroni

*To complete dish from freezer*
fresh white breadcrumbs
$\frac{1}{2}$ oz. butter

*Serves 2–3*

**TO MAKE:** Slice the leeks and celery and
wash them carefully. Put in a pan with the
milk, add salt and pepper, cover and simmer
until the vegetables are tender, about 30 min.
Strain off the cooking liquid and set it aside.

Melt the butter, stir in the flour; remove
from the heat, stir in the cooking liquid and
cheese. Return to the heat and cook until
thickened. Meanwhile, cook the macaroni in
boiling salted water until tender, drain and
place it in a buttered foil dish. Cover it with
the leeks and celery and pour the sauce over.
Cool.

**TO PACK AND FREEZE:** See notes at
beginning of chapter.

**TO USE:** Unwrap and reheat from frozen in
the oven at 400°F (mark 6) for about 45 min.
Sprinkle the breadcrumbs on top and dot with
butter half way through the cooking time.

# Chicory with cheese and ham

4 heads of chicory, trimmed
lemon juice
$1\frac{1}{2}$ oz. butter
$1\frac{1}{2}$ oz. flour
$\frac{1}{2}$ pt. milk
2 oz. cheese, grated
salt and pepper
grated nutmeg
4 slices cooked ham

*To complete dish from freezer*
grated cheese
fresh white breadcrumbs
butter

*Serves 2*

**TO MAKE:** Put the chicory into boiling
salted water to which a little lemon juice has
been added and cook slowly for 25 min.
Drain well and set aside $\frac{1}{2}$ pt. of the cooking
liquid. Melt the butter, stir in the flour,
cook for 1–2 min. then remove from the heat
and stir in the milk and $\frac{1}{2}$ pt. cooking liquid.
Return to the heat and cook until thickened.
Add the cheese and season with salt, pepper
and nutmeg.

Wrap each head of chicory in a slice of
ham and put in an individual foil container.
Pour over the sauce and cool quickly.

**TO PACK AND FREEZE:** See notes at
beginning of chapter.

**TO USE:** Unwrap the dishes and reheat
from frozen in the oven at 400°F (mark 6) for
about 45 min. After 30 min. sprinkle with a

*Sweet and sour sprouts* (page 152); *Whole bean provençale* (page 151);
*Sunshine carrots* (page 150); *Spicy cauliflower* (page 154)

*Mushroom ring*

mixture of grated cheese and breadcrumbs and dot with butter.

# Mushroom ring

(*see picture above*)

1 small onion, skinned
2 oz. bacon, rinded
8 oz. button mushrooms
1 oz. butter
salt and freshly ground black pepper
grated nutmeg
2 oz. flour
$\frac{1}{4}$ pt. milk
2 tbsps. cream or top of the milk
2 eggs, beaten
finely chopped parsley

*To complete dish from freezer*
bacon rolls
whole sauté mushrooms

*Serves 4*

**TO MAKE:** Chop the onion, bacon and mushrooms and fry them in butter for 5 min. Add salt, pepper and nutmeg. Gently stir in the flour, add the milk and continue to stir until the mixture comes to the boil. Remove the pan from the heat, and add the cream, beaten eggs and parsley. Turn the mixture into a greased 7-in. ring mould and cook in the oven at 350°F (mark 4) for about 30 min., or until firm. When cold, turn the ring out on to a foil plate.

**TO PACK AND FREEZE:** See notes at beginning of chapter.

**TO USE:** Leave the wrapping in position and thaw the ring for 8–12 hr. in the refrigerator. Remove freezer wrappings, cover loosely with foil, place on a baking sheet and reheat the ring in the oven at 350°F (mark 4) for 30 min. Serve garnished with bacon rolls and mushrooms, accompanied by a green salad.

*Gooseberry ice cream flan* (page 209); *Summer sponge pudding* (page 206); *Arctic mousse* (page 217); *Golden apple charlotte* (page 201)

# Aubergine in breadcrumbs

(*see picture below*)

2 aubergines
salt
flour
1 egg, beaten
fresh white breadcrumbs

*Serves 4*

TO MAKE: Wipe the aubergine, slice thinly and lay on a flat tray or plate. Sprinkle with salt and leave for about 1 hr. to extract the bitter juices. Wipe dry with absorbent paper. Dip each slice in flour, and coat with egg and breadcrumbs. Fry in deep fat until golden. Drain and cool.

TO PACK AND FREEZE: See notes at beginning of chapter.

TO USE: Unwrap and place in a single layer on a baking sheet. Reheat from frozen in the oven at 350°F (mark 4) for 30 min.

# Aubergine au gratin

2 medium sized aubergines
½ oz. butter
1 small onion, skinned and finely chopped
14-oz. can tomatoes
½ level tsp. dried basil
salt and freshly ground black pepper
6 rashers back bacon, rinded and cut into
    large pieces
cooking oil

*To complete dish from freezer*
1 pt. cheese sauce, frozen separately
    (see below)

*Serves 3–4*

TO MAKE: Wash, dry and thinly slice the

*Aubergine in breadcrumbs*

aubergines. Spread the slices out on a large tray or plate, sprinkle with salt and leave for 1 hr., to extract the bitter juices. Meanwhile melt the butter and sauté the onion until transparent. Add the canned tomatoes to the pan with their juice and season with the basil, salt and pepper. Allow to simmer gently for about 10 min. Place the bacon in a pan without fat and heat gently until the fat runs; sauté until it begins to crisp.

Dry the aubergine slices with absorbent paper. Heat 3 tbsps. oil in the pan and sauté half the aubergine, until beginning to brown on both sides. Remove from the pan, drain and divide between 2 deep foil containers, making a layer in the base of each. Cook the rest of the aubergine. Divide the tomato mixture evenly and spoon it over the aubergine. Scatter the bacon over the tomatoes. Top with the remaining aubergine slices. Allow to cool.

**TO PACK AND FREEZE:** See notes at beginning of chapter.

**TO USE:** Remove polythene wrapping and cook, covered, in the oven at 375°F (mark 5) for $\frac{3}{4}$–1 hr. Spoon the thawed cheese sauce over the aubergines after 30 min. Sprinkle a little extra grated cheese on top if wished. Serve with cold meats and roasts.

*Maximum storage time:* 1 month.

**CHEESE SAUCE**

1½ oz. butter
1½ oz. flour
1 pt. milk
4 oz. cheese, grated
1 tbsp. dry white wine
salt and freshly ground black pepper
½ level tsp. made mustard

**TO MAKE:** Melt the butter, remove from the heat and stir in the flour. Gradually blend in the milk, bring the sauce to the boil and cook for a few min., stirring all the time. Add the cheese and continue to cook for a further 5 min., stirring. Add the wine.

Season with salt, pepper and mustard. Cool quickly.

**TO PACK AND FREEZE:** Pour into 2 waxed cartons or pre-formed polythene bags. Freeze, seal, label and return to the freezer.

**TO USE:** Loosen the block of frozen sauce from its container by immersing in hot water for a few moments. Slip the block of frozen sauce into a pan and heat gently, stirring as the sauce thaws. When completely thawed, bring the sauce to the boil and boil for about 3 min., beating well.

# Duchesse potatoes

(*see colour picture facing page 32*)

**6 lb. old potatoes, peeled and boiled**
**2 oz. butter**
**1 egg**
**1 level tsp. salt**
**freshly ground black pepper**
**¼ level tsp. grated nutmeg**

*Serves 10*

**TO MAKE:** Mash the potatoes, add the remaining ingredients (no milk) and beat well. Line a baking sheet with non-stick paper. Using a forcing bag fitted with a large star vegetable nozzle, pipe on to the sheet about 20 pyramids of potato, with a base of about 2 in.

**TO PACK AND FREEZE:** Freeze uncovered until firm. Remove from the freezer, slide the potato pyramids off the sheet on to foil plates or into a rigid container. Cover with foil or polythene, seal and label.

**TO USE:** Grease some baking sheets or line them with kitchen foil, transfer the frozen duchesse potato portions to the sheets, brush lightly with egg glaze and put into a cold oven. Cook at 400°F (mark 6) for 20–30 min., or until heated and lightly browned.

*Note:* Milk has been omitted from the recipe because if it is included, the mixture 'weeps' slightly when defrosted owing to the large amount of water present.

# 8 Hot Puddings

Many puddings and pies freeze well and it is a tremendous time-saver to be able to mix and bake in bulk. Most baked and steamed puddings are best frozen raw in the dish or basin in which they will be cooked, as any ingredients put in the base of the dish ready for turning out as a topping may not freeze solid (this applies particularly to jams, syrup and some fruits). If you prefer to add topping ingredients when you remove the pudding from the freezer, pack in a foil-lined dish and, when the pudding is frozen, remove the pre-formed shape from the dish; this method also enables you to keep your pudding and pie dishes in use all the time. Most pastry dishes can be frozen either before or after cooking.

Sauces for puddings are best frozen separately. Freeze in pre-formed polythene bags or waxed cartons, as described on page 61.

## Basic sponge pudding (for steaming or baking)

1 lb. butter or margarine
1 lb caster sugar
8 eggs
1–1½ lb. self-raising flour, sifted
8 tbsps. milk, optional

*Makes four 1½-pt. puddings*

**TO MAKE:** Cream the fat and sugar together until light and fluffy. Add the eggs one at a time, beating well after each one. A little of the flour may be added to prevent curdling as the eggs are added. Lightly fold in the remaining flour. If the larger measure of flour is used, add the milk as well. Divide the mixture between 4 greased or foil-lined basins or pie dishes (see packing instructions below). As plain pudding, this quantity will make four 1½-pt. puddings; if bulky extra ingredients are to be added (see variations below), larger basins will be required. Remember that there should always be enough room left in the basin for the pudding to rise when cooked.

**TO PACK AND FREEZE: Either** Use greased foil containers or greased ovenproof basins or pie dishes. Fill with the mixture, cover tightly with foil, overwrap with polythene and seal before freezing. When required, unwrap the basin and cook as instructed. **Or** Line ovenproof basins or dishes with foil, pressing out as many of the creases as possible. Fill with the mixture and fold the foil lightly over the top. Freeze until firm, remove the pre-formed shape from the basin, overwrap with polythene, seal and return to the freezer. When required, remove the wrappings and return the frozen pudding to the dish in which it was shaped.

Puddings for steaming should be covered with a pleated circle of greased greaseproof paper and then a tight foil covering, pressed into the sides of the basin to keep the steam out. When puddings are frozen in the basin they are to be cooked in, this can be done before freezing and just checked before putting into the steamer. When a pre-formed pack is used, the covering is put in place after the pudding has been removed from the freezer and replaced in its basin.

TO USE: Sponge puddings may be baked or steamed. **To bake**, unwrap and cook from frozen in the oven at 350°F (mark 4) for $1\frac{3}{4}$–2 hr. Turn out and serve with custard or cream. **To steam** Either use a steamer, with the base half filled with water, or stand the securely covered pudding basin in a large saucepan, with water to come half way up the basin, and bring to the boil. Steam a $1\frac{1}{2}$-pt. pudding from frozen for $2\frac{1}{2}$ hr. from the time the water boils. Turn out and serve with custard, cream or a sauce.

## PRUNE AND PEACH BAKE

$\frac{1}{4}$ quantity basic sponge pudding recipe
7-oz. can sliced peaches
15-oz. can prunes, drained and stoned

*Serves 4*

TO MAKE: Drain the peaches, reserving 2 tbsps. juice. Place the fruits at the bottom of a greased or foil-lined $1\frac{1}{2}$–2-pt. pie dish and spoon the reserved juice over. Cover evenly with the sponge mixture.

TO PACK AND FREEZE: See basic recipe.

TO USE: Unwrap and place the dish on a baking sheet. Cook from frozen in the oven at 350°F (mark 4) for $1\frac{3}{4}$–2 hr. Turn out and serve hot with custard or cream.

## FIGGY LEMON BAKE

$\frac{1}{4}$ quantity basic sponge pudding recipe
1 lemon
15-oz. can figs

*Serves 4*

TO MAKE: Grate the rind from the lemon and add to the sponge. Peel the lemon, discarding all traces of white pith. Cut the flesh into slices approx. $\frac{1}{4}$ in. thick and

discard the pips. Cut each slice into 4. Drain the figs and reserve 2 tbsps. juice. Place the figs, reserved juice and lemon flesh in the base of a greased $1\frac{1}{2}$–2-pt. ovenproof pie dish. Cover evenly with the sponge mixture.

TO PACK AND FREEZE: See basic recipe.

TO USE: Unwrap and place the dish on a baking sheet. Cook from frozen in the oven at 350°F (mark 4) for $1\frac{3}{4}$–2 hr. Turn out and serve hot with custard or cream.

## STEAMED PEACH AND PINEAPPLE SPONGE PUDDING

$\frac{1}{4}$ quantity basic sponge pudding recipe
2 tbsps. golden syrup
15-oz. can pineapple rings, drained
7-oz. can sliced peaches, drained

*Serves 4*

TO MAKE: Spoon the syrup into the base of a $1\frac{1}{2}$–2-pt. greased or foil-lined basin. Add a ring of pineapple in the centre. Cover with a little of the sponge mixture and continue to fill the basin with layers of sponge and a mixture of the peaches and pineapple. End with a fruit layer.

TO PACK AND FREEZE: See basic recipe.

TO USE: Unwrap and place the basin in a steamer or saucepan of cold water. Bring to the boil and steam for $2\frac{1}{2}$ hr., from the time the water boils. Turn out and serve hot with custard or cream.

## STEAMED BLACK CHERRY LAYER PUDDING

(*see picture overleaf*)

$\frac{1}{4}$ quantity basic sponge pudding recipe
2 tbsps. black cherry jam
15-oz. can black cherries, drained and stoned

*Serves 4*

TO MAKE: Spoon the jam into the base of a prepared $1\frac{1}{2}$-pt. pudding basin. Cover evenly with half the sponge mixture. Place half the stoned cherries over the sponge, add the remaining sponge and top with the rest of the cherries; these will give a cherry base when turned out.

*Steamed black cherry layer pudding*

**TO PACK AND FREEZE:** See basic recipe.

**TO USE:** Unwrap and place the basin in a steamer or saucepan of cold water. Bring to the boil and steam for $2\frac{1}{2}$ hr., from the time the water boils. Turn out and serve hot with custard or cream.

# Crunch-top pineapple pudding

2 oz. Demerara sugar
6 glacé cherries, halved
$1\frac{1}{2}$ oz. shelled walnut halves
15-oz. can pineapple rings or pieces
4 oz. margarine
4 oz. caster sugar
2 eggs, beaten
6 oz. self-raising flour
2 level tsps. cornflour

*Serves 4–5*

**TO MAKE:** Grease a $1\frac{1}{2}$-pt. oval pie dish and sprinkle the Demerara sugar in the base. Cover the sugar with cherries and walnuts. Drain the pineapple and make up the juice to $\frac{1}{2}$ pt. with water. Chop the pineapple, saving 2 rings or 10–12 pieces for the sauce. Make up a sponge mixture using the margarine, caster sugar, eggs and flour as for Basic Sponge Pudding (page 164), and then fold in the chopped pineapple. Spread evenly over the sugar and nuts.

Make the sauce by blending the cornflour with 2 tsps. pineapple juice. Heat the remaining juice in a small pan, stir in the cornflour and bring to the boil. Chop the remaining pineapple rings and add to the sauce. Cool quickly.

**TO PACK AND FREEZE:** Pour the cooled sauce into a preformed polythene bag. Freeze until solid, remove the bag from the pre-former, seal and label it and return it to the freezer. Wrap the pudding dish in foil and place it in a polythene bag. Seal, label and freeze.

TO USE: Unwrap the pudding, place the dish on a baking sheet and cook from frozen in the oven at 325°F (mark 3) for about $1\frac{3}{4}$ hr. When the pudding is nearly cooked unwrap the sauce, turn it into a saucepan and heat through gently. Turn the pudding out and serve the sauce separately.

# Hot marble sponge

4 oz. margarine
4 oz. caster sugar
2 eggs, beaten
6 oz. self-raising flour
2 tbsps. milk
2 lemons
1 level tbsp. cocoa powder
1 level tbsp. cornflour
6 level tbsps. sugar

*Serves 4–5*

TO MAKE: Grease a $1\frac{1}{2}$-pt. fluted ring mould. Place a small piece of foil over the hole to prevent water seeping in during steaming. (If you do this now it is ready for when the pudding is removed from the freezer.)

Cream the margarine and sugar until light and fluffy, add the eggs one at a time and beat well after each addition. It may be necessary to add a little of the flour with the eggs to prevent curdling. Fold in the flour and add the milk to give a mixture that will drop from the spoon. Put half the mixture into a second bowl. To one bowl, add the grated rind of 1 lemon and mix well; to the other add the cocoa. Place a spoonful of each mixture alternately into the mould to give a marbled effect. (The spoonsful will melt into each other during cooking so there is no need to smooth over too much.)

To make the sauce, pare the rind from the other lemon free of all white pith and cut it into fine strips. Squeeze the juice from both lemons and make up to $\frac{1}{2}$ pt. with water. Blend the cornflour with 1 tbsp. of the liquid and heat the remainder in a pan with the sugar and strips of lemon rind. Add this

to the blended cornflour, stirring, and return it to the pan. Bring to the boil and cook, stirring, for a few min. Cool.

TO PACK AND FREEZE: Pour the sauce into a preformed polythene bag, freeze, overwrap, seal and label. For the sponge, cover the top of the mould with foil and secure at the edges. (This will be ready for steaming when removed from the freezer as well as providing a protective wrapping now.) Place the mould in a polythene bag, seal, label and freeze.

TO USE: Remove the sponge mould from the polythene bag but keep the foil on. Check to see that the foil is secure at the edges and tie with string if not. Place in a steamer or saucepan with cold water, bring to the boil and cook for 2 hr. from the time the water begins to boil. Unwrap the sauce, place it in a saucepan and heat through when the pudding is almost cooked. Turn out the sponge and serve the sauce separately.

# Mixed fruit dumpling

6 oz. self-raising flour
$\frac{1}{2}$ level tsp. each ground cinnamon, ginger and mixed spice
$\frac{1}{4}$ level tsp. bicarbonate of soda
$\frac{1}{4}$ level tsp. cream of tartar
3 oz. shredded suet
2 oz. currants
2 oz. sultanas
2 oz. stoned raisins
2 oz. cooking apples, grated
$2\frac{1}{2}$ oz. light soft brown sugar
1 egg, beaten
$1\frac{1}{2}$ tbsps. black treacle
about 4 tbsps. milk

*Serves 6*

TO MAKE: Sift together the flour, spices, bicarbonate of soda and cream of tartar into a bowl. Add the suet, dried fruit, apple and sugar. Combine the egg and treacle, add to the dry ingredients with enough milk to make a soft, scone-like dough. Turn the mixture into a greased 2-pt. pudding basin. Cover with greased greaseproof paper folded into a pleat across the middle to allow for expansion during cooking.

**TO PACK AND FREEZE:** Top with pleated foil, secure with string and place the basin in a polythene bag. Seal, label and freeze.

**TO USE:** Remove the polythene bag. Place the covered basin in a steamer or large saucepan with cold water. Bring to the boil and cook for $2\frac{1}{4}$ hr. from the time the water begins to boil. Turn out and serve with brandy butter.

## Cottage plum upside-down pudding

1-lb. 4-oz. can golden plums
2 tbsps. golden syrup
3 oz. margarine
3 oz. granulated sugar
1 large egg
5 oz. self-raising flour
$\frac{1}{2}$ level tsp. ground ginger
1–2 tbsps. milk
3 level tsps. arrowroot
knob of butter

*Serves 4*

**TO MAKE:** Drain and reserve the juice from the plums. Arrange the plums in a buttered 2-pt. pie dish and drizzle the syrup over them. Cream the margarine and sugar thoroughly and beat in the egg. Lightly beat in the sifted flour and ginger with enough milk to give a stiff mixture that will just drop from the spoon. Cover the plums with the sponge mixture.

To make the sauce, blend the arrowroot with the plum juice and bring to the boil, stirring. Add a knob of butter and cool the sauce.

**TO PACK AND FREEZE:** Pour the cooled sauce into a pre-formed polythene bag; freeze, overwrap, seal and label. Wrap the pudding in its dish in foil and place in a polythene bag. Seal, label and freeze.

**TO USE:** Unwrap the pudding, place the dish on a baking sheet and bake from frozen in the oven at 350°F (mark 4) for about $1\frac{3}{4}$ hr. When the pudding is almost cooked, unwrap the frozen sauce and heat gently in a saucepan. Turn out the pudding and serve the sauce separately.

## Plum cobbler

*For cobbler topping*
6 oz. self-raising flour
2 oz. granulated sugar
2–3 oz. butter or margarine
1 egg, beaten
milk to mix

*To complete dish from freezer*
16-oz. can yellow plums
2 tbsps. milk
1 tbsp. granulated sugar

*Serves 4*

**TO MAKE:** Mix the flour and sugar in a bowl and rub in the butter with the finger-tips until the mixture resembles bread-crumbs. Add the egg and enough milk to give a fairly soft but manageable dough. Knead lightly on a floured board and roll out to $\frac{1}{2}$-in. thickness. Cut out 12 2-in. rounds with a pastry cutter.

**TO PACK AND FREEZE:** Open freeze the rounds on a baking sheet. Pack them in foil, with non-stick waxed paper or polythene film interleaved between each one. Place the pack in a polythene bag; seal and label.

**TO USE:** Drain the plums, turn them into a greased ovenproof pie dish or pie plate and spoon in half the juice. Place the unwrapped frozen 'cobblers' in a slightly overlapping ring round the edge of the dish, leaving a gap in the centre. Brush each with milk and dredge with sugar. Place on a baking sheet and bake in the oven at 400°F (mark 6) for about 1 hr. Serve hot with custard sauce.

## Oaty plum crumble

1-lb. 4-oz. can red plums
2 level tbsps. granulated sugar
4 oz. plain flour
$\frac{1}{2}$ level tsp. ground mixed spice
3 oz. margarine
$1\frac{1}{2}$ oz. rolled oats (or crushed breakfast cereal)
3 oz. Demerara sugar

*Serves 4*

**TO MAKE:** Turn the plums with half their juice into a 2-pt. ovenproof dish. Sprinkle with granulated sugar. Sift together the flour and spice and rub in the margarine lightly with the fingertips until the mixture resembles breadcrumbs. Add the rolled oats and 2 oz. Demerara sugar; mix well. Sprinkle the mixture over the plums, level the surface but do not pack down tightly. Scatter the remaining 1 oz. Demerara sugar on top.

**TO PACK AND FREEZE:** Cover the dish with foil and wrap in a polythene bag. Seal, label and freeze.

**TO USE:** Unwrap, place on a baking sheet and bake from frozen in the oven at 350°F (mark 4) for about $1\frac{1}{4}$ hr. Serve with cream or custard.

# Hot tipsy trifle

6 slices stale Madeira cake
3 tbsps. redcurrant jelly
4 tbsps. Cointreau
1-lb. 12-oz. can peach slices, drained
1 level tbsp. custard powder
1 pt. milk
3–4 level tbsps. caster sugar
2 egg yolks

*To complete dish from freezer*
2 frozen egg whites, thawed
2 oz. caster sugar
1 level tbsp. granulated sugar

*Serves 6*

**TO MAKE:** Spread the cake slices with the jelly and use to make a sponge base in a straight sided ovenproof dish approx. 3 in. deep by $7\frac{1}{2}$ in. diameter (a large soufflé dish will do). Pour the Cointreau over the sponge, then add the drained peaches. Mix the custard powder with 2 tbsps. of the milk and heat the remainder with the sugar in a saucepan. Add the hot milk to the blended custard powder, stir and return it to the heat. Bring to the boil and cook for 2–3 min., stirring. Remove from the heat, cool a little and beat in the egg yolks. Pour over the peaches. Cool.

**TO PACK AND FREEZE:** Cover the dish with foil and place in a polythene bag. Seal, label and freeze.

**TO USE:** Unwrap the trifle and thaw at room temperature for 5–6 hr. Whisk the thawed egg whites until stiff, fold in the caster sugar, pile on top of the trifle and sprinkle with granulated sugar. Place in the oven at 300°F (mark 2) for 30–40 min. until the meringue topping is golden brown. Alternatively the trifle can be put in the oven frozen, at 350°F (mark 4) for 1 hr; add the topping, turn the oven down and continue as above. Serve immediately.

# Rhubarb and apple snow

1 lb. cooking apples, peeled, cored and sliced
2 tbsps. Madeira wine
1 level tbsp. caster sugar
2 tbsps. bramble jelly
2 eggs, separated
3 tbsps. cake crumbs, preferably plain Madeira or sponge
1-lb. 3-oz. can rhubarb, drained

*To complete dish from freezer*
2 frozen egg whites, thawed
2 oz. caster sugar
1 level tbsp. granulated sugar

*Serves 6*

**TO MAKE:** Cook the apples gently in a saucepan with the Madeira and sugar until soft. Remove from the heat, add the bramble jelly and beat to a pulp or purée in a blender. Add the egg yolks and crumbs. Put the rhubarb in the base of a $1\frac{1}{2}$-pt. ovenproof dish or rigid foil container and pour the apple mixture over.

**TO PACK AND· FREEZE:** Cover the dish with foil and place in a polythene bag. Seal, label and freeze.

**TO USE:** Thaw the pudding, wrapped, at room temperature for 5 hr. Unwrap, place the dish on a baking sheet and cook in the oven at 350°F (mark 4) for 30–40 min. Whisk the thawed egg whites until stiff and fold in the caster sugar. Remove the dish from the oven, increase to the hottest temperature. Pile the meringue mixture on top of the apple, sprinkle with the granulated sugar and pop back into the hot oven for 2–3 min. until the meringue topping is brown. Serve at once.

# Cherry-walnut upside down pudding

*For topping*
1 oz. butter
2 oz. soft brown sugar
3 oz. glacé cherries, halved
3 oz. shelled walnuts, coarsely chopped
1 tbsp. coffee essence

*For sponge base*
4 oz. butter or margarine
4 oz. caster sugar
2 eggs
2 tbsps. coffee essence
6 oz. self-raising flour

*Serves 6*

**TO MAKE:** Lightly grease a 6-in. round cake tin. Melt the butter with the sugar for the topping in a saucepan and stir in the cherries, walnuts and coffee essence. Spread over the base of the tin. Cream the butter and sugar until light and fluffy; beat in the eggs one at a time and stir in the coffee essence. Lightly beat in the sifted flour and spoon the sponge evenly over the cherry and walnut mixture.

**TO PACK AND FREEZE:** Cover the tin with foil and wrap in a polythene bag. Seal, label and freeze.

**TO USE: Either** Unwrap and thaw for approx. 4 hr. at room temperature and bake in the oven at 350°F (mark 4) for 50–55 min. Or Unwrap and cook from frozen in the oven at 350°F (mark 4) for about 2 hr. To serve, turn out upside down on to a heated plate.

# Apricot and cherry pudding

6 oz. suet pastry (i.e. made with 6 oz. flour etc.)
15-oz. can apricot halves, drained
2 oz. glacé cherries, halved

*Serves 4*

**TO MAKE:** Use a 1½-pt. pudding basin, either greased or lined with foil, or a greased rigid foil basin to both freeze and cook in (see Basic Sponge Pudding, page 164).
   Knead the pastry on a floured surface until smooth. Roll it out into a circle at least 1 in. larger all round than the top of the basin. Cut a quarter out of the round and line the basin with the remaining portion, damping the cut edges, overlapping them and pressing well to seal. Fill the basin with the apricots and cherries. Roll out the remaining pastry to make a lid and place it over the basin. Damp the edges and press well together.

**TO PACK AND FREEZE:** See Basic Sponge Pudding recipe, page 164.

**TO USE:** Remove wrappings and return the pudding to the original basin if necessary. Cover the pudding with greaseproof paper and then cover the top of the basin with foil, securing it well round the rim with string. Put the basin in a steamer or large saucepan with cold water, bring to the boil and cook for 2–2½ hr. from the time the water begins to boil. Unwrap the pudding, turn it out and serve with hot custard.

# Ginger apple charlotte

1 lb. cooking apples, peeled, cored and sliced
4 oz. caster sugar
4 oz. fresh white breadcrumbs
grated rind of 1 lemon
½ level tsp. ground ginger
2 oz. butter, melted

*Serves 4*

**TO MAKE:** Blanch the apple slices in boiling water for 1–2 min., drain and cool. Mix together the sugar, breadcrumbs, lemon rind and ginger. Place alternate layers of crumbs and apples in a 1½-pt. ovenproof dish or rigid foil container and after each layer pour over a little of the melted butter. End with a crumb layer.

**TO PACK AND FREEZE:** Cover the dish with foil, place in a polythene bag, seal, label and freeze.

**TO USE:** Unwrap the dish, place it on a baking sheet and cook from frozen in the oven at 375°F (mark 5) for 1 hr. Increase the oven temperature to 425°F (mark 7) and continue to cook for a further 15 min. to brown the top. Serve warm, with whipped cream or custard.

*Baked fruit salad*

# Orange layer pudding

juice and grated rind of 1 orange
2 oz. butter
4 oz. caster sugar
2 eggs, separated
2 oz. self-raising flour
½ pt. milk

*Serves 4*

TO MAKE: Add the orange rind to the
butter and cream it with the sugar until pale
and fluffy. Add the egg yolks and beat well.
Stir in the flour, followed by the milk and
orange juice. Whisk the egg whites until
stiff and fold into the mixture. Pour into two
1-pt. buttered foil containers. Stand the
dishes in a shallow tin of water and cook in
the oven at 375°F (mark 5) for 45–50 min.,
when the top should be set firm. Cool.

TO PACK AND FREEZE: Open freeze the
pudding until firm, then wrap the dishes in
foil, place in polythene bags, seal, label and
return to the freezer.

TO USE: Unwrap the dishes and place them
on a baking sheet. Reheat from frozen in the
oven at 350°F (mark 4) for 1 hr.

*Note:* The pudding separates during cooking
into a custard layer with a sponge topping.

# Baked fruit salad

(*see picture above*)

¼ lb. dried apricots
¼ pt. white wine
15-oz. can sliced yellow peaches
juice of 1 lemon
6-fl. oz. can frozen Florida orange concentrate
2 oz. caster sugar
butter

*To complete dish from freezer*
3–4 bananas, peeled and thickly sliced
3 oz. seedless raisins

*Serves 4–6*

TO MAKE: Using scissors, snip the apricots
in half. Soak them overnight in the white
wine. Drain the syrup from the peaches into
a measuring jug, add the lemon juice and

make up to $\frac{3}{4}$ pt. with water. In a small saucepan, combine the peach syrup and the undiluted orange concentrate with the apricots, wine and sugar. Simmer over a low heat to dissolve the sugar. Bring to the boil, then simmer, covered, for 20 min. until the apricots are tender. Cool quickly and add peaches.

TO PACK AND FREEZE: Line a large container with a polythene bag and pour in the cooled fruit mixture. Freeze until solid, remove the container, seal and label the package and return it to the freezer.

TO COOK AND SERVE: Place the unwrapped fruit salad in a large, greased ovenproof dish and cook from frozen in the oven at 375°F (mark 5) for 30 min. Add the sliced bananas and raisins and bake for a further 30–40 min. Serve warm, with cream.

# Bread pudding

8 oz. white bread
$\frac{1}{2}$ pt. milk
4 oz. currants, raisins and sultanas, mixed
2 oz. chopped mixed peel
2 oz. shredded suet
2 oz. Demerara sugar
1$\frac{1}{2}$ level tsps. ground mixed spice
1 egg, beaten

*To complete dish from freezer*
grated nutmeg
caster sugar

*Serves 4–6*

TO MAKE: Remove the crusts from the bread, break it into small pieces and place in a basin. Pour the milk over and leave to soak for 30 min. Beat out the lumps, add the fruit, peel, suet, sugar and spice. Mix well. Add the egg, mix again and pour into a greased or foil-lined 1$\frac{1}{2}$-pt. pie dish or a greased rigid foil container.

TO PACK AND FREEZE: See Basic Sponge Pudding, page 164.

TO USE: Remove wrappings and replace the pudding in its original dish if necessary. Sprinkle with grated nutmeg. Place in the oven and bake, uncovered, from frozen at 350°F (mark 4) for about 2 hr. Dredge with caster sugar and serve hot.

# Bread and butter pudding

6 slices bread and butter, from a large loaf
2 oz. currants
1 oz. caster sugar
$\frac{3}{4}$ pt. milk
2 eggs, lightly beaten

*To complete dish from freezer*
grated nutmeg

*Serves 4*

TO MAKE: Remove the crusts from the bread and cut the slices into strips. In a foil-lined 1$\frac{1}{2}$-pt. ovenproof dish or similar sized rigid foil container, place the strips in layers buttered side up and sprinkle each layer with the currants and sugar. Whisk the eggs and milk together and strain this mixture over the bread.

TO PACK AND FREEZE: If the foil container has a lid this is all that is necessary; otherwise wrap the dish in polythene and place in the freezer. If using a foil-lined dish, remove the pre-formed package, wrap in polythene, seal and label.

TO USE: Remove wrappings and if necessary replace the pudding in its original dish. Sprinkle with grated nutmeg and bake from frozen in the oven at 325°F (mark 3) for 1 hr. Remove from the oven and put under a hot grill to brown for 2–3 min. Serve with cream.

# Hot marmalade bake

12 slices white bread, from a large loaf
3 oz. butter, softened
4 tbsps. ginger marmalade
1 level tbsp. caster sugar
1 egg
$\frac{1}{2}$ pt. milk

*To complete dish from freezer*
1 oz. margarine or butter
1 oz. Demerara sugar

*Serves 4*

TO MAKE: Remove the crusts from the bread and spread the slices generously with the butter then the marmalade. Make into 6 sandwiches and cut into 4. Grease an ovenproof dish or foil container, preferably a rectangle approx. 9 in. by 5 in. and place the pieces of sandwich overlapping in the dish. Sprinkle with the caster sugar, whisk the egg and milk together and pour over.

**TO PACK AND FREEZE:** Freeze the dish until the contents are firm, then wrap in foil, place in a polythene bag, seal and label. (If the foil container has a lid there is no need to overwrap.) Return to the freezer.

**TO USE:** Remove wrappings, place the dish on a baking sheet and cook, uncovered, from frozen in the oven at 350°F (mark 4) for 30 min. Dot with the margarine or butter, sprinkle with Demerara sugar and return to the oven for about a further 30 min. Serve hot with cream.

# Fruity peach lattice

**6 oz. shortcrust pastry (i.e. made with 6 oz. flour etc.)**
**12 oz. mincemeat**
**15-oz. can sliced peaches, drained**

*To complete dish from freezer*
**beaten egg**

*Serves 6*

**TO MAKE:** Roll out the pastry and use it to line a 7-in. fluted flan ring, placed on a baking sheet. Keep the pastry trimmings. Spread the base of the flan with 4 tbsps. mincemeat and cover with the peaches, arranged in a wheel pattern. Cover with the remaining mincemeat. Roll out the pastry trimmings, cut them into narrow strips and arrange in a lattice pattern across the mincemeat. Damp the ends of each strip and press firmly on to the pastry case.

**TO PACK AND FREEZE:** Open freeze the flan until firm. Remove the flan ring and the baking sheet carefully, wrap the flan in foil and place in a polythene bag. Label and seal.

**TO USE:** Unwrap and place the flan in the original flan ring on a baking sheet. Brush the pastry strips with beaten egg and cook in the oven from frozen at 375°F (mark 5) for about 1¼ hr. Remove flan ring and serve hot with cream.

# Peach and almond pinwheels

**6 oz. shortcrust pastry (i.e. made with 6 oz. flour etc.)**
**15-oz. can peaches, drained and chopped**
**4 level tbsps. ground almonds**
**2 oz. seedless raisins, chopped**
**¾ level tsp. ground mixed spice**
**1 level tbsp. soft brown sugar**

*To complete dish from freezer*
**beaten egg**
**sifted icing sugar**

*Serves 4*

**TO MAKE:** Roll the pastry out into a rectangle approx 6 in. by 8 in. Mix the peaches, almonds, raisins, spice and sugar and spread over the pastry. Damp one of the 6-in. edges and roll the pastry up from the other 6-in. edge. Seal the join firmly. Cut into 4 slices.

**TO PACK AND FREEZE:** Interleave each slice with non-stick or waxed paper or polythene film, reassemble into a roll and wrap in foil. Seal and label. Freeze.

**TO USE:** Unwrap and lay the frozen slices cut side down on a greased baking sheet. Brush the top cut edge with beaten egg and bake from frozen in the oven at 375°F (mark 5) for about 1 hr. Dust with icing sugar. Serve warm.

# Mincemeat roly-poly

**6 oz. self-raising flour**
**3 oz. suet**
**1 level tsp. salt**
**water to mix**
**6 tbsps. mincemeat**
**a little milk**

*To complete dish from freezer*
**beaten egg**

*Serves 6*

**TO MAKE:** Mix the flour, suet and salt together and add enough water to make a soft but manageable dough. Knead until smooth on a floured surface and roll out the dough to approx. 8 in. by 10 in. Spread the mincemeat over the pastry to within $\frac{1}{2}$ in. of the edges, damp the edges with milk and roll up the dough, working from the 8-in. side.

**TO PACK AND FREEZE:** Place the roll on a foil-lined baking sheet and open freeze until firm. Wrap it in the foil and place in a polythene bag. Seal, label and return to the freezer.

**TO USE: Either** Unwrap and thaw overnight in the refrigerator, place on a baking sheet, brush with beaten egg and bake in the oven at 400°F (mark 6) for 1 hr. **Or** Unwrap, place on a baking sheet, brush with beaten egg and bake from frozen in the oven at 375°F (mark 5) for about 2 hr. Serve hot with custard.

# Apple and redcurrant flan

6 oz. shortcrust pastry (i.e. made with
   6 oz. flour etc.)

*For the filling*
4 tbsps. redcurrant jelly
1 lb. cooking apples, peeled, cored and
   sliced
2 tbsps. white wine
1 level tbsp. caster sugar
1 egg, beaten

*For the topping*
2 oz. butter
2 oz. fresh white breadcrumbs
2 level tbsps. Demerara sugar

*Serves 6*

**TO MAKE:** Roll out the pastry and use it to line an 8-in. flan ring placed on a baking sheet. Spread the base of the flan with 2 tbsps. redcurrant jelly. Cook the apples in the wine with the sugar until soft, then beat them to a pulp or purée in a blender. Add the egg and 2 tbsps. redcurrant jelly and pour the mixture into the flan case. Melt the butter in a pan and gently fry the breadcrumbs with the Demerara sugar until golden, stirring frequently. Sprinkle over the apple filling.

**TO PACK AND FREEZE:** Open freeze the flan in the ring until firm, then remove the baking sheet and ring. Wrap the flan in foil and place it in a polythene bag. Seal, label and return it to the freezer.

**TO USE:** Unwrap, place the flan on a baking sheet and replace the flan ring. Bake in the oven from frozen at 375°F (mark 5) for about $1\frac{1}{2}$ hr. Serve hot with cream.

# Mandarin treacle tart

(*see picture opposite*)

6 oz. shortcrust pastry (i.e. made with
   6 oz. flour etc.)
3 oz. fresh white breadcrumbs
8 oz. golden syrup
grated rind and juice of 1 lemon
11-oz. can mandarin oranges, drained

*To complete dish from freezer*
icing sugar

*Serves 6*

**TO MAKE:** Roll out the pastry and use to line a 7-in. fluted flan ring on a baking sheet. Combine the breadcrumbs, syrup, lemon rind and juice and mandarin oranges. Pour into the case. Bake in the oven at 400°F (mark 6) for 50–60 min. Remove the flan ring and cool on a wire rack.

**TO PACK AND FREEZE:** Open freeze until firm then wrap in foil. Place in a polythene bag, seal, label and return to the freezer.

**TO USE: Either** Unwrap, place on a baking sheet and allow to stand at room temperature for 5–6 hr., then reheat in the oven at 350°F (mark 4) for about 30 min. **Or** Unwrap, place on a baking sheet and reheat from frozen in the oven at 375°F (mark 5) for about 1 hr. Serve warm, dredged with icing sugar.

# Black cherry and rhubarb pie

1-lb. 4-oz. can black cherries
16-oz. can rhubarb, drained
1 level tbsp. tapioca
$\frac{1}{4}$ level tsp. ground cinnamon
8 oz. shortcrust pastry (i.e. made with
   8 oz. flour etc.)
1 oz. butter

*To complete dish from freezer*
1 egg white
granulated sugar

*Serves 4–6*

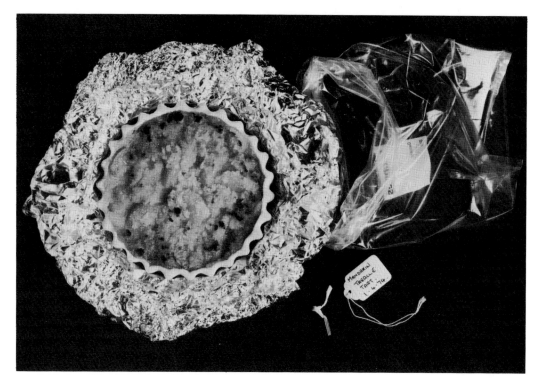

*Mandarin treacle tart*

**TO MAKE:** Drain the cherries, reserve the juice and stone the fruit. Mix the cherries and rhubarb with the tapioca, cinnamon and 6 tbsps. cherry juice. Leave to stand for 30 min. Roll out half the pastry and use it to line an $8\frac{1}{2}$-in. pie plate. Spoon in the filling and dot with butter. Roll out the remaining pastry and use to make a lid. Seal the edges well.

**TO PACK AND FREEZE:** Place the pie on a baking sheet (to keep it level) and open freeze. Remove from the baking sheet and wrap, complete with plate, in foil. Place in a polythene bag, seal, label and return to the freezer.

**TO USE:** Remove the wrappings, put the plate on a baking sheet and cook from frozen in the oven at 375°F (mark 5) for 30 min. Remove from the oven, brush with egg white, make a slit in the centre of the top crust and dust heavily with granulated sugar. Replace in the oven and cook for about a further 1 hr.

# Almond fruit puff

2 oz. butter
2 oz. caster sugar
$\frac{1}{2}$ tsp. almond essence
1 egg, beaten
$\frac{1}{2}$ oz. flour
2 oz. ground almonds
2 oz. Maraschino cherries, halved
8-oz. can each prunes, sliced peaches and
  apricot halves
1 small apple, peeled and chopped
1 lb. frozen puff pastry, thawed
1 egg, beaten

*To complete dish from freezer*
1 egg, beaten

*Serves 6*

**TO MAKE:** Cream the butter and sugar. Beat in the essence, egg, flour and almonds. Lightly fold in the cherries, well-drained canned fruits and apple. Roll out a quarter of the pastry thinly and trim to an 8-in. diameter circle. Place this on a baking sheet and prick with a fork. Brush a 1-in. rim with egg. Divide the remaining pastry into 2 portions, roll out one piece and trim into a 9-in. circle for the lid. Roll out the remainder in a strip 1 in. wide and about 14 in. long. Divide in half lengthwise. Lay the strips over the egg-glazed rim to form a 'wall'. Leave the pastry to relax for 30 min. in the refrigerator. Pile the fruit mixture within the 'wall'. Brush the 'wall' with egg and place the pastry lid on top. Press the edges together lightly. Roll out the trimmings thinly and cut in narrow strips; arrange these on top of the lid like spokes of a wheel.

**TO PACK AND FREEZE:** Open freeze until firm, then wrap in foil and place in a polythene bag. Seal, label and return to the freezer.

**TO USE:** Unwrap, place the puff on a baking sheet, brush with beaten egg and snip the top of the pastry to allow steam to escape. Bake from frozen in the oven at 450°F (mark 8) for 15 min. Reduce the oven temperature to 350°F (mark 4), cover the pastry with damp greaseproof paper and continue to cook for about 1 hr.

# Apricot cheese flan

6 oz. shortcrust pastry (i.e. made with
  6 oz. flour etc.)
1-lb. 3-oz. can apricot halves
4 oz. full fat soft cream cheese or curd
  cheese
3 oz. butter
3 oz. caster sugar
1 egg
4 oz. self-raising flour
2 level tsps. ground mixed spice

*To complete dish from freezer*
icing sugar

*Serves 6*

**TO MAKE:** Roll out the pastry and use it to line an 8-in. fluted flan ring on a baking sheet. Drain the apricot halves but reserve the juice. Beat the cheese until soft and smooth, cover the base of the pastry with apricots (cut side down) and dot with the cheese. Cream the butter and sugar until light and fluffy, then add the egg and beat well. Sift the flour with the mixed spice and

lightly beat into the butter and egg mixture, alternately with $1\frac{1}{2}$ tbsps. of the apricot juice. Pile the mixture on top of the cheese and smooth over.

**TO PACK AND FREEZE:** Open freeze until firm with the flan ring in position. Remove the ring, wrap in foil and place in a polythene bag. Seal, label and return to the freezer.

**TO USE:** Unwrap, place the flan on a baking sheet and put the flan ring firmly in place. Bake from frozen in the oven at 400°F (mark 6) for 15 min; reduce the oven temperature to 325°F (mark 3) and continue cooking for about 30 min. Dust with icing sugar and serve warm.

# Lemon cheese meringue

6 oz. shortcrust pastry (i.e. made with
  6 oz. flour etc.)

*For the filling*
2 oz. butter
3 oz. caster sugar
3 egg yolks
2 oz. plain flour
8 oz. cottage cheese, sieved
grated rind and juice of 2 lemons

*To complete dish from freezer*
3 frozen egg whites, thawed
2 oz. caster sugar
1 level tbsp. granulated sugar

*Serves 6*

**TO MAKE:** Roll out the pastry and use it to line an 8-in. flan ring on a baking sheet. Cream the butter and sugar until light and fluffy and beat in the egg yolks. Fold in the flour, cheese, lemon rind and juice. Turn into the pastry case.

**TO PACK AND FREEZE:** Open freeze the flan with the ring in position until firm. Remove the ring and wrap the flan in foil and place it in a polythene bag. Seal, label and return it to the freezer.

**TO USE:** Unwrap the flan, place it on a baking sheet and replace the flan ring. Thaw for 2 hr. at room temperature. Bake in the oven at 400°F (mark 6) for 35 min. Whisk the egg whites until stiff and fold in the caster sugar. Remove the flan from the oven; reduce

For all the wonderful benefits deep freezing brings it still means an expensive outlay on equipment. It's therefore sensible that once you buy a freezer you make the best and fullest use of it. At Alcan we've been involved in the freezer revolution from the start, developing a range of accessories with hundreds of practical home-freezing uses. Here are a few ideas using Alcan products that we've come-up with. We hope you'll try them for yourselves, and maybe send us your own suggestions.

## Alcan Wrap
### Keep the flavour out of the vegetables.

If you've never used Alcan Wrap you're in for a pleasant surprise. Wrap is a self-sealing, see-through, plastic that clings to itself or to a container's surface when pressed. Placed around food, Wrap forms a virtually air-tight parcel thus preventing smells from spreading. So you can safely put kippers next to cake in a deep-freeze! Wrap comes in handy when you need a lid for a foil container. It's also very good for wrapping cheese, particularly if you buy, say, a whole Cheddar and cut it into meal-sized chunks. Wrap comes in rolls 49' long and can be used in any Alcan Foil Dispenser.

Here's an interesting idea: frozen herbs. Just seal the herbs in Wrap, pop them in the deep-freeze ready later to be defrosted and added to soups, stews, casseroles or sauces.

## Alcan Foil

Alcan Foil has one great advantage: it moulds to fit any shape. Not only that, but it prevents smells from spreading and, correctly used, forms an airtight package. After wrapping in foil it's advisable to put the package inside a polythene bag to prevent tearing. Alcan Foil comes in two different widths (12" and 20" approx.) and six lengths ranging from 15' to 300' approx.

### If you ever find a really good baker buy a lot.

Extra-Wide Foil is marvellous for wrapping and freezing bread. Ordinary loaves should be wrapped in foil then placed inside an Alcan polythene Big-Bag. You might like to try hot garlic bread wrapped in foil and popped in the oven to warm. A quick breakfast tip: frozen sliced bread can be toasted straight from the freezer.

### If you're serious, try tray freezing.

The best way of freezing fruits and vegetables is to keep them separate using a method known as open freezing. Taking cauliflower for example, lay individual florets on a baking tray covered with foil, fit the tray into the fast freezing compartment, freeze until solid then pack the florets in Alcan polythene Big-Bags or Freezer Trays.

## Alcan Cook-Bags

Alcan see-through, oven Cook-Bags aren't specially designed for use in freezers. You can, however, take beef from your freezer, put it in a floured Cook-Bag and place the bag directly in the oven adding 30 minutes to normal cooking time. Meat done like this cooks in its own juices inside the bag so you're left with a clean oven, lots of meaty gravy, and tasty results every time.

## Alcan Big-Bags

Alcan have a pack of 25 bags-on-a-roll called Big-Bags. They're called Big because they have special gussets which expand as you fill them. So they measure 7" x 13½" empty opening out to 11" wide. It's best to double-wrap foods in Big-Bags; the double layer gives added protection. You probably already use polythene bags. So here are a few freezing ideas we've come up with for you to try.

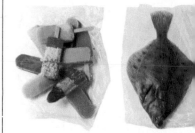

Whole fish, whales excepted, fit easily into a Big-Bag as do individual portions such as halibut or salmon steaks. (Do make sure fish is really fresh.) A refreshing idea is frozen lemon juice cubes. Squeeze fresh lemons, freeze juice in an ice-cube tray, then put the cubes in a polythene bag and wait for summer. Puréed tomato can also be done like this or you can simply pour the purée straight into the bag. Ice lollies stored in bulk will keep the kids happy all the year round.

Awkwardly shaped vegetables such as corn on the cob fit very well in bags. Do make sure that you pack them tightly. Soups, stocks, stews and Chili con Carne present no problem as far as shape is concerned: when cooled just pour them straight into polythene bags placed inside a rigid straight-sided container, seal, then freeze. Remove bag from container and store in freezer. Chips go with everything of course. They also go very well inside a Big-Bag. But if you fancy a change, what about Pizza Pie?

# out who your friends are

## Alcan Freezer Trays & Lids

One of the newest additions to Alcan's [ran]ge is Foil Freezer Trays with lids. These [var]y in size from a Two-Portion tray up to a [Par]ty Size container. For many fruits, [veg]etables and prepared meals, trays like [the]se are the best method available for [ho]me-freezing.

### [T]ake a tip.

[F]or your interest, here are a few general [hin]ts about freezing foods for packing in foil [tra]ys. Always freeze food at its freshest, [pre]ferably on the day you buy or pick it. [Af]ter blanching, cooling must be done very [qui]ckly. Use iced or running water.) Food [usu]ally expands when frozen so always leave [spa]ce over moist food packed in trays. Always [rem]ember to label your tray with freezing date, [con]tents and latest possible date for defrosting.

It's sometimes a good idea to freeze food [sli]ced or garnished in the way you intend [us]ing it. So why not add mint to peas before [free]zing? With many fruits you will want to [add] sugar to the layers, but do carefully [che]ck quantities first.

### [C]hocolate delights.

[I]f your family has a sweet tooth tempt them [wit]h this little number.

### [Ch]ocolate Cake

[Use a] 4-portion size Freezer Tray.

[... o]z. butter or margarine
[... o]z. soft brown sugar
[... la]rge egg
[... o]z. self-raising flour
[... le]vel tablesp. cocoa
[... ab]out 1 tablesp. milk

[... ] Cream butter and sugar together until [ligh]t and fluffy. Beat in egg. Fold in sieved [flo]ur and cocoa. Mix to a soft dropping [con]sistency with the milk. Turn mixture into [gre]ased four-portion Alcan Freezer Tray. [Le]vel mixture, hollowing centre slightly. Bake [in] a moderately hot oven (Gas No. 5 – 375°F.) [for] about 25 minutes. Cool. Cover with lid. [lab]el. Freeze quickly. When required, [all]ow cake to thaw, then decorate as liked.

## Alcan Dishes

1. 14 x 4½″ party dishes
2. 50 x 3″ tartlet cases
3. 7 Freezer Trays with lids, 2-portion size
4. 8 x 6¾″ plates
5. 34 x 3¼″ baking cases
6. 5 x 1 lb. pie dishes
7. 5 x 1 lb. pudding basins
8. 6 x 8″ plates
9. 5 Freezer Trays with lids, 3-portion size
10. 3 x 10″ plates
11. 3 x 2 lb. pie dishes
12. 2 Freezer Trays with lids, Family size
13. 4 Freezer Trays with lids, 4-portion size
14. 2 Freezer Trays with lids, Party size

### Sweetie pie.

Sweet red cherries freeze very well. Here's what you do: Remove stalks from the ripe cherries. Wash and stone if required. Layer the cherries in dry sugar (¼ lb. sugar to 1 lb. stoned cherries) in an Alcan Freezer Tray. Cherries will keep for up to 12 months. But why wait when you can use them to make a tasty pie?

## Feeding Five Thousand

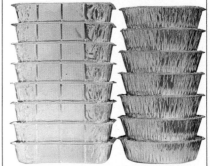

One of the great advantages of home freezing is that you can buy in bulk in season and save money. Here's a month-by-month seasonal buying guide for fruit and vegetables at their best.

| | |
|---|---|
| JAN. | **Fruit:** Seville oranges, cranberries, satsumas. **Veg:** Sprouts, celery, baby onions, aubergines, swedes, turnips, mushrooms. |
| FEB. | **Fruit:** Grapes. **Veg:** Parsnips, green peppers, swedes, turnips, Jerusalem artichokes, broccoli, cabbage, celery, mushrooms, aubergines. |
| MAR. | **Fruit:** Rhubarb. **Veg:** Aubergines, cabbage, mushrooms. |
| APR. | **Fruit:** Rhubarb. **Veg:** Cauliflower, aubergines, spinach. |
| MAY | **Fruit:** Apricots, rhubarb, gooseberries. **Veg:** Carrots, cauliflower, asparagus, spinach, aubergines. |
| JUNE | **Fruit:** Strawberries, apricots, cherries, pineapple, rhubarb, gooseberries. **Veg:** Peas, carrots, cauliflower, courgettes, new potatoes, asparagus, peppers, tomatoes, mushrooms. |
| JULY | **Fruit:** Raspberries, strawberries, apricots, currants, cherries, loganberries, cherries. **Veg:** Broad beans, French beans, cauliflower, tomatoes, new potatoes, peas. |
| AUG. | **Fruit:** Blackberries, cherries, currants, plums, greengages, peaches, seedless grapes, melons, pineapple. **Veg:** Runner beans, broad beans, French beans, spinach, tomatoes, marrow, carrots, artichokes, mushrooms, sweetcorn. |
| SEPT. | **Fruit:** Cooking apples, plums, blackberries, damsons, greengages, melons, pears, peaches, seedless grapes, strawberries. **Veg:** Runner beans, beetroot, courgettes, tomatoes, marrow, herbs, cauliflower, peppers, spinach, swedes, carrots, mushrooms, sweetcorn, artichokes. |
| OCT. | **Fruit:** Blackberries, cooking apples, pears, damsons, peaches, plums, strawberries. **Veg:** Courgettes, peppers, aubergines, leeks, baby onions, swedes, parsnips. |
| NOV. | **Fruit:** Cooking apples, pears, mandarins, tangerines, satsumas, cranberries. **Veg:** Parsnips, swedes, onions, carrots, sprouts, leeks, aubergines, celery. |
| DEC. | **Fruit:** Cranberries, mandarins, tangerines, satsumas. **Veg:** Sprouts, celery, swedes, turnips, baby onions, parsnips. |

To cope with your bulk buying you'll want plenty of trays and dishes. You'll find it worthwhile checking at your freezer centre to see if they stock the current range of Alcan Foil Catering Containers listed below.

1. 14 Square Trays with lids
2. 25 Rectangular Trays with lids
3. 50 Freezer Trays with lids
4. 40 2 lb. Pie Dishes
5. 25 Round Dishes with lids
6. 30 Rectangular Dishes with lids
7. 50 8″ Plates
8. 6 Serving Platters
9. 30 10″ Plates

*Peach and apple slice*

the temperature to 300°F (mark 2). Pile the meringue on top of the flan, making sure it covers the filling right to the edge but does not overlap (if it overlaps it is difficult to remove the flan ring). Sprinkle the granulated sugar over the top and return it to the oven for a further 20–25 min. until the topping is tinged with brown. Remove the flan ring and serve at once.

# Peach and apple slice

(*see picture above*)

8 oz. frozen puff pastry, thawed
beaten egg
¾ lb. cooking apples, peeled, cored and
   sliced
1–2 tbsps. water
1 oz. caster sugar
1 tbsp. currants
1 tbsp. sultanas
2 level tsps. ground cinnamon
15-oz. can sliced peaches, drained

*To complete dish from freezer*
icing sugar

*Serves 6*

TO MAKE: Roll the pastry out to approx. 9 in. square. Cut off a piece of pastry for the base approx 5 in. by 9 in. and place on a dampened baking sheet. Prick the base well. Brush the edges with beaten egg to make a 1-in. border of brushed pastry all round. From the remaining pastry cut 2 strips 1 in. by 9 in. and press these firmly on top of the long sides of the base. Cut 2 further pieces for the shorter sides and press them firmly into position. Brush the pastry borders with egg and bake in the oven at 425°F (mark 7) for 20–25 min. until golden brown.

Meanwhile place the apples, water and sugar in a pan and cook over a gentle heat until the apples are soft. Remove from the heat, beat well until broken down and stir in the currants, sultanas and cinnamon. Reserve 12 peach slices and chop the remaining slices into small pieces. Add the chopped peaches to the apple mixture. Remove the pastry from the oven. Pile the apple mixture in the centre and arrange the 12 remaining peach slices round the edge of the filling. Cool.

TO PACK AND FREEZE: Open freeze on a baking sheet until firm, then wrap the slice in foil and place it in a polythene bag. Seal, label and return to the freezer.

TO USE: Unwrap, place the slice on a baking sheet and reheat from frozen in the oven at 350°F (mark 4) for about 45 min. Dust with icing sugar and serve warm.

*Brandy cornets (page 235); Coffee éclairs (page 234);*
*Hungarian chocolate gâteau (page 221)*

# Ginger pear pie

8 oz. shortcrust pastry (i.e. made with
8 oz. flour etc.)

*For filling*
1-lb. 3-oz. can pear halves
4 pieces stem ginger, chopped
4 oz. stoned raisins
1 level tbsp. arrowroot
1 tbsp. ginger syrup (from stem ginger jar)

*To complete dish from freezer*
milk and caster sugar to glaze

*Serves 6*

**TO MAKE:** Roll out half the pastry and use
it to line a 10-in. foil pie plate. Drain the
pears, reserving $\frac{1}{4}$ pt. juice, and slice them
thickly. Put the pears, chopped ginger and
raisins into the lined pie plate. In a saucepan,
blend the arrowroot with 2 tbsps. of the
measured pear juice, add the rest of the juice
and the ginger syrup. Bring to the boil,
stirring until thickened and clear. Cool the
syrup and pour it over the fruit. Roll out the
remaining pastry to make a lid for the pie.
Damp the edges and press well together to
seal. Flute the edges of the pie.

**TO PACK AND FREEZE:** Open freeze until
firm, then wrap the pie in foil and place it in
a polythene bag. Seal, label and return it to
the freezer.

**TO USE:** Unwrap the pie and brush the top
with milk. Dredge heavily with sugar. Place
on a baking sheet and bake from frozen in
the oven at 375°F (mark 5) for about $1\frac{1}{2}$ hr.
Serve hot with custard or cream.

# Apricot and pineapple puffs

12 oz. frozen puff pastry, thawed
$7\frac{1}{2}$-oz. can apricot halves, drained and
chopped
7-oz. can pineapple slices, drained and
chopped
3 tbsps. apricot jam

*To complete dish from freezer*
1 egg white
1 level tbsp. granulated sugar

*Serves 4*

**TO MAKE:** Roll out the pastry to $\frac{1}{8}$-in.
thickness and cut into 8 squares, making 4
slightly larger than the others. Mix the
apricots and pineapple with half the jam
and spread the smaller squares to within
$\frac{1}{2}$-in. of the edge with the remaining half of
the jam. Pile the fruit mixture in the centre of
the smaller squares on top of the jam and
damp the $\frac{1}{2}$-in. border. Place the larger
squares on top of the filling, press the edges
together firmly and knock up.

**TO PACK AND FREEZE:** Open freeze the
puffs until firm. Pack them in foil and place
in a polythene bag. Seal, label and return to
the freezer.

**TO USE:** Unwrap and place the puffs on a
damp baking sheet. Brush them with egg
white and dust with sugar. Bake from
frozen in the oven at 400°F (mark 6) for
about 40 min., until puffed and golden
brown.

# DISHES USING FROZEN FRUIT FROM THE FREEZER

## Apple and blackberry lattice pie

1 lb. frozen sliced apples
4 oz. sugar
4 tbsps. water
8 oz. frozen blackberries
8 oz. shortcrust pastry (i.e. made with
    8 oz. flour etc.)

*Serves 4–6*

In a saucepan, cook together the apples, sugar and water until the apple is tender but not mushy. Leave until cold. Fold the blackberries through the apple mixture. Stand a $7\frac{1}{2}$-in. fluted flan ring on a baking sheet. Roll out the pastry and use it to line the ring, pressing it firmly into the flutes. Fill with the apple mixture. Roll out the pastry trimmings and cut them into strips, using a pastry wheel. Arrange these lattice-fashion over the fruit, using a little water to fix the strips. Bake in the oven at 400°F (mark 6) for about 45 min. until the pastry is light and golden. Serve warm with whipped cream.

## Cran-apple plate pie

$1\frac{1}{2}$ lb. frozen sliced apples
$\frac{1}{2}$ lb. frozen cranberries
4 oz. sugar
$\frac{1}{4}$ pt. water
8 oz. shortcrust pastry (i.e. made with
    8 oz. flour etc.)
milk and icing sugar to glaze

*Serves 6*

In a saucepan, cook together the apples, cranberries, sugar and water until the apples and cranberries are tender but not completely broken down; leave until cold. Roll out half the pastry and use it to line a 10-in.

pie plate. Spoon the cold fruit mixture on to the pastry, leaving the rim clear. Brush the rim with milk. Roll out the remaining pastry for a lid. Place it in position and press the rims together. Knock up the edges and make 2 slits in the lid to allow the steam to escape. Brush the pastry with milk and dredge heavily with icing sugar. Bake in the oven at 400°F (mark 6) for about 45 min. Serve hot with cream.

## Marshmallow soufflé pudding

12-oz. pkt. frozen gooseberries
3 oz. granulated sugar
4 oz. margarine
4 oz. caster sugar
2 eggs, beaten
6 oz. self-raising flour
grated rind and juice of 1 orange
12 marshmallows
1 tbsp. long-shredded coconut
lightly whipped cream

*Serves 4–6*

Place an $8\frac{1}{2}$-in. ovenproof glass or china flan dish on a baking sheet. Fix a band of foil (double thickness) round the outside of the dish to form a collar about 1 in. above the rim. Grease the dish and the foil. Place the frozen gooseberries in the base and add the granulated sugar. Make up a sponge mixture using the margarine, caster sugar, eggs and flour, adding the orange rind and juice with the eggs (see Basic Sponge Pudding, page 164). Use scissors to halve the marshmallows. Stir 18 pieces into the sponge mixture and spoon it evenly over the fruit. Top with the remaining marshmallows and the coconut. Bake in the oven at 350°F (mark 4) for about $1\frac{1}{4}$ hr. Remove the foil band and serve warm with lightly whipped cream.

# 9 Party Ice Creams and Desserts

There are several varieties of both cream and water ices but the most common types are:

*Cream ices:* These are seldom made entirely of cream, but any of the following mixtures give good results: equal parts cream and custard (egg custard, made with yolks only); cream and fruit purée; cream and egg whites. The cream may be replaced by unsweetened evaporated milk and flavouring and colouring ingredients are added as required.

*Water ices:* The foundation is a sugar and water syrup, usually flavoured with fruit juice or purée; wine or liqueur is frequently added.

*Sherbets:* A true sherbet is a water ice with whipped egg white added, to give a fluffy texture.

*Sorbets:* Semi-frozen ices, sometimes flavoured with liqueur. Because they are soft, they are not moulded but served in tall glasses or goblets.

*Bombes:* Iced puddings frozen in special bombe shaped moulds; bombes may be made of a single ice cream mixture known as a parfait, or of two or more. They are often elaborately decorated after unmoulding.

## Handy hints for making ice cream

To obtain the best results when making your own ice cream, use a rich mixture with plenty of flavouring. A flavour that tastes quite strong at room temperature will become much less so when frozen. See that the mixture is well sweetened, too, as this helps to accentuate the flavour of the frozen ice. But don't go to the other extreme and over-sweeten, as too much sugar will prevent the mixture freezing properly (too much alcohol used as flavouring will also prevent freezing). Colouring is not affected by freezing, so use sparingly, a drop at a time. Where recipes use double cream, whipping cream may be substituted. Use the same quantity of cream and when the recipe requires the cream to be whipped it should just leave a trail in the bowl. Cream which is whipped until stiff is difficult to incorporate into the mixture.

## To pack and freeze

Home-made ice cream generally has a firmer texture than the commercial ice cream, which is more highly aerated. Home-made ice cream may be made in any shallow rigid container, such as a 10-in. by 4-in. ice-cube tray without partitions, a 7-in. sandwich tin or a shallow foil container, but it is not advisable to leave it in freezing trays for storage in the freezer, as the acid can react on the metal. It is better to line the ice trays with foil or waxed paper before the ice cream is finally frozen; alternatively, when the ice cream is almost frozen, but still malleable, scoop it into rigid polythene containers or clean, used bought ice cream boxes. If using the first method, wrap the ice cream in foil, overwrap, seal and label; for the second method, simply put the lid on, seal with freezer tape and label.

### To use home-made ices
To use ice cream from the freezer it is necessary to let it 'come to', i.e. to soften a little, and timing is very much a question of personal judgement. With some recipes, just 10 min. at room temperature is enough, but to get even thawing, the ice-cube compartment of an ordinary refrigerator is the better place, provided you know that it is set at about 20°F (−7°C). Put the ice cream or sorbet from the freezer into the compartment and leave for 4–6 hr. Or, if this compartment proves too cold, leave it in the body of the refrigerator for about 2 hr.

### Commercial ice cream
Ice cream is best kept away from the coldest part of the freezer, so that it doesn't become excessively hard. When taking ice cream from a bulk container, a special scoop is very practical; using what is known as a 'size 16' scoop (giving a 3-fl. oz. portion) you should get about 52 helpings from 1 gallon. Scoop evenly across the can, and replace the lid at once. It does help to warm the scoop before use, but don't make it too hot or you will melt the surrounding ice cream.

### To serve
Serve in glass dishes, goblets or stemmed glasses and top with a delicious sauce (see page 199 and Sauces chapter), whipped cream and nuts, fruit, or serve plain with a wafer.

# CREAM ICES

## Basic (vanilla) ice cream

½ pt. milk
1 vanilla pod
1 whole egg
2 egg yolks
3 oz. caster sugar
½ pt. double or whipping cream

*Makes about 1 pt.: serves 4–6*

TO MAKE: Bring the milk and vanilla pod almost to the boil and infuse off the heat for about 15 min.; remove the pod. Cream the whole egg, yolks and sugar together until pale. Stir in the flavoured milk and strain the mixture through a sieve back into the saucepan. Heat the custard very slowly, stirring all the time, until it just coats the back of a spoon. Pour it into a basin and leave to cool.

Pour the cold custard into a container and freeze until mushy. Remove from the tray and beat well. Lightly whip the cream and fold it through the mixture. Continue freezing until mushy; beat it a second time and freeze again until firm.

TO PACK AND USE: See notes at beginning of chapter.

# Rum and raisin ice cream

4 oz. raisins
4 tbsps. rum
4 eggs, separated
4 oz. icing sugar
½ pt. double cream

*Serves 6–8*

TO MAKE: Soak the raisins in the rum for about 30 min. Whisk the egg yolks and sugar together until thick and fluffy. Beat the cream until thick and fold it into the egg mixture. Add the rum-soaked raisins. Whisk the egg whites until stiff and fold into the mixture. Pour into a container, freeze for about 1 hr., then stir the mixture to re-distribute the raisins. Seal the container with freezer tape and label.

TO PACK AND USE: See notes at beginning of chapter.

# Fruit ice cream

small can evaporated milk
8 oz. fresh raspberries or a 15-oz. can, drained, or other fruit such as strawberries, pears, bananas
2 oz. sugar (if fresh fruit is used)

*Serves 4*

TO MAKE: Chill the unopened can of milk for several hours. Sieve the fruit to make ¼ pt. purée; add the sugar if required. Open the can of milk and whip until stiff. Fold in the purée. Pour into a container and freeze.

TO PACK AND USE: See notes at beginning of chapter.

# Orange ice cream

½ pt. double cream
½ pt. orange juice
caster sugar

*Serves 4*

TO MAKE: Whip the cream until it leaves a trail when the whisk is lifted. Stir in ¼ pt. orange juice and add the remaining ¼ pt. a little at a time. Add sufficient sugar to make the mixture taste slightly over-sweet. Pour into a container and freeze.

TO PACK AND USE: See notes at beginning of chapter.

# Rich chocolate ice cream

½ pt. milk
3 oz. sugar
2 eggs, beaten
3 oz. plain chocolate, melted
½ pt. double cream

*Serves 4*

TO MAKE: Heat the milk and sugar and pour on to the eggs, stirring. Return the mixture to the saucepan and cook over gentle heat, stirring all the time until the custard thickens; strain. Add the melted chocolate and allow to cool. Whip the cream until it leaves a trail when the whisk is lifted then fold in the mixture. Pour into a container and freeze.

TO PACK AND USE: See notes at beginning of chapter.

# Christmas ice cream

¼ pt. milk
½ tsp. vanilla essence
1 egg
1½ oz. caster sugar
2 tbsps. mincemeat
1 tbsp. brandy
¼ pt. double cream

*Serves 4*

TO MAKE: Heat the milk and essence, cream the egg and sugar together and pour on the milk. Strain through a sieve back into the saucepan and heat gently to thicken, stirring all the time. Cool. Add the mincemeat and brandy. Whip the cream lightly and fold it into the mixture. Pour into a container and freeze until firm.

**TO PACK AND USE:** See notes at beginning of chapter.

# Chestnut ice cream

¼ pt. milk
1 egg
1½ oz. caster sugar
8¾-oz. can chestnut spread
¼ pt. double cream

*Serves 4–6*

**TO MAKE:** Heat the milk, cream the egg and sugar together and pour on the milk. Strain through a sieve back into the pan and heat slowly to thicken, stirring all the time. Stir in the chestnut spread and mix well. Cool. Whip the cream until it leaves a trail when the whisk is lifted and fold it into the chestnut mixture. Pour into a container and freeze until firm.

**TO PACK AND USE:** See notes at beginning of chapter.

# Caramel ice cream

3½ oz. caster sugar
3 tbsps. boiling water
¼ pt. milk
1 egg
1 egg yolk
¼ pt. double or whipping cream, lightly
   whipped

*Serves 4*

**TO MAKE:** Put 2 oz. sugar in a heavy pan and heat it very gently until it dissolves; continue to heat until it browns very slightly. Remove from the heat and pour on 3 tbsps. of boiling water, very slowly. Return to the heat, simmer to dissolve the caramel, add the milk and stir. Beat the whole egg and yolk with 1½ oz. sugar, add the caramel milk and strain into a saucepan. Cook very gently until the custard coats the back of a spoon. Pour into a basin and cool. Transfer the

cooled custard to a container. Freeze until mushy. Beat well and fold in the lightly whipped cream. Freeze again until firm.

**TO PACK AND USE:** See notes at beginning of chapter.

# Choc-de-menthe

1 recipe basic vanilla ice cream (page 181)
4 tsps. crème de menthe
green colouring
bar of flake chocolate

*Serves 4–6*

**TO MAKE:** Make up the basic vanilla ice cream recipe, omitting the vanilla pod and adding 4 tsps. crème de menthe and a few drops of green colouring. When the mixture is half frozen, turn into a chilled bowl and whisk. Crumble a small bar of flake chocolate and fold in. Return to the tray and freeze.

**TO PACK AND USE:** See notes at beginning of chapter.

# Berry ice cream

2 level tsps. powdered gelatine
2 tbsps. water
15-oz. can strawberries or raspberries,
   puréed, or 1 lb. fresh fruit, puréed
1–3 oz. soft brown sugar
1½ tbsps. lemon juice
¼ pt. single cream
2 oz. digestive biscuit crumbs
½ pt. double cream

*Serves 8*

**TO MAKE:** Dissolve the gelatine in the water in a basin placed over a pan of hot water. Pour the fruit purée into a bowl, add sugar to taste and the lemon juice. Stir in the dissolved gelatine, single cream and crushed biscuit crumbs. Two-thirds fill a container and freeze until fairly stiff. Beat again and fold in the whipped double cream. Freeze again.

**TO PACK AND USE:** See notes at beginning of chapter.

# Toasted almond ice

½ pt. milk
3 oz. sugar
pinch of salt
4 egg yolks
¼ pt. double cream
vanilla essence
2 oz. crushed praline (see below)

*To complete ice cream from freezer*
2 egg whites
3 level tbsps. icing sugar
1 tbsp. chopped toasted almonds

*Serves 4*

TO MAKE: Boil the milk. Cream the sugar,
salt and egg yolks and stir in the milk. Cook
over a very gentle heat, stirring constantly,
until thick, then allow to cool. Pour into a
container and freeze for 20–30 min. Turn the
mixture out, whip until creamy and add the
cream, a little vanilla essence and the
crushed praline. Pour the mixture back into
the freezing container and freeze until firm,
stirring it after the first 30 min.

TO PACK: See notes at beginning of chapter.

TO USE: Serve with a sauce made by whipping the egg whites and icing sugar together
until just stiff enough to drop from a spoon,
then adding the chopped nuts.

### PRALINE

2 oz. caster sugar
2 oz. chopped toasted almonds

TO MAKE: Heat the sugar in a pan until a
deep amber colour. Add the chopped
almonds and pour the mixture on to a
greased tin or slab. When it is cool, pound in
a mortar or crush with a rolling pin.

# Honey ice cream

1 lb. fresh or frozen thawed raspberries
¼ pt. double cream
¼ pt. plain yoghurt
2 tbsps. lemon juice
10 level tbsps. clear honey
pinch of salt
3 egg whites

*Serves 8*

TO MAKE: Sieve the raspberries to give ½ pt.
purée. In a bowl, blend together the raspberry purée, cream, yoghurt, lemon juice,
honey and salt. Turn the mixture into a
container and freeze until firm. Turn out
into a bowl and beat until smooth. Stiffly
whisk the egg whites and fold in. Return the
mixture to the container and freeze.

TO PACK AND USE: See notes at beginning
of chapter.

# Lemon ice cream

½ pt. double cream
2 eggs, beaten
grated rind and juice of 2 lemons
8–10 oz. caster sugar
½ pt. milk

*Serves 6*

TO MAKE: Beat together the cream and eggs
until smooth. Add the lemon rind, the juice,
sugar and milk and mix thoroughly. Pour
into container and freeze for about 2 hr. Do
not stir while freezing.

TO PACK AND USE: See notes at beginning
of chapter.

# Apple and ginger ice cream

¼ pt. milk
1 egg yolk
1½ oz. caster sugar
3 pieces stem ginger, chopped
2 tbsps. ginger syrup
¼ pt. apple purée
edible green colouring
¼ pt. double cream

*Serves 4–6*

TO MAKE: Heat the milk gently but do not
allow to boil. Cream the egg yolk with the
sugar and pour on the milk. Strain through
a sieve back into the saucepan and heat,
slowly stirring all the time until it thickens.
Cool. Add the chopped ginger and syrup to
the apple purée and add enough colouring
to give a rich green appearance. (The colour
will become pale once the custard and cream

are added.) Combine the apple mixture with the custard. Whip the cream lightly and fold it into the apple mixture. Turn into a container and freeze until firm.

**TO PACK AND USE:** See notes at beginning of chapter.

# Banana ice cream

4 bananas
2 oz. icing sugar
2 tbsps. orange juice
2 tbsps. lemon juice
½ pt. whipping cream
edible yellow colouring

*Serves 6*

**TO MAKE:** Purée the bananas, sugar, orange and lemon juice in a blender. Whip the cream until it leaves a trail when the whisk is lifted and lightly fold into the banana purée. Colour if you wish (left without colouring the banana ice cream looks rather muddy). Turn into a container and freeze.

**TO PACK AND USE:** See notes at beginning of chapter.

# Apricot ice cream

1 lb. fresh ripe apricots
water
6 oz. caster sugar
5 egg yolks
½ pt. double cream

*Serves 8*

**TO MAKE:** Halve and stone the apricots. Cook until soft in the minimum of water, then sieve. Dissolve the sugar in ¼ pt. water, bring to the boil and boil to a syrupy consistency (217°F). Whisk the yolks in a deep bowl and whilst still whisking pour in the slightly cooled syrup. Whisk until thick, frothy and cool. When the egg mixture is cold, whip the cream until it leaves a trail when the whisk is lifted and fold into the egg mixture with the fruit purée. Turn into a container and freeze.

**TO PACK AND USE:** See notes at beginning of chapter.

# Grape ice cream

4 oz. caster sugar
6 tbsps. water
5 egg yolks
1 lb. white grapes, peeled and pipped
2 tbsps. lemon juice
½ pt. double cream

*Serves 8*

**TO MAKE:** Dissolve the sugar in the water, bring to the boil and boil to a syrupy consistency (217°F). Whisk the yolks in a deep bowl until pale and while still whisking pour in the slightly cooled syrup; whisk until frothy, thick and cool. When the egg mixture is cold, purée in a blender with the grapes and lemon juice. Whip the cream until it leaves a trail when the whisk is lifted and fold into the purée. Turn into a container and freeze.

**TO PACK AND USE:** See notes at beginning of chapter.

# Butterscotch ice cream

2 oz. butter
2 oz. golden syrup
2 oz. dark soft brown sugar
¼ pt. milk
½ tsp. vanilla essence
1 egg
1½ oz. caster sugar
¼ pt. double cream

*Serves 4–5*

**TO MAKE:** Melt the butter with the syrup and brown sugar in a saucepan. Cool. Heat the milk and essence in a saucepan until just warm. Beat together the egg and sugar and stir in the milk. Strain the mixture through a sieve back into the saucepan and heat gently, stirring all the time until the custard thickens slightly. Add the syrup mixture and cool. Whip the cream lightly until it just leaves a trail when the whisk is lifted and fold it into the mixture. Pour into a container and freeze.

**TO PACK AND USE:** See notes at beginning of chapter.

*Strawberry yoghurt ice*

# Strawberry yoghurt ice

(*see picture above*)

15-oz. can strawberries
½ oz. powdered gelatine
1 pt. natural yoghurt
4 oz. caster sugar
1 tbsp. lemon juice
colouring if required

*Serves 4*

**TO MAKE:** Drain the can of strawberries, sprinkle the gelatine onto 4 tbsps. of the syrup in a basin and dissolve over hot water. Mix the remaining syrup, yoghurt, sugar and lemon juice together. Fold in the gelatine and chill until the consistency of unbeaten egg white. Fold in the whole fruit, turn into a container and freeze.

**TO PACK AND USE:** See notes at beginning of chapter.

# After-dinner ice cream

½ pt. thick custard
8 oz. full fat cream cheese
2 tbsps. lemon juice
8 tbsps. evaporated milk, whipped
1 oz. shelled walnuts, chopped

*Serves 6*

**TO MAKE:** Leave the custard to cool. Blend the cheese with the lemon juice and stir in the whipped evaporated milk. Fold in the walnuts and finally stir in the cooled custard. Pour into a container and freeze.

TO PACK AND USE: See notes at beginning of chapter.

## Drambuie dream

½ pt. milk
2 egg yolks
2 oz. caster sugar
1½ tbsps. clear dark honey
4 tbsps. Drambuie
1 tbsp. orange juice
¼ pt. double cream

*Serves 6*

TO MAKE: Heat the milk gently. Cream the yolks and sugar together. Pour on the milk and stir well. Return mixture to the pan and heat slowly to thicken, stirring all the time. Add the honey and stir well. Remove from heat, add the Drambuie and orange juice. Cool. Whip the cream until it just leaves a trail when the whisk is lifted and fold it into the cooled mixture. Pour into a container and freeze until firm.

TO PACK AND USE: See notes at beginning of chapter.

## Rhubarb and ginger ice cream

1-lb. 3-oz. can rhubarb, drained and puréed
2 pieces stem ginger, finely chopped
2 tbsps. ginger syrup (from the stem ginger jar)
½ level tsp. ground ginger
edible pink colouring
½ pt. double cream, whipped

*Serves 4–5*

TO MAKE: Mix together the purée, chopped ginger, ginger syrup and ground ginger and add sufficient colouring to give a deep pink tone. (This will become paler when the cream is added.) Fold in the whipped cream and pour it into a container to freeze until firm. **Rhubarb and ginger water ice** can be made using the same recipe but omitting the cream.

TO PACK: See notes at beginning of chapter.

TO USE: Serve both ice cream and water ice straight from the freezer.

# WATER ICES

BASIC SUGAR SYRUP FOR SORBET AND WATER ICES

4 oz. sugar
½ pt. water

Dissolve the sugar in the water over a low heat. Bring to the boil and boil gently for 10 min. Cool.

## Port and lemon sorbet

¼ pt. fresh lemon juice
4 tbsps. port
1 recipe quantity basic sugar syrup (above)
1 egg white

*Serves 4*

TO MAKE: Combine the lemon juice, port and syrup and pour into a container to freeze until mushy. Whisk the egg white until stiff but not dry, turn the half-frozen mixture into a bowl and fold in the egg white. Return to the container and freeze until firm.

TO PACK AND USE: See notes at beginning of chapter.

# Gooseberry sorbet

2 lb. gooseberries, topped and tailed
¼ pt. water
8 oz. caster sugar
2 tbsps. lemon juice
edible green colouring (optional)
2 egg whites

*Serves 6–8*

**TO MAKE:** Place the gooseberries in a saucepan with the water, bring to the boil, cover and cook until soft. Stir in the sugar and cool. Put through a sieve into a freezer container, stir in the lemon juice and colouring if desired. Freeze until mushy. Whisk the egg whites until stiff and fold into the half-frozen mixture. Turn into a container and freeze.

*Note:* When fresh fruit is not available use a can but omit the water and sugar, or use frozen fruit.

**TO PACK AND USE:** See notes at beginning of chapter.

*Martini cooler*

# Martini cooler

(*see picture above right*)

1 fresh grapefruit, peeled or a 15-oz. can grapefruit segments, drained
4 tbsps. dry Martini
1 recipe quantity basic sugar syrup (page 187)

*Serves 4*

**TO MAKE:** Put the grapefruit flesh in a blender to make ½ pt. purée. Stir in the Martini and add the syrup. Pour into a container and freeze until firm.

**TO PACK:** See notes at beginning of chapter.

**TO USE:** Use straight from the freezer.

# Pineapple sorbet

2¾-lb. pineapple
6 tbsps. lemon juice
4 tbsps. orange juice
¼ pt. water
11 oz. caster sugar

*Serves 6*

**TO MAKE:** Peel the pineapple, remove the core and discard. Pulp the flesh in a blender. Put the pineapple pulp through a sieve to give ¾ pt. purée. Mix the purée with the lemon and orange juice, water and sugar, stirring until the sugar has completely dissolved. Pour into containers and freeze without stirring until crystals have formed but the spoon still goes in.

**TO PACK AND USE:** See notes at beginning of chapter. This recipe looks particularly attractive when the sorbet is returned to the empty pineapple shell for serving. (To preserve the shell, wrap it in foil and freeze with or without the completed sorbet.)

# Apricot brandy water ice

15-oz. can apricot halves
½ recipe quantity basic sugar syrup (page 187)
5 tbsps. brandy
3 tsps. lemon juice

*Serves 4–6*

**TO MAKE:** Drain the apricot halves and keep ¼ pt. juice. Purée the apricots. Combine together the purée, reserved juice, sugar syrup, brandy and lemon juice. Pour into a container and freeze until firm.

TO PACK AND USE: See notes at beginning of chapter.

## Orange sherbet

4 oz. caster sugar
½ pt. water
1 tbsp. lemon juice
grated rind of 1 orange
grated rind of 1 lemon
juice of 3 oranges and 1 lemon
1 egg white

*Serves 4*

TO MAKE: Dissolve the sugar in the water over a low heat, bring to the boil and boil gently for 10 min. Add 1 tbsp. lemon juice. Put the grated fruit rinds in a basin, pour the boiling syrup over and leave until cold. Add the mixed fruit juices and strain into a container. Freeze the mixture until mushy then turn it into a bowl, whisk the egg white and fold it in, mixing thoroughly. Return to the container and freeze.

TO PACK AND USE: See notes at beginning of chapter.

## Rum and blackcurrant sorbet

1-lb. can blackcurrants, puréed
1 recipe quantity sugar syrup (page 187)
4 tbsps. rum
2 egg whites

*Serves 4–6*

TO MAKE: Sieve the puréed blackcurrants to remove the pips. Combine the blackcurrants, syrup and rum together and mix well. Pour into a container and freeze until mushy. Turn into a bowl and beat with a fork. Whisk the egg whites until stiff and fold them into the blackcurrant mixture. Return to the container and freeze.

TO PACK: See notes at beginning of chapter.

TO USE: This recipe is quite soft and can be used straight from the freezer.

## Black cherry sorbet

15-oz. can black cherries, stoned and puréed
2 tbsps. Kirsch
1 tbsp. lemon juice
1 egg white

*Serves 4*

TO MAKE: Combine the cherry purée, Kirsch and lemon juice. Pour into a container and freeze until mushy. Whisk the egg white, turn the cherry mixture into a bowl, and fold the stiffly whisked egg white into the mixture. Return to the container and freeze until firm.

TO PACK AND USE: See notes at beginning of chapter.

## Lemon sherbet

8 oz. caster sugar
1 pt. water
thinly pared rind and juice of 3 lemons
1 egg white

*Serves 4*

TO MAKE: Dissolve the sugar in the water over a low heat, add the thinly pared lemon rind, free of all white pith, and boil gently for 10 min., leave to cool. Add the lemon juice and strain the mixture into a container. Freeze until mushy, then turn into a bowl, whisk the egg white and fold it in, mixing thoroughly. Return to the container and freeze.

TO PACK AND USE: See notes at beginning of chapter.

## Avocado cream sherbet

(*see picture overleaf*)

1 oz. powdered gelatine
½ pt. milk
3 large avocados, stoned, peeled and diced
½ level tsp. salt
grated rind of 1 lemon and 1 orange
juice of 2 lemons and 2 oranges
1 pt. double cream, whipped
8 oz. caster sugar

*Serves 10*

*Avocado cream
sherbet*

TO MAKE: Sprinkle the gelatine over the milk in a saucepan, heat gently to dissolve. Place the avocado, salt, rinds and juices in a blender and operate on high speed until smooth. Fold in the whipped cream. Add the sugar to the gelatine and milk and stir slowly into the avocado mixture. Turn into a container and freeze until semi-hard; return to the blender and blend until smooth. Return to the container and freeze again.

TO PACK AND USE: See notes at beginning of chapter.

## SIX SIMPLE WAYS TO MAKE ICE CREAM SUPER!

**Chocolate cups:** Brush melted chocolate into paper cake cases. Leave to set upside-down on non-stick paper, then peel off the cases and fill the chocolate cups with ice cream.

**Sponge Ice Flan:** Fill a sponge flan case with scoops of ice cream, then top with a sauce and chopped nuts.

**Fruit Sundaes:** Purée drained canned fruits—black cherries (stoned), raspberries and strawberries—and spoon over ice cream in sundae glasses. Top with chopped nuts.

**Jelly Sundaes:** Serve scoops of ice cream layered with coloured jelly and fruit. Top with a swirl of cream and drizzled jam.

**Tutti-Frutti:** Mix raisins, nuts and cherries into softened plain vanilla or dairy ice cream. Re-freeze rapidly before serving.

**Individual Alaskas:** Slice a Swiss roll, place a scoop of vanilla ice cream on each slice, top with fruit and completely cover with meringue. Flash-bake at 450°F (mark 8) for 2–3 min.

# DESSERTS USING ICE CREAM

## Mountain glory

1-lb. 13-oz. can white peaches, chilled
2 tbsps. Madeira wine
16 profiterole buns (page 240)
about 16 scoops of vanilla ice cream
3 tbsps. apricot jam
grated chocolate

*Serves 6*

Turn the peaches and their juice into a glass serving dish, stir in the Madeira wine and chill. Fill each profiterole with a little ice cream (use a small knife to push a little nut of ice cream into the centre). Pile scoops of ice cream on top of the peaches and then pile the profiteroles on top in a pyramid shape. Heat the apricot jam until liquid (do not sieve) and trickle over the profiteroles; finish with the grated chocolate. Serve immediately.

## Chocolate cream meringue

*For meringue*
4 egg whites
4 oz. caster sugar
4 oz. granulated sugar

*For cream filling*
2 oz. chocolate
2 oz. cornflour
2 oz. sugar
2 eggs, beaten
$\frac{1}{2}$ pt. milk
$\frac{1}{2}$ tsp. vanilla essence

*To finish*
12 scoops chocolate ice cream
$\frac{1}{4}$ pt. double cream, whipped

*Serves 6*

Cover 2 baking sheets with non-stick paper and draw a 7-in. circle on each. Whisk the egg whites until light, fluffy and dry and gradually whisk in the caster sugar a spoonful at a time. Fold in the granulated sugar. Spoon the mixture into a forcing bag fitted with a large star vegetable nozzle. Using one of the drawn circles as a guide, pipe the meringue to fill the circle starting from the centre and working outwards. Using the other circle, pipe a ring of meringue over the actual line drawn and a second ring inside that. With the remaining meringue and the space left on the paper around each circle pipe 9 individual whirls. Dry the meringues in the oven for about 4 hr. at the lowest setting. When dry, remove from the oven and cool slightly. Then turn the meringues over and gently peel off the paper. Cool on a wire rack.

Melt the chocolate in a small basin over a pan of hot water. Cream together the cornflour, sugar, eggs and 2 tbsps. of the measured milk. Heat the remaining milk then pour it on to the cornflour, return to the pan and add the vanilla. Bring to the boil slowly, stirring all the time. When the custard has thickened combine it with the melted chocolate. Cover and leave until cold.

Spread the chocolate cream over the flat meringue base and place the meringue ring on the top. Fill the centre with the scoops of chocolate ice cream and place the meringue whirls half on the meringue border and half on the ice cream. Secure with the whipped cream and serve immediately.

## After Eight gâteau

(*see picture overleaf*)

small pkt. After Eight Chocolate Mints
2 7-in. chocolate sponge layers
$\frac{1}{2}$ pt. double cream, whipped
1 small 4-in. square block ice cream

*Serves 6*

Chop 4 of the mints into small pieces. Sandwich the chocolate sponge layers together with some of the cream and the chopped mints. Spread the sides of the sandwich with some more of the cream. At this stage place the cake on the serving dish. Place the After Eight Mints close together around the edge of the cake, pressing them against the cream. With a sharp knife cut a 4-in. square out of the top layer of sponge (the size of the ice cream block) and insert the ice cream. Arrange more mints to decorate the top of the ice cream and pipe the remaining cream in small whirls around the base of the ice cream. Serve immediately.

## Iced apricot slice

7-oz. pkt. frozen puff pastry, thawed
15-oz. can apricot halves, drained
2 tbsps. apricot jam
1 tbsp. brandy
17-fl. oz. block vanilla ice cream
$\frac{1}{4}$ pt. double cream, whipped
1 oz. shelled walnuts, chopped

*Serves 4–6*

Roll out the pastry to a rectangle approx. 6 in. by 10 in. and divide in half (3 in. by 10 in.). Place each piece of pastry on a dampened baking sheet, prick well and bake at 450°F (mark 8) for 10–15 min. until brown. Cool on a wire rack. Chop all but 6 of the apricot halves. Heat the apricot jam with the brandy. Place the pastry slices on a flat tray or baking sheet and brush with the apricot brandy glaze. Transfer one slice to a serving dish. Slice half the ice cream block and use it to cover the pastry. Place the second slice of pastry on top of the ice cream. Cover with the chopped apricots and slice the remaining ice cream to cover. Cover the ice cream with piped cream and decorate with the chopped nuts and remaining apricots. Serve at once.

## Chocolate peach gâteau

*For chocolate sponge*
4 oz. margarine or butter
4 oz. caster sugar
2 eggs
3 oz. self-raising flour
1 oz. cocoa powder

*For topping*
3 tbsps. bramble jelly
1 oz. nibbed almonds
15-oz. can peach halves, drained
chocolate ice cream
grated plain chocolate

*Serves 6*

Make up the sponge mixture following the basic sponge pudding recipe (page 164). Turn into a lightly greased and floured tin measuring 11 in. by 7 in.; bake in the oven at 375°F (mark 5) for 25–30 min. Cool on a wire rack. Spread the sponge with the bramble jelly and sprinkle over the almonds. Cut the peach halves in half again and arrange over the almonds. Spoon the ice cream over to cover and top with grated chocolate. Serve at once with cream.

# Crunchy top

2 oz. butter
2 oz. sugar
1 oz. cornflakes
1 oz. desiccated coconut
1 oz. shelled walnuts, chopped
2 tbsps. honey
17-fl. oz. block ice cream

*Serves 6*

Melt the butter, add the sugar and dissolve over a low heat. Add the other ingredients (except the ice cream) and mix well. Place the ice cream block on a chilled plate. Press the crunchy topping on to the ice cream block. Return the dessert to the freezing compartment until ready to serve. Cut into squares or slices to serve.

# Chocolate caramel flan

24 caramel wafer fingers, halved
4 oz. butter
3 oz. golden syrup
1 oz. cooking chocolate, grated
1 oz. caster sugar
$\frac{1}{2}$ lb. digestive biscuits, crushed
16–20 scoops chocolate ice cream
$\frac{1}{4}$ pt. caramel sauce (see page 199)
1 oz. shelled walnuts, chopped

*Serves 6*

Place an 8-in. flan ring on a flat serving plate and line with the wafers. Melt the butter and stir in the syrup, chocolate, caster sugar and crushed biscuits. Mix well together. Press the mixture into the flan ring and leave to set in the refrigerator. Fill with chocolate ice cream and pour over the caramel sauce.

Sprinkle with the chopped walnuts. Serve at once.

# Iced lemon meringue pie

*For meringue*
3 egg whites
3 oz. caster sugar
3 oz. granulated sugar

*For lemon cream*
6 tbsps. lemon curd
2 tbsps. lemon juice
$\frac{1}{4}$ pt. milk
2 egg yolks
1 level tbsp. cornflour
$\frac{1}{2}$ tbsp. water

*For topping*
9–10 scoops vanilla ice cream
angelica leaves
strips of lemon rind

*Serves 6*

Line 2 baking sheets with greaseproof or non-stick paper and draw an 8-in. circle on each. Whisk the egg whites until fluffy and dry and gradually whisk in the caster sugar a little at a time. Fold in the granulated sugar. Spoon into a forcing bag and pipe on to baking sheets making one 8-in. round base and one 8-in. circle. Bake in the oven at 250°F (mark $\frac{1}{4}$) for about 4 hr. until dry. Cool. Mix all the ingredients for the lemon cream together and heat in a saucepan slowly until thickened, stirring all the time. Allow to cool slightly then spread over the meringue base. Place the circle on top and fill with the ice cream. Decorate with angelica leaves and the lemon strips. Serve at once.

# Peach brandy gâteau

1 recipe quantity vanilla ice cream
  (page 181) or a 17-fl. oz. block ice cream
2 15-oz. cans sliced peaches
4 tbsps. brandy
2 8-in. fatless sponge layers
$\frac{1}{4}$ pt. double or whipping cream
macaroon sticks

*Serves 8*

Soften the ice cream until of a stiff spreading

193

consistency. Line an 8-in. round cake tin with non-stick paper and spoon in the ice cream. Allow to harden in the freezer. Meanwhile drain the peaches but reserve 2 tbsps. syrup. Soak the drained peaches in the reserved syrup and the brandy. Arrange all but 10 peach slices over the ice cream layer, spoon the juice over and chill until firm. To serve, take the ice cream layer from the freezer, sandwich it between the sponge cakes and leave in the refrigerator at normal temperature to soften a little. Before serving, lightly whip the cream until it holds its shape. Pile on top of the gâteau and decorate with the reserved peach slices and macaroon sticks.

# Omelette en surprise

2 eggs, separated
1 level tbsp. sugar
$\frac{1}{2}$ oz. unsalted butter
2 scoops dairy vanilla ice cream

*Serves 2*

Whisk the egg yolks with the sugar and whip the whites separately until stiff. Fold the yolks and sugar into the whites. Melt the butter in an omelette pan and pour in the egg mixture. Cook for 2–3 min. until golden on the underside, then flash under the grill to brown the top surface. Place the ice cream in the centre and fold the omelette over. Serve immediately on a warm plate.

# Nutty butterscotch pies

*For pastry cases*
3 oz. plain flour
2 oz. butter
1 oz. caster sugar

*For honey sauce*
2 oz. butter
$1\frac{1}{2}$ level tsps. cornflour
5 oz. thin honey

*To finish*
2 oz. plain chocolate
1 recipe quantity butterscotch ice cream
   (page 185)
2 oz. shelled walnuts, chopped

*Serves 4*

Use the fingertips to knead the flour, butter and sugar to a manageable dough. Roll out the dough thinly and use to line four $3\frac{1}{2}$-in. shallow patty pans. Bake blind in the oven at 375°F (mark 5) for 15 min. until light golden brown. Cool. To make the honey sauce, melt the butter, stir in the cornflour and gradually add the honey. Bring to the boil and cook for 1–2 min. Cool. Melt the chocolate in a small bowl over hot water. Dip the edges of the pastry cases into the soft chocolate and leave to set. Pile scoops of ice cream into the pastry cases and top with honey sauce and chopped walnuts. Serve at once.

# Fruit-and-nut sundae

4 oz. dried apricots
2 level tbsps. caster sugar
lemon juice
1 oz. seedless raisins
1 oz. flaked almonds
Raspberry Ripple Family Sweet
whipped cream for decoration, optional

*Serves 4*

Soak the apricots overnight in water to cover—if large cut in half. Drain. Cover with fresh water and cook gently for 15–20 min. until soft. Drain and sieve. Add the sugar to the purée and leave until cold. Sharpen with lemon juice and fold the raisins and almonds through the apricot purée. Spoon Raspberry Ripple ice cream into 4 sundae glasses. Top with the apricot mixture. Decorate each with a whirl of whipped cream, if wished. Serve at once.

# Choco-rum flan

3 oz. chocolate
3 oz. cornflakes
3 oz. sponge cake crumbs
3 tbsps. rum
8–10 scoops chocolate ice cream
grated chocolate

*Serves 6*

Melt the chocolate in a basin over hot water and stir in the cornflakes. Stir gently until the cornflakes are evenly coated then turn them into an 8-in. flan ring placed on a flat serving plate. Chill until firm. Mix the cake crumbs with rum and spread over the centre of the cornflake base. Spoon the chocolate ice cream over the rum mixture and sprinkle with coarsely grated chocolate. Serve at once.

# Melon and ice cream balls

1 honeydew melon
17-fl. oz. vanilla ice cream
stem ginger sauce (page 199)

*Serves 4*

Cut the melon in half, remove the seeds and scoop out as many balls as possible from each half, using a melon baller. Make the same number of ice cream balls the same size; put them in a single layer on non-stick paper or foil and quickly put them back into the freezer to retain their shape. Just before serving, pile the melon and ice cream balls into glasses and pour over the warm ginger sauce.

*Note:* Alternatively, use a smaller ogen melon to serve 2 people and serve the balls in the empty melon shells.

# Rhubarb and ginger sundae

4 oz. plain chocolate
2 level tbsps. golden syrup
4 tbsps. evaporated milk
6 oz. stem ginger, drained and chopped
1 recipe quantity rhubarb and ginger
   ice cream (page 187)
nibbed almonds

*Serves 4*

Melt the chocolate in a basin over hot water, stir in the syrup and evaporated milk. Stir until well blended then cool a little. Divide the chopped ginger between 4 stemmed glasses and top with scoops of ice cream. Pour a little sauce over each serving and sprinkle with the almonds. Serve at once.

# Lemon candy Alaska

3 oz. butter
7 oz. caster sugar
2 oz. cornflakes, crushed
1 recipe quantity vanilla ice cream
   (page 181)
9 sour lemon sweets, crushed
2 egg whites
juice of $\frac{1}{2}$ lemon

*Serves 4*

Melt the butter in a saucepan, stir in 3 oz. sugar, heat gently until dissolved and then toss in the cornflakes. Use this mixture to line a 7-in. pie plate, and chill. Soften the ice cream in a basin, beat in the sour lemon sweets and re-freeze. Whisk the egg whites until stiff, add 2 oz. sugar and whisk again until stiff. Fold in the remaining 2 oz. sugar. Scoop the lemon ice cream into the chilled pie shell, sprinkling it with lemon juice. Pile the meringue over the ice cream, covering it completely, and place in the oven at 450°F (mark 8) for about 5 min. until light brown. Serve at once.

# Tulip fruit sundaes

2½ oz. plain flour
2½ oz. icing sugar
1 egg, beaten
1 large lemon, greased
15-oz. can fruit cocktail, drained
17-fl. oz. block vanilla ice cream
   (6 scoops)
2½ fl. oz. double cream, whipped

*Serves 6*

Sift the flour and icing sugar into a bowl. Add the egg and mix well. Grease a baking sheet and spread out 1 tbsp. of the mixture into a round approx. 4-in. diameter. (You should get 2 on a sheet.) Bake in the oven at 350°F (mark 4) for 5 min. until just brown at the edges. Carefully lift the round from the baking sheet with a palette knife whilst still warm and mould it round the lemon, keeping one end 'open' to form a tulip shape. Remove the lemon. Allow the tulip cases to cool. Repeat, making 4 more tulips, but as the shaping has to be done quickly it is advisable to bake only 2 rounds at a time.

Half-fill the 6 cases with fruit cocktail, top with a scoop of ice cream and decorate with whirls of cream.

## Surprise soufflé

15-oz. can apricots, drained
2 tbsps. brandy
3 eggs, separated
1 level tbsp. caster sugar
3 drops vanilla essence
17-fl. oz. block ice cream
icing sugar
large quantity of ice cubes (sufficient to
    pile in a roasting tin)

*Serves 4*

Soak the apricots in the brandy. Whisk the egg yolks, sugar and vanilla until pale and thick. Whisk the egg whites until stiff and fold into the yolk mixture. Place the ice cream in the bottom of a $2\frac{1}{2}$–3 pt. ovenproof soufflé dish and spoon the apricots over. Pile the soufflé mixture on top and dust with icing sugar. Stand the soufflé dish in a roasting tin packed with ice cubes, making sure there is plenty of ice both underneath and around the sides of the dish. Bake at 450°F (mark 8) for 5–7 min. Serve at once.

## Iced raspberry tartlets

$\frac{1}{2}$ oz. flour
$\frac{1}{2}$ oz. sugar
1 egg, beaten
4 tbsps. milk
vanilla essence
2 tbsps. redcurrant jelly
12 raspberries
6 shortcrust or pâte sucrée tartlet cases
6 scoops ice cream

*Serves 6*

Mix the flour, sugar and egg together in a saucepan, stir in the milk and vanilla essence. Heat the cream gently to cook and thicken it, then cool. Warm the redcurrant jelly and fold in the raspberries. Leave to cool. Spoon some of the cream filling into the base of each tartlet case, cover with a scoop of ice cream and top with the raspberry mixture. Serve immediately.

## Iced fruit crispies

3 level tbsps. golden syrup
1 oz. soft brown sugar
1 oz. butter
2 oz. marshmallows
2 oz. rice crispies
17-fl. oz. block vanilla ice cream
14-oz. can crushed pineapple, drained
2 oz. chocolate, melted

*Serves 6–8*

Gently heat the syrup, sugar and butter in a saucepan and stir together. When the butter has melted, remove from the heat, add the marshmallows and allow them to dissolve. Add the rice crispies and stir well. Turn the mixture on to a greased baking sheet and press into an oblong shape approx. 8 in. by 6 in. and $\frac{1}{2}$ in. thick. Chill. Cover the chilled base with the ice cream, cut into slices, and top with the crushed pineapple. Trickle the melted chocolate over the top and cut into squares. Serve at once.

## Banana rum crunch

$3\frac{1}{2}$ oz. plain flour
$\frac{1}{2}$ oz. cocoa powder
$1\frac{1}{2}$ oz. margarine
1 oz. lard
1 oz. sugar
water to mix
2 bananas
2 tbsps. rum
16 scoops ice cream (banana, chocolate or
    vanilla)
whipped cream

*Serves 8*

Sift the flour and cocoa together into a bowl, rub in the fat with the fingertips until it resembles crumbs, add the sugar and mix together with the water to give a pastry dough. Roll out the dough to a 7-in. round and place on a baking sheet. Bake in the oven at 400°F (mark 6) for 20–25 min. Remove from the oven and cut into 8 wedges while still warm. Allow to cool on a wire rack. Slice the bananas and leave soaking in the rum. When the base is cold, cover each piece with two scoops of ice cream and decorate with the bananas and whipped cream. Serve at once.

## Ice cream trifle

8 trifle sponges
15-oz. can raspberries
3 tbsps. sherry (optional)
8 large scoops ice cream
$\frac{3}{4}$ pt. custard, chilled

*For meringue*
2 egg whites
2 oz. caster sugar
2 oz. granulated sugar

*Serves 6–8*

Arrange the trifle sponges in the base and up the sides of a chilled $2\frac{1}{2}$-pt. ovenproof dish. Drain the raspberries and combine the juice with the sherry. Moisten the sponges with the juice. Whisk the egg whites until fluffy and dry, whisk in the caster sugar a spoonful at a time, then fold in the granulated sugar. Place the raspberries over the sponges, followed by the ice cream, and custard. Pile the meringue over the custard. Bake in the oven at 450°F (mark 8) for 3–5 min. until brown. Serve at once.

## Jamaican Alaska

1 bought ginger bar cake
2–3 pieces stem ginger, chopped
4 large scoops vanilla ice cream

*For meringue*
2 egg whites
2 oz. caster sugar
2 oz. granulated sugar

*Serves 4*

Cut a hollow from the centre of the ginger cake and put the chopped ginger in the base of the hollow. Whisk the egg whites until fluffy and dry, whisk in the caster sugar a spoonful at a time, then fold in the granulated sugar. Place the cake on an ovenproof serving plate. Fill the hollow with scoops of ice cream and top with the meringue mixture, making sure that all the cake and ice cream is covered. Bake in the oven at 450°F (mark 8) until golden. Serve at once.

# CHILDREN'S FAVOURITES

## Mickey mouse

(*see picture overleaf*)

1 individual ice cream brick
1 pear half
2 currants
2 flaked almonds (or 1 almond, halved)
1 piece angelica
a little apricot glaze

*Serves 1*

Place the ice cream brick on a chilled plate. Place the pear half on the ice cream, cut side down. Place the currants in the pear at the narrow end to make the eyes, use the almonds for ears, angelica for the tail and finish with a little glaze over the back.

## Orange baskets

(*see picture overleaf*)

1 thick slice of ice cream (same size as biscuit but 1-in. thick)
3 Nice biscuits
a little apricot jam
7 mandarin segments
1 strip angelica

*Serves 1*

Place the ice cream on a chilled plate. Press 2 of the Nice biscuits along each long side of the ice cream and break the third biscuit in half to go along the shorter sides. Heat the jam in a small pan. Fill the cavity with the mandarins and brush with glaze. Bend the

*Children's favourites: Sunflower; Orange baskets; Mickey mouse; Dice*

angelica strip over the mandarins to form the handle. Secure the ends of the handle by pushing them into the ice cream.

## Sunflower

(*see picture above*)

1 pineapple ring
1 digestive biscuit
1 scoop ice cream
6 mandarin segments
6 angelica diamond shapes

*Serves 1*

Place the pineapple ring on the digestive biscuit and top with the scoop of ice cream. Arrange the mandarin segments on the pineapple around the ice cream, to form petals, and place a piece of angelica between each mandarin.

## Butterfly sundaes

15½-oz. can pineapple pieces, drained
½ pt. strawberry jelly, chopped
ice cream
2 glacé cherries, chopped in half
8 Matchmakers

*Serves 4*

Reserve 8 pineapple pieces. Layer the chopped jelly, pineapple and ice cream in sundae glasses, finishing with a scoop of ice cream. To make the butterfly, place half a cherry on the ice cream and break one Matchmaker stick in half to form the antennae sticking out from the cherry. Place one Matchmaker at the other end of the cherry to make a tail and put a piece of pineapple each side of the cherry to form the wings.

## Dice

(*see picture opposite*)

2-in. square block of ice cream
6 After Eight Mints
a little butter icing, or jam
Smarties

*Serves 1*

Cover the 6 sides of the ice cream block with the After Eight Mints. Spread a little butter icing or jam on the Smarties and stick on to the Mints to turn them into dice.

Children adore their favourite sweets on top of their ice cream. Use Smarties, jelly tots, choc drops, crushed chocolate flake or buttered popcorn. Serve scoops of different flavoured ice creams in cornets and sprinkle with chocolate vermicelli or hundreds and thousands. Roughly chop Crunchy Bars or Matchmakers and serve over ice cream.

# QUICK SAUCES FOR ICE CREAM

## Stem ginger sauce

3 oz. Demerara sugar
$\frac{1}{2}$ pt. water
pinch ground cinnamon
thinly pared rind and juice of 1 lemon
2 level tsps. arrowroot
2 oz. stem ginger, chopped
1 tbsp. ginger syrup (from stem ginger jar)

Dissolve the sugar in the water in a saucepan. Add the cinnamon and lemon rind and simmer for 5 min. Remove the rind. Blend the arrowroot with the lemon juice and add it to the sugar syrup. Add the remaining ingredients, bring to the boil and allow to thicken. Serve warm.

## Quicky mint sauce

$3\frac{1}{2}$-oz. pkt. peppermint creams
2 tbsps. milk

Warm the mints and milk together in a saucepan until the mints have melted. Beat till smooth and serve warm.

## Spicy peach sauce

15-oz. can peaches
2 tbsps. lemon juice
$\frac{1}{2}$ level tsp. ground mixed spice

Purée all the ingredients together in a blender or pass through a sieve. Heat gently and serve warm.

## Honey almond sauce

1 level tsp. arrowroot
1 tbsp. lemon juice
4 tbsps. clear honey
$\frac{1}{4}$ pt. water
$\frac{1}{2}$ oz. flaked almonds, toasted

Blend the arrowroot with the lemon juice in a saucepan; add the honey, water and almonds. Bring to the boil, stirring, and cook until thickened. Serve warm.

## Caramel sauce

8 oz. caster sugar
$\frac{1}{2}$ pt. boiling water

Dissolve the sugar in a pan over a gentle heat. Shake the pan occasionally and increase the heat to brown the sugar. Slowly add the boiling water and boil for 6 min. Cool slightly but serve while still warm.

## Peanut sauce

2 oz. dark soft brown sugar
$\frac{1}{2}$ pt. water
4 level tbsps. crunchy peanut butter
2 level tsps. arrowroot
4 tbsps. milk

*Serve bought ice cream with a sauce or fruit*

Place the sugar and water in saucepan and heat gently until the sugar dissolves. Whisk in the peanut butter. Blend the arrowroot with the milk and stir into the peanut syrup. Bring to the boil and stir until thickened. Serve warm.

## Coffee sauce

2 oz. Demerara sugar
½ pt. strong black coffee
2 level tsps. arrowroot
4 tsps. water
2 tbsps. cream (optional)

Dissolve the sugar in the coffee over a low heat. Blend the arrowroot with the water, add to the coffee and heat gently. Remove from the heat and if desired add the cream. Serve warm.

## Lemon cheese sauce

3 oz. full fat soft cream cheese
1 egg
1 tbsp. lemon juice
1 level tsp. grated lemon rind
1 level tbsp. caster sugar
2 tbsps. sultanas

Blend all the ingredients together and spoon over the ice cream.

# 10 Fruits out of Season

When preparing fruits for the freezer, make sure you have sufficient syrup available. This should be made in advance, and the best plan is to make it a day ahead and allow it to chill overnight in the refrigerator, since it has to be used cold. As a rough guide, for every 1 lb. fruit you need $\frac{1}{2}$ pt. syrup (made according to the individual directions). Dissolve the sugar in the water, bring to the boil, remove from the heat, strain if necessary and leave to cool, lightly covered. Pour the syrup over the fruit or place the fruit in a container with the syrup. Light-weight fruits which tend to rise in liquids can be held below the surface by means of a dampened and crumpled piece of non-absorbent paper.

Incidentally, when you come to serve fruit frozen in syrup, open it only just before serving; keep stone fruits submerged in the syrup as long as possible.

## Apple shortcakes

$\frac{3}{4}$ lb. frozen unsweetened apple slices, thawed
juice and finely grated rind of 1 lemon
8 oz. self-raising flour
1 level tsp. baking powder
3 oz. butter or margarine
1 oz. caster sugar
cold milk
2 level tbsps. caster sugar
1 level tsp. ground cinnamon
clear honey
cream

*Makes 8*

Toss the apple slices in the lemon juice. Sift together the flour and baking powder, rub in the fat and add the grated lemon rind and caster sugar. Blend to a softish dough with cold milk. Roll out the dough and use it to line 8 shallow individual Yorkshire pudding tins, about 4-in. diameter (top measurement). Arrange the apple slices over the shortcake base, blend together the 2 tbsps. sugar and the cinnamon and sprinkle over the apples.

Bake in the oven at 400°F (mark 6) for about 25 min. While still warm, brush the shortcakes with honey and serve with cream.

## Golden apple charlotte

(*see colour picture facing page 161*)

2 oz. golden syrup
2 oz. thick-cut marmalade
$\frac{3}{4}$ lb. frozen unsweetened apple slices
4 oz. white bread, crusts removed
2 oz. margarine or butter
2 level tbsps. granulated sugar
top of the milk or single cream, optional

*Serves 4*

Warm the syrup and marmalade in a saucepan, add the apples and stir. Cook gently over a medium heat until the apples are soft but not too broken up. Meanwhile, cut the bread into even $\frac{1}{2}$-in. cubes. Melt the fat in a frying pan, add the cubes of bread and fry gently until evenly brown, turning the cubes occasionally. Stir in the sugar. Turn the apple mixture into 4 small warmed dishes and top with fried bread cubes. If you wish, drizzle each dish with top of the milk or single cream. Alternatively, serve in a single large dish with single cream.

# Apple meringue

6-oz. jam Swiss roll
grated rind and juice of 1 lemon or orange
$\frac{3}{4}$ pt. frozen, unsweetened, thick apple
   purée, thawed
2 egg whites
4 oz. caster sugar

*Serves 4*

Slice the Swiss roll and arrange the slices in the base of a 2-pt. soufflé dish. Spoon over the grated rind and juice. Spread the apple over the sponge. Whisk the egg whites until stiff, whisk in half the sugar, until the meringue is stiff again, then fold in all but 1 level tbsp. of the remaining sugar. Pile the meringue over the apple, pull up in peaks and dredge with the rest of the sugar. Cook in the oven at 300°F (mark 2) for about 30 min. Serve warm.

*Note:* If you prefer, use trifle sponge cakes sandwiched with jam instead of the Swiss roll.

# Belgian Torte

8 oz. butter
4 oz. caster sugar
2 tbsps. corn oil
few drops vanilla essence
1 egg, beaten
1 lb. plain flour
1 level tsp. baking powder
$\frac{1}{4}$ level tsp. salt
2 lb. frozen, unsweetened apple slices
2 tbsps. water
icing sugar to dredge

*Serves 6*

Cream the butter and 2 oz. sugar together until fluffy and pale in colour. Add the oil, vanilla essence and egg. Work in the flour, sifted with the baking powder and salt. Form into a dough and chill.

Meanwhile, make the apple marmalade; stew the apples with the remaining 2 oz. caster sugar and the water until a really thick purée is obtained, that will hold its shape when lifted with a spoon. Leave to cool.

Coarsely grate the pastry and place half of it in a lightly-greased, $8\frac{1}{2}$-in. diameter, loose-bottomed, spring-release cake tin. Press down lightly. Spread the apple marmalade over and cover it with the remainder of the grated pastry—do not press down this time. Bake in the oven at 350°F (mark 4) for about $1\frac{1}{4}$ hr. Leave to cool until lukewarm before turning out. Dredge heavily with icing sugar when cold.

# French apple flan

(*see picture opposite*)

*For pâte sucrée*
4 oz. plain flour
pinch salt
2 oz. caster sugar
2 oz. butter, at room temperature
2 egg yolks

*For filling*
$1\frac{1}{4}$ lb. frozen, unsweetened apple slices,
   thawed
sugar to taste
melted butter
apricot jam, sieved

*Serves 6*

Sift together the flour and salt on to a working surface. Make a well in the centre and into it put the sugar, butter and egg yolks. Using the fingertips of one hand, pinch and work the sugar, butter and egg yolks until well blended. Gradually work in all the flour and knead until smooth. Put the paste in a cool place for at least 1 hr.

Cook $\frac{3}{4}$ lb. apples in a saucepan with about 1 level tbsp. sugar and a little water until soft and thick. Sieve and allow to cool. Roll out the pâte sucrée and use it to line a $7\frac{3}{4}$-in. diameter, loose-bottomed, French fluted flan tin. Prick with a fork and bake blind in the oven at 375°F (mark 5) for 15–20 min. Remove from the oven and fill the flan with the apple purée. Arrange the remaining apple slices in overlapping circles on top of the purée. Brush with melted butter and return to the oven at 400°F (mark 6) until the apples are tinged brown. Protect the pastry edges from overbrowning with foil, if necessary. Remove from the oven and brush with warm, sieved

*French
apple flan*

apricot jam. Leave until cold and serve with pouring cream.

## Arctic apple slice

2 oz. granulated sugar
½ pt. water
12 frozen apple rings
1 Arctic roll
¼ pt. double cream, whipped
toasted hazelnuts, to decorate

*Serves 4–6*

Dissolve the sugar in the water in a large saucepan. Add the apple rings and simmer for 10–15 min. until tender. Drain and cool. Cut the Arctic roll into 13 slices. Place an apple ring between each slice and press the slices well together to form a roll again. Decorate with whipped cream and hazelnuts. Serve immediately.

## Date toffee apples

6 frozen, whole, peeled and cored apples
4 oz. stoned dates
2 oz. Demerara sugar
2 oz. granulated sugar

*Serves 6*

Place the frozen apples in a large saucepan, cover with water and simmer gently for 20–25 min. until soft but not collapsed. Drain. Chop the dates, mix with the Demerara sugar and use to stuff the apples. Place them on a baking sheet and bake in the oven at 350°F (mark 4) for 15 min. Dissolve the granulated sugar slowly in a heavy based saucepan, increase the heat and allow the sugar to caramellise. Pour the caramel over the apples and chill. Serve with whipped cream.

## Apple meringue bombes

4 oz. granulated sugar
1 pt. water
6 frozen, whole, peeled and cored apples
2 oz. butter
6 rounds sliced bread, cut to the size of
   the apples
2 oz. sultanas
¼ level tsp. ground cinnamon
2 egg whites
4 oz. caster sugar

*Serves 6*

Dissolve the granulated sugar in the water. Add the apples, cover and simmer gently for 20–25 min. Drain. Melt the butter and fry the rounds of bread on both sides until crisp. Place the apples on the fried bread in an ovenproof dish. Mix the sultanas and cinnamon together and stuff the apples with this mixture. Stiffly whisk the egg whites, add half the caster sugar and continue whisking. Fold in the remaining sugar. Cover the apples with the meringue mixture and bake in the oven at 400°F (mark 6) for 3–4 min. until golden. Serve at once.

# Blackberry and apple condé

1 pt. milk
3 oz. pudding rice
1 oz. caster sugar
$\frac{1}{4}$ pt. single cream
$\frac{3}{4}$ lb. frozen, unsweetened apple slices
$\frac{1}{2}$ lb. frozen blackberries
2 tbsps. water
1 oz. butter
2 oz. light soft brown sugar
$\frac{1}{4}$ pt. double cream

*Serves 6–8*

Bring the milk to the boil with the rice and caster sugar. Turn into a casserole, cover with a lid and cook in the oven at 300°F (mark 2) for about 1 hr., until the rice is tender. Allow to cool then stir in 4 tbsps. single cream and turn it into a $2\frac{1}{2}$-pt. glass serving dish. Place the apples, blackberries, water, butter and brown sugar in a saucepan. Simmer gently for 20–30 min., until tender. Cool and spoon over the rice. Chill. Whisk together the remaining single and double cream. Spoon over the condé before serving.

# Blackberry and apple soufflé

3 eggs, separated
3 oz. caster sugar
3 tbsps. lemon juice
3 tbsps. water
1 level tbsp. powdered gelatine
1 pt. frozen blackberry and apple purée, thawed
4 tbsps. double cream, whipped
extra whipped cream, for decoration

*Serves 6*

Tie a band of greaseproof paper round the outside of a 5-in. ($1\frac{1}{2}$-pt.) soufflé dish. Place the egg yolks, sugar and lemon juice in a bowl, stand it over a pan of hot water and whisk until thick and creamy. Remove from the heat and continue to whisk occasionally until the mixture is cool. Put the 3 tbsps. water into a small basin and sprinkle the gelatine over it. Stand the basin in a pan of hot water and heat gently until the gelatine is dissolved. Allow it to cool slightly, then pour it into the egg mixture in a steady stream, stirring the mixture all the time. Stir in the purée and cream until evenly mixed. Whisk the egg whites in a clean bowl until stiff and fold these into the mixture. Pour it into the prepared soufflé dish and leave to set. Decorate with whipped cream before serving.

# Blackberry Marnier flan

1 lb. frozen blackberries
2 oz. sugar
2 tbsps. milk
2 level tbsps. custard powder
1 level tbsp. caster sugar
$\frac{1}{2}$ pt. soured cream
1 tbsp. orange juice
2 tbsps. Grand Marnier
8-in. baked sponge flan case
2 level tsps. arrowroot
$\frac{1}{2}$ level tsp. grated orange rind

*Serves 8*

Place the frozen blackberries in a saucepan with the sugar and heat gently until the sugar dissolves. Cover and simmer the fruit for about 20–25 min. until tender. Drain and reserve the juice. Blend together the milk, custard powder and sugar. Add the soured cream and orange juice and heat the mixture gently in a saucepan, stirring continuously until thickened. Remove from the heat and add 1 tbsp. Grand Marnier. Cool a little and pour the custard into the flan case.

In a small pan, blend the arrowroot with a little of the blackberry juice, add the rest and the orange rind and bring to the boil. Remove from the heat, add the remaining 1 tbsp. Grand Marnier. Arrange the blackberries over the custard and spoon the cool glaze over the fruit. Chill slightly before serving.

# Crunchy butter crumbcake

6 oz. butter
8 oz. fresh white breadcrumbs
4 oz. Demerara sugar
2 oz. nibbed almonds
1 pt. frozen blackberry and apple purée, thawed
whipped cream, for decoration
few extra nibbed almonds, toasted, for decoration

*Serves 6–8*

Melt the butter in a large frying pan. Add the breadcrumbs and brown gently until all the butter is absorbed. Remove from the heat, add the sugar and almonds and mix well. Layer the crumbs and purée alternately in a glass serving dish, finishing with crumbs. Decorate with whipped cream and toasted almonds.

# Banana and mandarin chartreuse

1 lime jelly tablet
1 lemon jelly tablet
1 orange jelly tablet
water
juice of 1 orange
$1\frac{1}{2}$ lb. frozen mandarins in syrup, thawed
2 bananas, peeled
juice of 1 lemon

*Serves 6–8*

Place the broken up jelly tablets in a large bowl. Pour over 1 pt. boiling water and stir until dissolved. Pour the orange juice into a measuring jug, add the syrup from the mandarins and make up to 1 pt. with cold water. Strain before adding to the hot liquid. Cool. Spoon in enough jelly to cover the base of a 3-pt. fancy jelly mould. Allow to set. Thinly slice the bananas and dip the slices in lemon juice. If the mandarins are plump and large, slice them across horizontally. Dip the fruit in a little jelly, then arrange alternately in the base of the mould. Leave to set. Cover the fruit with more jelly. Leave to set before coating the sides of the mould with jelly. To aid the setting, rotate the mould at an angle over a shallow dish containing iced water and ice cubes. Arrange the orange and banana alternately up the sides of the mould and allow to set. Layer the sides and middle alternately with layers of fruit, then jelly. Chill until firm for at least 2 hr. Turn out on to a flat serving plate and serve with pouring cream.

# Blackcurrant cheese flan

$\frac{3}{4}$ lb. frozen blackcurrants
$1\frac{1}{2}$ level tbsps. cornflour
2 oz. sugar
7-in. frozen baked shortcrust flan case
6 oz. soft cream cheese
1 large egg
1 oz. sugar
chopped walnuts

*Serves 6*

Place the blackcurrants in a saucepan with a minimum of water and simmer until tender. Drain and reserve 4 tbsps. juice. Blend the cornflour with the reserved juice, then heat the fruit, sugar and blended cornflour in a saucepan, stirring all the time until thick. Place the flan case on a baking sheet and pour in the fruit filling. Beat the cheese, egg and sugar together and spoon over the blackcurrants. Bake in the oven at 350°F (mark 4) for about 30 min. Cool on a wire rack. When cold, top the flan with chopped nuts.

# Blackcurrant whip

1 lb. frozen unsweetened blackcurrants
water
4 oz. sugar
$\frac{1}{2}$ oz. powdered gelatine
$\frac{1}{4}$ pt. evaporated milk, chilled
1 egg, separated
whipped cream, to decorate

*Serves 8*

Place the blackcurrants, $\frac{1}{4}$ pt. water and sugar in a saucepan. Heat gently until the sugar dissolves, cover and simmer for about 20 min. until the fruit is tender. Cool. Sieve or purée the fruit in a blender. Sprinkle the gelatine over $\frac{1}{4}$ pt. water in a small basin; stand it in a pan with a little hot water and dissolve the gelatine over a low heat. Stir it into the fruit purée. Whisk the evaporated milk in a bowl until thick and frothy. Mix the egg yolk into the fruit mixture and fold in the whipped evaporated milk. Whisk the egg white stiffly and fold into the mixture. Turn into a glass serving dish and leave to set. Serve decorated with whipped cream.

# Rum and black chiffon

1½ lb. frozen, unsweetened blackcurrants
6 oz. granulated sugar
3 eggs, separated
3 tbsps. rum

*Serves 6–8*

Place the frozen blackcurrants in a saucepan with the sugar. Heat until the sugar dissolves then simmer gently for about 20 min. until the fruit is tender. Cool. Place two-thirds of the blackcurrants in the base of a glass serving dish. Purée the remaining one-third of the blackcurrants in a blender. Place the egg yolks in a basin over hot water and whisk with the purée until thick and creamy. Remove from the heat. Add the rum and continue beating for a few min. Whisk the egg whites until stiff and fold them into the fruit mixture. Pour the chiffon over the whole fruit. Chill and serve.

# Cranberry and apple cheese mousse

½ lb. frozen cranberries
½ lb. frozen apple slices
5 oz. caster sugar
water
edible pink colouring
powdered gelatine
grated rind and juice of 1 orange
grated rind and juice of 1 lemon
12 oz. cottage cheese
¼ pt. soured cream
2 eggs, separated
¼ pt. double cream

*Serves 6–8*

Place the cranberries and apple slices in a saucepan with 2 oz. sugar. Heat gently until the fruit is soft. Drain, make up the juice to ¼ pt. with water and add a few drops of pink colouring. Dissolve 1½ level tsps. gelatine in the fruit syrup, stir it into the fruit and pour

into a 3½-pt. ring mould. Chill to set.

Put the rinds and juices from the orange and lemon with 2 oz. caster sugar, the cottage cheese, soured cream and egg yolks in a blender. Sprinkle 1½ level tbsps. gelatine over 3 tbsps. water in a small basin, place over a pan with hot water and heat gently until the gelatine dissolves. Add to the blender, switch on for a few sec., then turn the mixture into a bowl. Whip the cream until the whisk leaves a trail when lifted. Fold it into the cheese mixture. Whisk the egg whites stiffly, add the remaining 1 oz. sugar and whisk again. Fold into the bulk mixture. Pour the cheese mousse over the cranberry and apple jelly in the mould. Chill until set, then turn out on to flat serving plate.

# Summer sponge pudding

(*see colour picture facing page 161*)

¾ lb. frozen bilberries
¾ lb. frozen raspberries
1 lb. frozen apple slices
2 oz. sugar
2 pkts. boudoir biscuits (28 fingers)
¼ pt. milk
1 level tbsp. powdered gelatine
6 fl. oz. double cream
1 tbsp. milk
crystallised rose petals, for decoration

*Serves 8*

Place the fruit and sugar in a saucepan, heat gently until the sugar dissolves and simmer the fruit until tender. Drain and reserve the juice. Dip the boudoir biscuits, one at a time, into the milk just long enough to soften them sufficiently to mould around the inside of a 2½-pt. pudding basin. Line the base first, then work round the sides. Keep back 4–5 fingers for the top.

Sprinkle the gelatine over the fruit juice and heat gently over a pan of hot water until it dissolves. Mix the gelatine with the fruit and allow to cool before pouring into the lined basin. Cover with the remaining boudoir biscuits soaked in milk. Turn down the ends of any protruding biscuits. Cover with a piece of greaseproof paper and a small weighted saucer. Leave overnight in the refrigerator. Just before serving, turn the pudding out on to a serving plate and mask

*Bilberry flans*

with cream whipped with milk. Decorate with crystallised rose petals.

# Bilberry flans

(*see picture above*)

4 oz. butter
2 oz. caster sugar
6 oz. plain flour
5 level tbsps. redcurrant jelly
2 tbsps. lemon juice
1 level tbsp. arrowroot
½ lb. frozen bilberries, thawed
¼ pt. double cream
1 tbsp. milk

*Makes 12*

Cream the butter until soft, add the sugar and beat well. Gradually work in the flour using the fingertips. Chill the dough, if necessary, to a rolling consistency. Roll out the dough and stamp out 12 plain 3-in. rounds to line 12 2½-in. bun tins. Prick and then bake in the oven at 325°F (mark 3) for about 25 min. Cool on a wire rack.

Heat the redcurrant jelly in a small pan. Blend together the lemon juice and arrowroot. Drain the bilberries, reserve the juice and add it with the arrowroot to the jelly. Bring to the boil, stirring. Add the bilberries

and coat with the glaze. Leave until cold. Whip the cream with the milk until it just holds its shape and divide between the pastry shells. Top with bilberries and serve on the same day.

# Liqueured cherry crêpes

6 oz. butter, softened
2 oz. caster sugar
2 tbsps. Kirsch
2 tbsps. lemon juice
12 oz. frozen cherries, thawed and
    stoned
8 7-in. frozen pancakes, partly thawed
1 tbsp. brandy

*Serves 6–8*

Cream together 4 oz. butter, the sugar, Kirsch and lemon juice. Melt the remaining butter in a large frying pan. Heat the cherries in the butter. Divide the creamed butter mixture into 8, spreading it over the pancakes, and fold each into 4. Place the pancakes in the frying pan with the cherries and heat through gently for about 15 min., allowing the butter inside to melt and make a sauce. Just before serving add the brandy, ignite it and heat for a further 1 min. Serve quickly.

207

*Note:* If the cherries were stoned before freezing do not thaw before heating in the butter.

# Liqueur cherry brûlée

(*see picture opposite*)

1 lb. frozen unsweetened cherries, thawed and stoned
8 oz. caster sugar
2 tbsps. Kirsch
$\frac{1}{4}$ pt. double cream
$\frac{1}{4}$ pt. single cream

*Serves 6–8*

Place the cherries in the base of a flameproof serving dish. Sprinkle with 2 oz. sugar and the Kirsch. Whip the creams together, spread them on top of the cherries and chill for 2 hr. Sprinkle 6 oz. sugar over the cream, place the dish under a pre-heated grill and allow the sugar to caramellise. Chill the brûlée again until required.

# Top crust cinnamon cherry pie

*For pastry*
6 oz. plain flour
pinch of salt
$3\frac{1}{2}$ oz. margarine
2 level tsps. caster sugar
1 egg yolk
water

*For filling*
1 lb. frozen unsweetened cherries, thawed and stoned
water
2 oz. sugar
4 level tsps. arrowroot
4 level tbsps. redcurrant jelly
$\frac{1}{4}$ level tsp. ground cinnamon
1 tbsp. Kirsch, optional
milk and granulated sugar for glazing

*Serves 6*

Sift the flour and salt into a bowl. Rub in the margarine with the fingertips and add the sugar. Blend the egg yolk with 2 tbsps. water and add to the flour, working to a manageable dough with a round-bladed knife; use more water if necessary. Place the

*Liqueur cherry brûlée*

cherries in a pan with $\frac{1}{4}$ pt. water and the sugar and cook gently until they begin to soften. Strain off the juice and make up to $\frac{1}{2}$ pt. with water.

Blend the juice with the arrowroot, add the redcurrant jelly and cinnamon and boil gently, stirring until the sauce is clear. Add the Kirsch, if you wish. Stir the cherries into the sauce and turn them into a 2-pt. pie dish. Add a pie funnel and leave to cool. Roll out the pastry and use it to top the pie. Knock up the edges, brush with milk and dust with granulated sugar. Place on a baking sheet and bake in the oven at 400°F (mark 6) for about 30 min. Serve cold with cream or ice cream.

# Gooseberry cream mould

(*see picture opposite*)

$\frac{1}{2}$ pt. frozen unsweetened gooseberry purée, thawed
$\frac{1}{2}$ pt. cold thick custard
3 tbsps. water
3 level tsps. powdered gelatine
$\frac{1}{4}$ pt. double cream
$\frac{1}{4}$ lb. frozen gooseberries, poached

*Serves 4*

Combine the fruit pureé and custard. Spoon the water into a small basin, sprinkle the gelatine over it, place the basin in a saucepan with hot water and heat gently until the gelatine dissolves. Stir into the custard mixture. Whip the cream until it just leaves a trail when the whisk is lifted and fold it into the mixture. Transfer it to a dampened 1½-pt. mould and leave until firm. Turn out on to a flat serving plate and decorate with a border of whole gooseberries.

# Gooseberry crisp

1½ lb. frozen unsweetened gooseberries, thawed
4 oz. sugar
grated rind of 1 orange
a little water
2 oz. butter
2 oz. Demerara sugar
2 oz. cornflakes

*Serves 6*

Place the gooseberries, sugar, orange rind and water in a casserole. Melt the butter in a saucepan, add the sugar and stir in the cornflakes. Turn on to the top of the gooseberries and cook in the oven at 350°F (mark 4) for 30–40 min. until golden and crisp. Serve in the dish in which it was cooked.

# Gooseberry ice cream flan

(*see colour picture facing page 161*)

*For crunch crust*
4 oz. cornflakes
2 oz. butter
2 oz. Demerara or light soft brown sugar

*For filling*
2 oz. caster or granulated sugar
½ pt. water
½ lb. frozen unsweetened gooseberries, thawed
1 level tbsp. cornflour
juice of 1 orange
17-fl. oz. block Cornish ice cream

*Serves 6*

Roughly crush the cornflakes. Melt the butter, stir in the sugar and heat, without boiling, until the sugar has dissolved. Stir in the cornflakes and heat gently for about 3 min. Spoon the cornflake mixture into a 9-in. fluted porcelain flan dish. Press it into the sides and base of the dish to form a shell. Chill.

In a saucepan, dissolve the sugar in the water over a low heat. Add the gooseberries and simmer until just tender. Blend the

*Gooseberry cream mould*

cornflour with the orange juice. Stir in a little gooseberry juice and return to the pan; bring to the boil and simmer for 1–2 min. To serve, scoop the ice cream with a tbsp. and pack into the chilled crust and spoon the warm sauce over just before serving. Serve in wedges.

## Gooseberry fool

¼ pt. double or whipping cream
¼ pt. thick custard
¾ pt. frozen sweetened gooseberry purée, thawed
edible green colouring
grated chocolate for decoration

*Serves 4*

Whip the cream until it just holds its shape. Fold the custard and then the cream into the gooseberry purée. Add a few drops of colouring to give a pale green. Chill. To serve, divide the fool between 4 stemmed glasses and decorate with grated chocolate.

## Gooseberry soufflé

4 large eggs, separated
4 oz. caster sugar
½ pt. frozen unsweetened gooseberry purée, thawed
water
1 tbsp. orange liqueur
½ oz. powdered gelatine
edible green colouring
¼ pt. single cream
¼ pt. double cream
poached gooseberries, or ratafias soaked in orange juice and orange liqueur
whipped cream, for decoration

*Serves 8*

Tie a paper collar round the outside of a 7-in. (2-pt.) soufflé dish and place a 1-lb. jam jar in the centre. Whisk the egg yolks and sugar over hot water until thick. Add the gooseberry purée and continue to whisk until beginning to thicken again. Remove from the heat. Place 2 tbsps. water and the liqueur in a small basin, sprinkle the gelatine over and dissolve it by placing in a pan of hot water. Stir the dissolved gelatine into the purée mixture and add a few drops of green colouring. Allow the mixture to cool until

just beginning to set, whisking it occasionally.

Meanwhile, whisk the creams together until they are the same consistency as the gooseberry mixture and whisk the egg whites. Lightly fold the cream through the gooseberry mixture, followed by the egg whites. Turn the soufflé into the prepared dish and leave to set.

To serve, pour a little warm water into the jam jar, ease it out carefully and fill the cavity with drained, sweetened poached fruit or with ratafias soaked in orange juice and orange liqueur. Ease the paper band away with a warm knife. Decorate with whipped cream.

## Grapefruit and ginger compote

6 oz. caster sugar
¼ pt. white wine
2 lb. frozen unsweetened grapefruit segments
3 oz. stem ginger, chopped
3 tbsps. ginger syrup
2 tbsps. orange squash

*Serves 8–10*

Heat the sugar in the white wine in a large saucepan. When the sugar is dissolved, add the frozen grapefruit segments. Cover and simmer for 20–30 min. until the fruit is well thawed. Stir in the remaining ingredients. Transfer to a serving dish and chill before serving.

## Loganberry tartlets

4 oz. plain flour
pinch salt
½ level tsp. baking powder
2 oz. caster sugar
3 oz. butter
1 egg yolk
12 oz. frozen loganberries, thawed
2 tbsps. water
2 level tsps. arrowroot
2 oz. sugar
¼ pt. soured cream
grated nutmeg

*Makes 12*

Sift together the flour, salt and baking powder. Stir in the sugar and rub in the butter with the finger tips. Add the egg yolk and work together to give a smooth dough. Chill if too soft to roll out. Thinly roll out the dough and use it to line 12 2½-in. diameter bun tins; prick them well. Bake blind in the oven at 400°F (mark 6) for 10 min., then cool on a wire rack.

Drain the juice from the loganberries. In a pan, blend the water and arrowroot, add 2 oz. sugar and the loganberry juice and bring to the boil, stirring until clear. Stir in the fruit. When cold, divide the fruit between the pastry cases. Just before serving top with the soured cream and dust with grated nutmeg.

# Melon and orange jelly

6-fl. oz. can frozen orange juice concentrate
water
4 level tsps. powdered gelatine
1 lb. frozen melon balls, thawed

*Serves 8*

Make the orange juice up to 1 pt. with water. Place 3 tbsps. orange juice into a small bowl, sprinkle over the gelatine, place the bowl in a saucepan with a little hot water and warm gently over the heat, until dissolved. Stir into the rest of the orange juice. Add 8 oz. chopped melon balls to the jelly and pour into a 2-pt. ring mould. Allow to set. When set, turn out on to a flat plate, decorate with the remaining melon balls in the centre of the ring.

# Peach chiffon pie

6 oz. digestive biscuits
2 oz. walnut pieces
4 oz. butter, melted
1 pt. fruit cocktail jelly tablet
water
½ pt. frozen peach purée, thawed
1 tbsp. lemon juice
1 egg white
1 level tbsp. caster sugar
¼ pt. double cream
1 tbsp. milk

*Serves 6–8*

Place the biscuits in a polythene bag and crush to fine crumbs with a rolling pin. Chop the walnuts very finely. In a bowl combine the biscuits, walnuts and melted butter. Line a 10-in. loose-bottomed, French fluted, flan tin, with the crumb mixture, then using the back of a spoon press into place to form a shell. Make the jelly up to 1 pt. as directed on the packet. Combine with the purée and lemon juice and leave until beginning to set. Stiffly whisk the egg white, then continue to whisk whilst adding the sugar. Fold the meringue through the half-set peach mixture and turn into the biscuit case. Chill. To serve, whip the cream and milk until it just holds its shape, swirl over the filling. Chill.

# Plum whip

1 lb. frozen, stoned plums
4 oz. granulated sugar
12-fl. oz. black cherry yoghurt
2 egg whites

*Serves 6–8*

Place the plums and sugar in a saucepan and heat gently until the sugar dissolves. Simmer the fruit for about 20 min. until soft. Cool and then purée the fruit in a blender or pass through a sieve. Stir in the yoghurt. Stiffly whisk the egg whites and fold into the fruit mixture. Pour into individual glasses and serve.

# Plum and almond flummery

½ lb. frozen unsweetened, stoned plums, thawed
2 level tbsps. arrowroot
½ pt. milk
2–3 oz. sugar
1 oz. ground almonds
¼ tsp. almond essence
1 egg yolk
toasted almonds, to decorate

*Serves 4*

Place the plums in a serving dish. Mix the arrowroot with a little milk. Add the remaining milk, sugar to taste, ground almonds and essence. Pour into a saucepan and bring to the boil, stirring all the time.

Reduce the heat and simmer gently for a few min. until mixture thickens. Remove from the heat, cool a little, and stir in the egg yolk. Pour over the plums. Sprinkle with toasted almonds and serve warm.

# Plum crumble slice

7½-oz. pkt. frozen shortcrust pastry, thawed
2 oz. butter
3 oz. plain flour
½ oz. rolled oats
3 oz. brown sugar
1 lb. frozen sweetened, stoned plums, thawed

*Serves 8*

Roll out the pastry and use it to line a 7-in. by 11-in. by 1-in. deep cake tin. Melt the butter. In a bowl blend together the flour, oats, and brown sugar. Pour over the butter and fork it through. Layer the plums on the pastry, top with the crumble topping and bake in the oven at 400°F (mark 6) for about 30 min. Cool and cut into squares to serve.

# Cheesy plum dessert

8 oz. frozen stoned plums, thawed
1½ level tbsps. caster sugar
8 oz. cream cheese
2 tbsps. condensed milk
1 egg yolk
½ tsp. vanilla essence
flaked almonds, toasted, to decorate

*Serves 4*

Divide the plums between 4 sundae glasses and sweeten to taste. Blend the remaining ingredients together in a bowl or blender until smooth. Pour over the plums. Decorate with toasted, flaked almonds.

# Royal raspberry slices

8 slices stale white bread, crusts removed
red wine
8 tbsps. frozen raspberry purée, thawed
butter for frying
beaten egg
cinnamon sugar
whipped cream

*Serves 4*

Cut each slice of bread in half, dip in wine and sandwich with fruit purée. Melt the butter in a frying pan. Dip the raspberry sandwiches in beaten egg and fry sandwiches for 1–2 min. on each side, turning them with a fish slice or spatula. Remove from the pan. Sprinkle with cinnamon sugar, top with whipped cream and serve immediately.

# Orange jelly with raspberries

6-fl. oz. can frozen orange juice concentrate
water
1 level tbsp. powdered gelatine
½ lb. frozen raspberries
2 egg whites
1 oz. caster sugar

*Serves 6*

Make the orange juice up to 1 pt. with water. Spoon 2 tbsps. orange juice into a small basin, sprinkle over the gelatine, place the basin in a saucepan containing a little water and heat until the gelatine dissolves. Allow it to cool a little and stir into the rest of the orange juice. Add the raspberries to the jelly and leave until on the point of setting (the frozen raspberries act as ice cubes). Whisk the egg whites until stiff, then gradually add the sugar and fold into the jelly. Turn into a glass dish or individual glasses and leave to set.

# Iced raspberry soufflé

2 level tsps. cornflour
½ pt. milk
3 oz. caster sugar
2 eggs, beaten
½ lb. frozen unsweetened raspberries, thawed
6 fl. oz. double cream
chocolate cornets (see below)
fresh raspberries, optional
mint sprigs

*Makes 4*

Choose 4 small straight-sided soufflé dishes measuring 3-in. across the top and holding 7 tbsps. water. Other dishes of a similar type and size could also be adapted. Cut strips of non-stick paper long enough to fit round the outside of each dish and overlap slightly, and wide enough to extend 1 in. above the rim. Secure the paper with paperclips, pulling it tightly into place round the dishes. Put the dishes on a flat baking sheet. Blend the cornflour with 2 tbsps. cold milk. In a small pan, dissolve the sugar in the remaining milk over a low heat. Pour on to the blended cornflour, stirring, then pour this mixture slowly on to the beaten eggs, whisking. Return to the pan and cook, stirring, for a few min. without boiling, until the mixture thickens. Allow the sauce to cool. Sieve the raspberries to remove the seeds and stir into the cornflour sauce. Leave covered until cold. Whip the cream until it just holds its shape, and is the same consistency as the cornflour sauce, then fold the sauce evenly through the whipped cream. Spoon into the prepared soufflé dishes so that the mixture comes about $\frac{1}{2}$ in. above rim.

Freeze until firm but not solid. Just before serving remove the paper collar and decorate with chocolate cornets, fresh raspberries and mint sprigs.

Chocolate Cornets: Cut non-stick paper into 4-in. squares then cut each into 2 triangles. With the base towards you fold the right-hand corner up to the top point of the triangle and then turn round to take in the other point. Slightly overlap the points and secure with a staple or fold the tips over. Melt 2–3 oz. chocolate in a small basin over a pan of hot, not boiling, water. Spoon a little of the chocolate, cool but still flowing, in the centre of each paper cornet, turn the cornet to evenly coat the sides. Chill, recoat if necessary. When set carefully peel away the paper. Store in a cool place.

# Raspberry napolitain

1 lb. frozen puff pastry, thawed
$\frac{1}{4}$ lb. raspberry jam, sieved
$\frac{3}{4}$ pt. double cream
3 egg whites
1 oz. caster sugar
1 lb. frozen raspberries, thawed
icing sugar

*Serves 6–8*

Roll the pastry into 3 wafer thin 10-in.

rounds. Lift on to baking sheets, using a rolling pin. Prick well and leave the pastry to relax for 15 min. in a cool place then bake in the oven at 425°F (mark 7) for about 10 min. or until golden brown. Cool. With scissors, trim each disc to the same size and crush and retain the trimmings. Spread the sieved jam over one surface of 2 of the rounds. Whisk together the cream, egg whites and sugar until they hold their shape. Place one round, jam side uppermost, on a flat serving plate. Spread a film of cream over the jam. Put two spoonsful of raspberries to one side for decoration and place half the remaining raspberries in the centre of the round. Use half the cream to make a collar level with the berries. Repeat with second round. Place the third round on top and press lightly. Level off the cream edges if necessary. Dust with icing sugar and spoon a band of pastry trimmings round the top edge. Arrange a spoonful of berries in the centre. Refrigerate for up to 1 hr. before serving.

# Raspberry cream pie

6 oz. shortcrust pastry (i.e. made with
    6 oz. flour etc.)
1 level tbsp. cornflour
$\frac{1}{2}$ pt. milk
2 eggs, separated
5 oz. caster sugar
finely grated rind of 1 lemon
icing sugar
$\frac{1}{2}$ lb. frozen raspberries, thawed
6 fl. oz. double cream

*Serves 6–8*

Roll out the pastry and use it to line a $9\frac{3}{4}$-in. ovenproof pie plate. Bake blind in the oven at 400°F (mark 6) for about 20 min. Meanwhile, blend the cornflour with a little milk. Bring the rest of the milk to the boil and pour it on the cornflour, stirring; return to the pan and cook for 1–2 min. stirring. Remove from the heat and add the beaten egg yolks, blended with 1 oz. sugar and the lemon rind; beat well. Again return to the heat and bring almost to boiling point, stirring. Pour into a bowl and dust with icing sugar to prevent a skin forming.

Whisk the egg whites stiffly, gradually

add 4 oz. caster sugar, whisking all the time. Line a baking sheet with non-stick paper and mark an 8-in. round on it (use a pan lid as a guide). Spoon the meringue into the circle. Cook in the oven at 350°F (mark 4) for about 30 min. until tipped with brown and crisp. Cool and carefully remove the paper. Layer the raspberries in the pastry case. Fold the stiffly whipped cream into the cold custard and turn into the flan case. Top with the cold meringue disc. Eat on the day of making.

# Raspberry and almond tranche

*For pastry*
6 oz. plain flour
pinch salt
3 oz. butter at room temperature
1½ oz. caster sugar
1 egg, beaten
¼ tsp. vanilla essence

*For almond cream*
3 oz. butter
3 oz. caster sugar
1 large egg
3 oz. ground almonds
½ oz. plain flour
¼ tsp. almond essence

12 oz. frozen raspberries
2–3 tbsps. redcurrant jelly

*Serves 6–8*

Sift the flour and salt on to a work surface. Make a well in centre, add the butter and sugar and cream with the fingertips. Add the beaten egg and vanilla essence then gradually work in the flour using the fingertips. Knead lightly and chill whilst making the filling. Cream the butter and sugar and beat in the egg. Stir in the ground almonds, flour and almond essence. Roll out the pastry and use it to line a tranche frame 14 in. by 4½ in. by 1 in., placed on a foil-lined baking sheet. Crimp the pastry edge. Spread almond cream in the base and top with the frozen raspberries. Bake in the oven at 400°F (mark 6) for 25 min., reduce the oven temperature to 350°F (mark 4), and cook for about a further 15 min. Cover with foil if the pastry edges start to overbrown. Warm the redcurrant jelly and brush it over the fruit. Serve warm or cold.

# Redcurrant Griestorte

3 large eggs, separated
4 oz. caster sugar
grated rind and juice of ½ lemon
2 oz. fine semolina
1 level tbsp. ground almonds
6 fl. oz. double cream
1 tbsp. milk
¼ lb. frozen unsweetened redcurrants, partly thawed
icing sugar

*Serves 6*

Grease and line an 8-in. by 12-in. Swiss roll tin with non-stick paper to extend above the sides. Grease the paper, sprinkle with caster sugar and a dusting of flour. Whisk the egg yolks with the sugar until thick and pale. Whisk in the lemon juice. Stir in the rind, semolina and almonds mixed together. Whisk the egg whites until stiff and fold the yolk mixture through the whites. Turn into the prepared tin, level the surface and bake in the oven at 350°F (mark 4) for about 30 min., until puffed and pale gold. Turn the Griestorte carefully on to a sheet of non-stick paper dusted with caster sugar. Trim the long edges and roll up loosely with the paper inside. Cool on wire rack.

To finish, whisk together the cream and milk until it just holds its shape. Unroll the Griestorte—don't worry if it cracks a little—spread the cream over but not quite to the edges, sprinkle with currants. Roll up, pressing lightly into shape. Dust with icing sugar. Eat on the day of making.

# Baked strawberry meringue

1 lb. frozen unsweetened strawberries, thawed
2 oz. caster sugar
3 egg whites
5 level tbsps. icing sugar

*Serves 4*

Place the strawberries in an ovenproof serving dish. Sprinkle with sugar. Whisk the egg whites stiffly and fold in the sifted icing sugar. Spoon over the strawberries. Bake in the oven at 400°F (mark 6) for a few min., until golden brown. Serve immediately.

# Fruit and cream puffs

6 frozen vol-au-vent cases
½ lb. frozen sweetened strawberries,
   thawed
whipped cream, to decorate

*Serves 6*

Bake the vol-au-vent cases as directed on
the pkt. When cool, fill with strawberries
and top with rosettes of cream.

# Strawberry cream gâteau

(*see picture below*)

¾ lb. frozen strawberries, thawed
2 oz. plain chocolate
¼ pt. double cream
a little sugar
1 frozen 8-in. Victoria sandwich cake or
   whisked sponge, thawed

*Serves 8*

Reserve 12 whole strawberries, wipe them
dry and place on non-stick paper. Melt the
chocolate on a plate over a pan of hot water
and, using a teaspoon, drizzle a little chocolate over the 12 strawberries. Whip the
cream with sugar to taste. Split the cake,
sandwich with half the cream and the strawberries. Cover the top of the cake with the
remaining cream, and drizzle any leftover

chocolate over the top. Decorate with the
chocolate strawberries.

# Strawberry mille feuilles

13-oz. pkt. frozen puff pastry, thawed
½ lb. frozen unsweetened strawberries,
   thawed
1 level tbsp. caster sugar
½ pt. double cream
¼ pt. single cream
½ lb. strawberry jam
1 tbsp. lemon juice

*Serves 6–8*

Roll out the pastry to a rectangle about
16 in. by 13 in. Trim ½ in. off the edges to
neaten it. Mark into 3 down the length and
cut into equal pieces about 5 in. by 12 in.
Prick the surface well. Place the pieces on a
damp baking sheet and chill for 30 min.
Bake in the oven at 425°F (mark 7) for 10–12
min. Cool on a wire rack.

   Leave 4 strawberries whole and cut the
remainder into small pieces. Sprinkle sugar
over and leave for 10 min. Whip the creams
together until they hold their shape. In a
bowl combine the jam and lemon juice. To
assemble the mille feuilles, spread 5 tbsps.
whipped cream over 1 pastry layer. Halve
the jam and lightly spread it on top of the
cream. Scatter with half the berries. Repeat

*Strawberry
cream gâteau*

with a second layer then put one on top of the other, filling side uppermost. Position the last pastry layer on top, spread a little of the remaining cream over and place the bulk in a forcing bag fitted with a large star vegetable nozzle. Pipe the cream as a decoration and add the reserved strawberries, cut in half. Serve on a flat plate or board.

# Strawberry almond tartlets

4 oz. butter
2 oz. caster sugar
1 egg yolk
6 oz. plain flour
2 oz. ground almonds

*For filling*
8 Petit-Suisse cheeses
2 level tsps. caster sugar
1 lb. frozen unsweetened strawberries,
    thawed and sliced
3 tbsps. redcurrant jelly

*Makes 8*

Cream together the butter and sugar. Beat in the egg yolk, then work in the flour and ground almonds. Chill until of a manageable consistency—about 1 hr. Divide into 8 portions. Roll the pastry out thinly on a lightly floured surface and use to line 8 loose-based, 4-in. diameter, fluted French flan tins. Bake blind at 375°F (mark 5) for about 15 min. until lightly coloured. Cool on a wire rack. Cream the Petits-Suisses with the caster sugar and spread it over the base of each pastry case. Top with overlapping, sliced strawberries. Warm the redcurrant jelly and brush it over the strawberries to glaze them.

# Strawberry foam

1 lb. frozen unsweetened strawberries,
    thawed
2 egg whites
4 oz. icing sugar, sifted
few drops edible red colouring

*Serves 6*

Turn the berries into a deep bowl and mash. Add the egg whites, sugar and colouring and whisk with an electric mixer

for about 10 min. until thick and frothy. Turn into glasses, piling the mixture well up. Chill for up to 6 hr. before serving. Serve with single cream.

# Strawberry shortcake

8 oz. plain flour
1 level tsp. cream of tartar
$\frac{1}{2}$ level tsp. bicarbonate of soda
pinch of salt
2 oz. butter or margarine
$1\frac{1}{2}$ oz. caster sugar
1 egg, beaten
3–4 tbsps. milk
$\frac{1}{2}$ lb. frozen unsweetened strawberries,
    thawed
$\frac{1}{4}$ pt. double cream
1 tbsp. milk
butter

*Serves 6–8*

Sift together the flour, cream of tartar, bicarbonate of soda and salt into a bowl. Rub in the fat with the fingertips and stir in the sugar. Make a well in the centre and add the beaten egg and enough milk to give a soft but manageable dough. Knead lightly on a floured surface, then shape out into a 7-in. round. Place on a baking sheet, dust lightly with flour and bake in the oven at 425°F (mark 7) for about 20 min. Cool a little on a wire rack covered with a clean tea towel.

Slice the strawberries thickly. Whisk together the cream and milk until it holds its shape. Split the warm shortcake and lightly butter. Spread half the cream over the base, then add the fruit and the rest of the cream. Replace the top and serve at once.

# Strawberry and pineapple Romanoff

1 lb. frozen unsweetened strawberries,
    thawed
$\frac{1}{2}$ lb. frozen pineapple chunks, thawed
sugar to taste
$\frac{1}{4}$ pt. double cream
17-fl. oz. block Cornish ice cream,
    softened
3 tbsps. Cointreau

*Serves 8*

Reserve a few whole strawberries and pineapple chunks for decoration and place the remaining fruit in a serving dish. Sprinkle with sugar. Lightly whip the cream. Beat the softened ice cream and stir in the cream and Cointreau. Pour this over the fruit. Decorate with the reserved fruit and return to the freezer until ready to serve. Do not re-freeze completely but serve the same day.

# Pineapple snow

1 lb. frozen pineapple in syrup, thawed
2 eggs, separated
1 level tbsp. cornflour
grated rind $\frac{1}{2}$ lemon
whipped cream, for decoration

*Serves 6–8*

Purée the pineapple in a blender. Blend the egg yolks, cornflour and lemon rind into the pineapple purée and turn the mixture into a saucepan. Heat gently for a few min. until the mixture thickens, then allow to cool. Whisk the egg whites stiffly and fold into the pineapple mixture. Serve in individual sundae glasses, decorated with whipped cream.

# Arctic mousse

(*see colour picture facing page 161*)

$\frac{1}{2}$ pt. frozen thick fruit purée, thawed
  (e.g. raspberry or blackcurrant)
4 egg yolks
sugar to taste
$\frac{1}{2}$ oz. powdered gelatine
3 tbsps. water
$\frac{1}{4}$ pt. double cream
1 Arctic roll
2 egg whites

*Serves 6*

Place the purée, yolks and sugar to taste in a large basin over hot water and whisk until thick and creamy. Remove from the heat. Sprinkle the gelatine on to the water in a small basin, place in a pan of hot water and heat very gently to dissolve. Blend into the fruit mixture and cool until beginning to set. Whip the cream to the same consistency as the fruit mixture and fold in, lightly and

evenly. Cut the Arctic roll into thin slices and use to line an 8-in. loose-based cake tin. Whisk the egg whites stiffly, fold into the mousse, then turn into the lined cake tin. Freeze until firm. To serve, unmould and slice.

# Sugared fruit fool

$\frac{1}{2}$ pt. frozen sweetened fruit purée, thawed
$\frac{1}{4}$ pt. custard
$\frac{1}{4}$ pt. natural yoghurt
Demerara sugar

*Serves 4*

Mix the fruit purée and custard together and fold in the yoghurt gently, so as to leave a marbling of white. Turn the fool into individual serving dishes. Sprinkle with Demerara sugar and serve chilled.

# Russian kissel

$1\frac{1}{2}$ level tbsps. cornflour
1 pt. frozen unsweetened fruit purée,
  thawed
sugar to taste

*Serves 4*

Blend the cornflour with the fruit purée and add sugar to taste. Turn into a saucepan, bring to the boil, stirring, and heat gently for a few min. until the mixture thickens. Pour into a serving dish and chill. Serve with whipped cream.

# Fruit blancmange

$\frac{1}{2}$ pt. milk
3 level tbsps. cornflour
$\frac{1}{2}$ pt. frozen fruit purée, thawed
  (e.g. blackberries, raspberries etc.)
sugar to taste
knob of butter
whipped cream, to decorate

*Serves 6*

Blend the milk and cornflour. Place in a saucepan, bring to boil then reduce the heat. Add the remaining ingredients. Simmer gently for few min. until the mixture thickens. Pour into a dampened 1-pt. mould and allow to set. Decorate with rosettes of whipped cream.

# 11 Cakes, Tea Breads and Biscuits

Baked goods freeze extremely well. Whether you are thinking of plain biscuits and cookies, tea breads and buns, plain cakes or decorative gâteaux, all freeze and thaw equally well. Remember that custards tend to separate, so do not freeze cakes with this type of filling—otherwise most fillings and icings will freeze; butter cream and fudge type decorations are perhaps the best to use.

### To pack and freeze cakes and tea breads
Always cool baked goods before starting to freeze. Plain cakes and tea breads should be wrapped whole in polythene bags and sealed and labelled before freezing. This type of cake can be sliced while frozen or partly thawed. If you wish to pack several cakes together, such as sponge layers, separate each layer with a double thickness of polythene film or waxed paper.

Decorative iced cakes cannot be cut while frozen, so it will save thawing time to slice them before freezing. Protect the cut edges with polythene film or waxed paper and re-assemble the portions in the shape of the original cake. Open freeze the cake on a baking sheet then place it in a rigid plastic or foil container with a lid that will protect it during storage, and seal it. For a special occasion the cake will, of course, look better if displayed whole, and for this you should allow the extra thawing time.

Small uniced cakes may be frozen and packed in polythene bags; iced ones need packing in a single layer in a rigid container with a lid.

### To pack and freeze biscuits and cookies
Biscuits and cookies can be frozen before or after baking. It saves storage space to freeze unbaked dough before shaping—just mix the dough, form it into a convenient shape such as a roll or a brick, open freeze until firm, then wrap in a polythene bag and seal. Alternatively, pipe or shape the unbaked dough on a baking sheet, open freeze for 1–2 hr. until firm, then wrap the shapes in a polythene bag and seal.

Baked biscuits and cookies are open frozen on the baking sheet or in the baking tins, then removed from the sheet, wrapped and sealed in a polythene bag or, for extra protection, a rigid container.

## To use
Unwrap any iced cakes or biscuits before thawing, to prevent the wrapping sticking to the icing. Leave plain cakes, breads and baked cookies or biscuits in their wrapping, but remove any outer wrappings.

## Thawing times

| | |
|---|---|
| Flans and sponge layers | $1\frac{1}{2}$–2 hr. at room temperature |
| Large plain cakes and | |
|    gâteaux (whole) | 4–6 hr. ,,  ,,  ,, |
|    (cut into portions) | $2\frac{1}{2}$ hr. ,,  ,,  ,, |
| Swiss rolls | $2\frac{1}{2}$ hr. ,,  ,,  ,, |
| Small cakes | 1 hr. ,,  ,,  ,, |
| Cookies and biscuits | 15 min. ,,  ,,  ,, |
| | (longer if filled) |
| Cakes decorated with | |
|    fresh cream | 6–7 hr. in the refrigerator |

Undecorated cookies or biscuits that need to be extra crisp may be refreshed in the oven on a baking sheet for about 5 min. at 375°F (mark 5).

## To use unbaked cookie and biscuit dough
Shaped dough can be baked without thawing; simply add a few min. extra time to the baking time given in the recipe. Refrigerator cookie dough can be sliced as needed while still frozen and any surplus returned to the freezer. Uncooked dough that was not shaped before freezing should be thawed at room temperature until easy to handle and baked the same as dough that has not been frozen.

## Storage times
Cooked cakes, teabreads and biscuits all store well for about 6 months. Uncooked biscuit dough can be kept for a similar period. Icings and frostings, however, tend to start deteriorating sooner and it is not recommended that iced cakes be stored for longer than 2 months.

## Note on fats used in cakes
Occasionally we specify that butter should be used in a recipe because it will affect the flavour if margarine is used. Where this is not so and the alternative 'butter or margarine' is given, this refers to the traditional, firm, block-type margarine. The soft margarine now sold in tubs should not be used unless specifically stated, as it gives a mixture of a different consistency.

# Cherry and raisin cake

12 oz. self-raising flour
pinch of salt
6 oz. margarine
6 oz. glacé cherries
4 oz. stoned raisins
2 oz. desiccated coconut
6 oz. caster sugar
2 large eggs
$\frac{1}{4}$ pt. milk, approx.

TO MAKE: Grease a $7\frac{1}{2}$–8-in. round cake tin. Sift the flour and salt into a bowl. Rub in the margarine with the fingertips until the mixture resembles fine breadcrumbs. Halve the cherries and raisins, using scissors; toss them in the coconut and stir them into the rubbed-in mixture with the caster sugar. Whisk the eggs and milk together and stir into the mixture; beat lightly. Turn the mixture into the prepared tin and level the surface. Bake in the oven at 350°F (mark 4) for about $1\frac{1}{2}$ hr. until well risen and golden brown. Leave in the tin for 15 min. before turning out on to a wire rack to cool.

TO PACK AND FREEZE: See notes at beginning of chapter.

TO USE: Thaw for about 4 hr.—see notes.

# Tyrol cake

8 oz. plain flour
1 level tsp. ground cinnamon
$3\frac{1}{2}$ oz. butter
2 oz. caster sugar
2 oz. currants
2 oz. sultanas
1 level tsp. bicarbonate of soda
$\frac{1}{4}$ pt. milk, approx.
3 level tbsps. clear honey, warmed

TO MAKE: Grease and flour a 6-in. round cake tin. Sift the flour and cinnamon into a basin and rub in the butter with the fingertips until the mixture resembles fine breadcrumbs. Stir in the sugar and dried fruit. Dissolve the bicarbonate of soda in the milk, add it to the warmed honey and stir well. Add the liquid to the dry ingredients, adding more milk if necessary to give a firm dropping consistency. Turn the mixture into the prepared tin and level the top. Bake in the oven at 325°F (mark 3) for $1\frac{3}{4}$–2 hr. Cool for

a few min. in the tin, then turn out on to a wire rack.

TO PACK AND FREEZE: See notes at beginning of chapter.

TO USE: Thaw for about 4 hr.—see notes.

# Golden syrup cake

6 oz. margarine
6 oz. golden syrup
3 large eggs, beaten
6 oz. self-raising flour
pinch of salt
grated rind of 1 lemon
2 slices candied peel

TO MAKE: Line a 7-in. round cake tin with greased greaseproof paper. Cream the margarine and syrup together until light and fluffy, then beat in the eggs. Sift the flour and salt over the surface and fold them into the creamed mixture with the lemon rind, using a metal spoon. Turn the mixture into the prepared tin and smooth the surface. Bake it in the oven at 350°F (mark 4) for 50 min., quickly open the oven door, arrange the peel on top of the cake and continue to cook for about a further 10 min. Turn out and cool the cake on a wire rack.

TO PACK AND FREEZE: See notes at beginning of chapter.

TO USE: Thaw for 3–4 hr.—see notes.

# Carrot cake

4 eggs, separated
6 tsps. warm water
8 oz. caster sugar
4 oz. plain flour
2 level tsps. baking powder
$\frac{1}{2}$ lb. tender carrots, peeled and finely grated
$\frac{1}{2}$ lb. ground hazelnuts
1 tsp. rum essence
few drops lemon juice

TO MAKE: Line a $9\frac{3}{4}$-in. by 5-in. loaf tin (top measurement) with foil and grease the foil. In a bowl, beat the egg yolks with the water until frothy, then beat in 6 oz. sugar and continue to whisk until thick and creamy. Whisk the egg whites in a second bowl until stiff, add the remaining 2 oz. sugar and whisk again. Fold the whites into the yolks, then sift the flour and baking powder together over the surface of the mixture in the bowl and fold it in with a metal spoon. Next fold in the carrots and ground hazelnuts and add the essence and lemon juice. Turn the mixture into the prepared tin and bake in the oven at 400°F (mark 6) for about 50 min. Turn out and leave to cool on a wire rack.

**TO PACK AND FREEZE:** See notes at beginning of chapter.

**TO USE:** Unwrap the cake, place it on a wire rack and leave to thaw at room temperature for 3–4 hr.

# Cinnamon date cake

2 oz. plain flour
3 level tsps. ground cinnamon
8 oz. margarine
6 oz. caster sugar
3 level tbsps. clear honey
3 large eggs, beaten
6 oz. self-raising flour
grated rind of ½ lemon
1 tbsp. milk
3 oz. stoned dates, finely chopped
1 oz. halved, shelled walnuts

**TO MAKE:** Line an 8-in. square cake tin with greased greaseproof paper. Sift the plain flour and 1 level tsp. cinnamon into a bowl, rub in 2 oz. margarine and stir in 3 oz. caster sugar. This is the topping mixture. Cream the remaining margarine with the rest of the sugar and the honey until fluffy. Beat in the egg a little at a time. Sift the self-raising flour and the rest of the cinnamon over the creamed mixture and fold it in with a metal spoon, then mix in the lemon rind, milk and dates. Turn the mixture into the prepared tin. Smooth the surface, spoon over the topping and arrange halved nuts on the top. Bake in the oven at 375°F (mark 5) for about 1 hr. 10 min. or until well risen, firm and golden. Leave the cake in the tin for a while, then turn it out on to a wire rack to cool.

**TO PACK AND FREEZE:** See notes at beginning of chapter.

**TO USE:** Thaw for about 4 hr.—see notes.

# Hungarian chocolate gâteau

(*see colour picture facing page 177*)

6 eggs
6 oz. caster sugar
1 tbsp. coffee essence
1 oz. cocoa powder
4 oz. plain flour
1 oz. cornflour
1 oz. ground almonds
3 oz. butter (preferably unsalted), melted

*For filling and decoration*
9 oz. caster sugar
12 tbsps. water
6 egg yolks, beaten
12–14 oz. butter, creamed
2–3 tbsps. coffee essence
2 oz. shelled walnuts, chopped

*To complete cake from freezer*
chocolate cobwebs (see below)
sugared coffee beans

*Serves 8–10*

**TO MAKE:** Line two 9½-in. straight-sided sandwich tins with greased greaseproof paper. Whisk the eggs and sugar in a large deep bowl over a pan of hot water until thick and the mixture retains the impression of the whisk for a moment when lifted. Whisk in the coffee essence. Remove from heat and continue whisking until cold. Sift together the cocoa, flour, cornflour and almonds. Sift half the dry ingredients a second time and fold carefully into the whisked mixture, using a metal spoon. Then pour in the melted but not oily butter, round the side of the bowl, and fold it in with the remaining dry ingredients. Divide the mixture between the tins and bake at 375°F (mark 5) for about 30 min. Turn out and cool on wire rack.

Meanwhile make a coffee crème au beurre. Put the sugar in a heavy based saucepan. Add the water and dissolve the sugar over gentle heat. When dissolved, bring it to boiling point and bubble for 2–3 min. to reach 225°F. Pour the syrup gradually, in a thin stream, on to the beaten egg yolks, whisking all the time until thick and cold. Gradually beat the creamed butter into the egg yolk mixture and flavour it with 2–3 tbsps. coffee essence.

Sandwich the cakes with coffee crème au beurre, spread the sides as well and coat with chopped walnuts, pressing them in well with a palette knife. Cover the top of the cake with more crème au beurre and decorate with piping.

**TO PACK AND FREEZE:** See notes at beginning of chapter.

**TO USE:** Thaw according to notes, then complete the decoration of the thawed gâteau with chocolate cobwebs and sugared coffee beans.

### CHOCOLATE COBWEBS

Put 2 oz. plain chocolate cake covering into a basin over hot water and allow it to melt. Pour the liquid chocolate into a paper forcing bag fitted with a No. 2 icing nozzle. On waxed paper, draw a pencil line round a medium ($3\frac{3}{4}$-in.) boat-shaped pastry cutter. Pipe chocolate around the outline and fill in with a wriggly, continuous trellis work. Chill until set, then peel away the paper. These may be made a few days ahead and stored in an airtight tin.

# Apricot wine cake

(*see picture opposite*)

4 oz. butter
4 oz. caster sugar
2 eggs, beaten
4 fl. oz. dry white wine
2 oz. dried apricots, roughly chopped
8 oz. plain flour
1 level tsp. bicarbonate of soda
$\frac{1}{4}$ level tsp. grated nutmeg

*To complete cake from freezer*
Either: 3 oz. butter
       2 level tbsps. clear honey
       1 tsp. lemon juice
Or:    whipped double cream
       15-oz. can apricots, drained and
       puréed
icing sugar

**TO MAKE:** Line a 7-in. round cake tin with greased greaseproof paper. Cream together

the butter and sugar until light and creamy. Add the eggs gradually, beating well, then add the wine and apricots. Sift the flour, bicarbonate of soda and grated nutmeg over the creamed mixture and fold them in with a metal spoon. Turn the mixture into the prepared tin and bake in the oven at 375°F (mark 5) for about 45 min., until risen and firm in the centre. Turn out and cool on a wire rack.

**TO PACK AND FREEZE:** Split the cake in half, wrap the layers separately in foil and overwrap them together in a polythene bag. Seal, label and freeze.

**TO USE:** Remove the overwrapping and thaw the cake for about 2 hr. at room temperature. Meanwhile make the filling. Cream the butter until soft; beat in the honey and lemon juice. Sandwich the layers with this filling. Alternatively, sandwich the layers with cream and apricot purée. Dredge the top of the cake with icing sugar.

# Coffee gâteau

8 oz. butter
8 oz. caster sugar
4 eggs
2 heaped tsps. instant coffee, dissolved in a
   little boiling water
8 oz. self-raising flour, sifted

*For crème au beurre*
6 oz. caster sugar
8 tbsps. water
4 egg yolks, beaten
10 oz. butter (preferably unsalted)
2–3 tbsps. coffee essence to flavour

**TO MAKE:** Grease and line three 7-in. sandwich tins. Cream the fat and caster sugar until pale, soft and fluffy. Gradually beat in the eggs and coffee, beating well after each addition. Fold in the flour and then divide evenly between the three tins, levelling the tops with a knife. Bake in the oven at 375°F (mark 5) for about 35 min., until golden and beginning to shrink from the sides of the tin. Turn the cakes out on to a wire rack to cool. Make the crème au beurre. Place the sugar in a heavy based saucepan, add the water and leave over a

*Apricot wine cake*

very low heat to dissolve the sugar, without boiling. When completely dissolved, bring to boiling point and boil steadily for 2–3 min. to 225°F. Pour the syrup in a thin stream on to the egg yolks, whisking all the time. Continue to whisk until the mixture is thick and cold. Gradually add to the creamed butter and flavour with the coffee essence.

Sandwich the cakes together with a little of the crème au beurre, and spread the remainder over the sides and top of the cake; draw the blade of a small round bladed knife over the surface of the crème au beurre, to decorate.

**TO PACK AND FREEZE:** See notes at beginning of chapter.

**TO USE:** Unwrap the cake and leave at room temperature for about 6 hr. to thaw.

## Chocolate crunch cake

8 oz. digestive biscuits
2 oz. seedless raisins
3 oz. margarine or butter
2 tbsps. golden syrup
1 level tbsp. caster sugar
1 level tbsp. cocoa powder

*To complete cake from freezer*
2 oz. chocolate, grated

**TO MAKE:** Roughly crush the biscuits with a rolling pin and mix the crumbs with the raisins. Put the margarine or butter, syrup, sugar and cocoa in a saucepan and melt them over a low heat. Stir the mixture into the crumbs and raisins, mix well then pack into a greased 8-in. flan ring placed on a flat serving plate or baking sheet. Leave to set.

**TO PACK AND FREEZE:** Open freeze until firm, then remove from the flan ring, wrap in foil, overwrap, seal and label. Return to the freezer.

**TO USE:** Unwrap and thaw at room temperature for 1 hr. Sprinkle with grated chocolate to serve.

## Austrian cheesecake

5 oz. butter
5 oz. caster sugar
5 oz. curd cheese
5 eggs, separated
grated rind and juice of 1 lemon
5 oz. ground walnuts
1 oz. self-raising flour

**TO MAKE:** Line an 8-in. round cake tin with greased greaseproof paper. Cream together the butter, sugar, cheese, egg yolks, grated rind and juice of the lemon and fold in the nuts. Stir in the flour and then the stiffly whisked egg whites. Turn the mixture into the prepared tin and level the surface. Bake in the oven at 425°F (mark 7) for 20 min.; reduce the oven temperature to 350°F (mark 4) and continue cooking for about a further 1 hr. The cake should be firm to the touch. Turn out and cool on a wire rack.

**TO PACK AND FREEZE:** See notes at beginning of chapter.

**TO USE:** See notes.

# Plum cake with orange rum butter

(*see colour picture facing page 176*)

6 oz. soft tub margarine
6 oz. light soft brown sugar
3 large eggs
1 tbsp. golden syrup
3 tbsps. blackcurrant jam
12 oz. seedless raisins
6 oz. currants
2 oz. chopped mixed peel
2 pieces crystallised ginger, chopped
2 oz. shelled walnuts, chopped
2 oz. ground almonds
2 tbsps. rum
8 oz. self-raising flour
1 level tsp. baking powder
1 level tsp. ground mixed spice

*For orange rum butter*
4 oz. soft tub margarine
12 oz. icing sugar, sifted
1½ tbsps. rum
1 tbsp. orange juice
½ level tsp. ground cinnamon

**TO MAKE:** Grease and flour a 3-pt. ring mould. Place the first 12 ingredients in a large bowl and sift the flour, baking powder and mixed spice over the top. Beat well with a wooden spoon for about 3 min. until all the ingredients are thoroughly combined. Turn the mixture into the prepared mould, bake in the oven at 400°F (mark 6) for 15 min., then reduce the oven temperature to 300°F (mark 1–2) for a further 2½–2¾ hr. Allow to cool slightly before removing from the tin to cool on a wire rack.

In a deep bowl, cream together all the

ingredients for the orange rum butter until soft and well blended.

**TO PACK AND FREEZE:** Pack and freeze the cake according to the notes at the beginning of the chapter. Place the orange rum butter in a rigid polythene container, seal, label and freeze.

**TO USE:** Remove the overwrapping from the cake and allow to mature for at least a day at room temperature. Remove the butter from the freezer with the cake and allow to thaw in the refrigerator. Serve the cake in slices, spread with the orange rum butter.

# Orange refrigerator cake

1 oz. butter
1 oz. flour
¼ pt. milk
1 egg
grated rind and juice of 1 small orange
juice of 1 lemon
6 oz. butter
6 oz. caster sugar
1 pkt. soft sponge fingers (about 24)
6 tbsps. dry sherry
edible orange colouring

*To complete cake from freezer*
¼ pt. double cream, whipped
mandarin oranges and angelica

*Serves 8*

**TO MAKE:** Melt the butter in a small pan, stir in the flour and cook for a few min. Remove from the heat, add the milk and beat well, then gradually add the egg. Beat in the grated rind and juice of the orange. Return the pan to a low heat and cook the sauce until thick without boiling. Remove from the heat and beat in the lemon juice. Cool the sauce whilst creaming the butter and sugar to a very light and fluffy consistency. Beat in the sauce a spoonful at a time (an electric mixer is ideal for this). Line an oblong 3-pt. plastic box with greaseproof paper. Dip the sponge fingers in sherry and line the base of the box with them. Cover with half the filling, and add another layer of fingers. Tint the remaining filling orange and layer it over the fingers. Finish with a final layer of fingers and chill.

**TO PACK AND FREEZE:** When set, lift the

*Coffee marble cake*

cake from the box. Wrap it in foil, overwrap with polythene, seal, label and freeze.

**TO USE:** Unwrap the cake, place it on a serving dish and thaw at room temperature for about 2 hr. Decorate with cream, fruit and angelica.

# Coffee marble cake

(*see picture above*)

8 oz. butter or margarine
8 oz. caster sugar
4 large eggs
8 oz. self-raising flour
2 tbsps. coffee essence

*To complete cake from freezer*
icing sugar

**TO MAKE:** Grease a 3½-pt. ring mould. Beat the fat until soft but not oily, add the sugar and cream them until light and fluffy. Add the eggs one at a time, beating between each addition. Lightly fold in the sifted flour. Divide the mixture into 2 portions and stir the coffee essence into 1 portion. Place the mixtures into the prepared mould in alternate spoonsful. Swirl through the mixture with a skewer, but do not overwork as this will spoil the marbled effect. Level the surface with a palette knife. Bake the cake in the oven at 350°F (mark 4) for about

45 min. until well risen and spongy to the touch. Turn out on to a wire rack to cool.

**TO PACK AND FREEZE:** See notes at beginning of chapter.

**TO USE:** Remove overwrap and thaw at room temperature for about 3 hr. Dust with icing sugar before serving.

# Chocolate fudge cake

3 oz. butter or margarine
3 oz. caster sugar
1 large egg
2 oz. golden syrup
5 oz. plain flour
1 oz. cocoa powder
½ level tsp. bicarbonate of soda
2 level tsps. baking powder
few drops vanilla essence
¼ pt. buttermilk

*For frosting*
2 oz. plain chocolate
3 tbsps. water
1 oz. butter
7 oz. icing sugar, sifted

**TO MAKE:** Line a 7–7½-in. double sandwich tin with greased greaseproof paper or raise the sides of a single sandwich tin with a greaseproof paper collar to give a total depth of about 2 in. Cream together the fat and sugar until light and fluffy. Beat in the egg and syrup. Sift together the flour, cocoa, bicarbonate of soda and baking powder and add alternately to the creamed mixture with the vanilla essence and buttermilk, beating lightly. Bake in the oven at 375°F (mark 5) for about 30 min. Turn out on to a wire rack to cool.

To make the frosting, melt the chocolate with the water in a small pan, taking care not to boil it. Remove from the heat and add the butter in small pieces. When melted, gradually beat in the icing sugar. Fasten a band of double greaseproof paper round the cake to come $\frac{3}{4}$–1 in. above the highest part of the cake; secure with a paper clip or pin. Spoon the frosting over the top when it is beginning to set and swirl it with a knife. Leave until firm but not hard before cutting.

**TO PACK AND FREEZE:** See notes at beginning of chapter.

**TO USE:** See notes.

# Oregon prune cake

4 oz. plain flour
1$\frac{1}{2}$ level tsps. baking powder
$\frac{1}{2}$ level tsp. salt
2 oz. margarine or butter
4 oz. caster sugar
1 egg, beaten
1 tbsp. milk

*For topping*
15$\frac{1}{2}$-oz. can prunes, drained and stoned
$\frac{1}{4}$ level tsp. ground cinnamon
1 level tbsp. sugar
2 oz. shelled walnuts, chopped
1 oz. butter, melted

**TO MAKE:** Grease a 7$\frac{1}{2}$-in. square cake tin. Sift the flour, baking powder and salt into a basin and rub in the fat with the fingertips until it resembles breadcrumbs. Add the remaining ingredients and mix to a soft consistency. Turn the mixture into the prepared tin. Press the prunes into the top of the cake. Mix the remaining ingredients for the topping and spread over the prunes. Bake in the oven at 375°F (mark 5) for 30–40 min. until firm. Turn out and cool on a wire rack.

**TO PACK AND FREEZE:** Wrap the cake in foil, overwrap in a polythene bag, seal, label and freeze.

**TO USE:** Unwrap the cake and thaw for 3–4 hr. at room temperature. Cut into squares to serve.

# Banana and honey loaf cake

12 oz. self-raising flour
$\frac{3}{4}$ level tsp. ground mixed spice
$\frac{3}{4}$ lb. bananas
6 oz. shelled walnut halves
6 oz. butter
6 oz. light soft brown sugar
3 eggs, beaten
3 level tbsps. honey
4 oz. glacé cherries, halved
4 oz. sultanas
4 oz. currants

**TO MAKE:** Grease and double line a 10$\frac{1}{2}$-in. by 6$\frac{1}{2}$-in. loaf tin (top measurement). Sift together the flour and spice. Mash the peeled bananas, reserve about 12 walnut halves and roughly chop the remainder. Cream together the butter and sugar and beat in the eggs, one at a time. Fold the sifted flour into the creamed mixture, alternately with the bananas and honey. Stir in the cherries, sultanas, currants and roughly chopped walnuts. Turn the mixture into the prepared tin. Slightly hollow the centre of the mixture and top with the reserved walnut halves. Bake in the oven at 350°F (mark 4) for 1 hr. If the cake starts to over-brown, cover with damp greaseproof paper. Reduce the oven temperature to 300°F (mark 2) and cook for about a further 1 hr. Turn out and cool on a wire rack.

**TO PACK AND FREEZE:** See notes at beginning of chapter.

**TO USE:** Remove overwrappings and leave wrapped in foil only at room temperature. Allow to mature for 2 days. Serve sliced, either plain or buttered.

# Apple and walnut teabread

4 oz. butter or margarine
4 oz. caster sugar
2 large eggs
1 level tbsp. honey or golden syrup
4 oz. sultanas
2 oz. shelled walnuts, chopped
8 oz. self-raising flour
pinch of salt
1 level tsp. ground mixed spice
1 medium cooking apple, peeled cored
   and chopped

TO MAKE: Line a loaf tin $8\frac{1}{2}$ in. by $4\frac{1}{2}$ in (top measurement) by $2\frac{1}{2}$ in. deep with greased greaseproof paper. Place all the ingredients in a large deep bowl and beat with a wooden spoon for about 2 min., until well combined. Turn the mixture into the prepared tin and level the surface. Bake the bread in the oven at 350°F (mark 4) for 1 hr. Reduce the oven temperature to 325°F (mark 3) and cook for about a further 20 min. Turn out and cool on a wire rack.

TO PACK AND FREEZE: See notes at beginning of chapter.

TO USE: Thaw at room temperature for 3–4 hr. Unwrap and slice as soon as possible for quicker thawing. Serve sliced and buttered.

# Malt fruit bread

(*see colour picture facing page 176*)

12 oz. plain flour
$\frac{1}{2}$ level tsp. bicarbonate of soda
1 level tsp. baking powder
2 level tbsps. Demerara sugar
4 level tbsps. golden syrup
4 level tbsps. malt extract
$\frac{1}{4}$ pt. milk
2 eggs, beaten
9 oz. sultanas, seedless raisins or chopped
   dates

TO MAKE: Grease a 9-in. by 5-in. loaf tin (top measurement). Sift together the flour, bicarbonate of soda and baking powder into a bowl and add the sugar. Warm the syrup, malt and milk together in a small pan—do not overheat, as this will cause curdling. Cool

the milk mixture and beat it with the eggs and fruit into the flour, until evenly mixed. Turn the dough into the prepared tin. Bake in the oven at 300°F (mark 2) for about $1\frac{1}{2}$ hr. Turn the loaf out and cool it on a wire rack.

TO PACK AND FREEZE: See notes at beginning of chapter.

TO USE: Remove the polythene bag and allow to mature at room temperature for at least 1 day, before serving sliced and buttered.

# Barm brack

7 oz. dark soft brown sugar
12 oz. mixed dried fruit
$\frac{3}{4}$ pt. cold, freshly made tea
1 egg, beaten
12 oz. self-raising flour

TO MAKE: Line an 8-in. round cake tin with greased greaseproof paper. Soak the sugar and fruit in the tea for at least 4 hr. (overnight would be ideal). Stir in the beaten egg and sifted flour. Turn the mixture into the prepared tin and bake in the oven at 350°F (mark 4) for about 1 hr., until well risen and a rich brown colour. Turn out and cool on a wire rack.

TO PACK AND FREEZE: See notes at beginning of chapter.

TO USE: Thaw at room temperature for about 3 hr. Serve sliced and buttered.

# Polish honey cake

4 oz. caster sugar
3 egg yolks
8 oz. self-raising flour
1 level tsp. ground cinnamon
$\frac{1}{4}$ level tsp. ground cloves
$\frac{1}{2}$ level tsp. grated nutmeg
$\frac{1}{2}$ pt. clear honey, warmed
3 oz. shelled walnuts, chopped
5 egg whites

TO MAKE: Line the base and grease and flour an 8-in. round cake tin. Whisk together the sugar and egg yolks until pale and creamy. Sift together the flour and spices and add them to the yolks with the honey and nuts. Stiffly whisk the egg whites and fold them into the mixture quickly but thoroughly. Turn it into the prepared tin and bake in the oven at 350°F (mark 4) for about 1 hr. Turn the cake out on to a wire rack to cool.

TO PACK AND FREEZE: See notes at beginning of chapter.

TO USE: Unwrap and thaw for about 4 hr. Serve sliced and buttered or sprinkled with icing sugar.

# Cherry buttermilk loaf

8 oz. self-raising flour
1½ level tsps. baking powder
¼ level tsp. ground mixed spice
¼ level tsp. ground ginger
4 oz. butter or margarine
4 oz. soft dark brown sugar
6 oz. glacé cherries, halved
½ tbsp. black treacle
1 egg, beaten
¼ pt. buttermilk

TO MAKE: Line a 9¾-in. by 5-in. loaf tin (top measurement) with greased greaseproof paper. Sift the flour, baking powder and spices into a basin. Rub the butter or margarine into the flour with the fingertips, until dry and crumbly. Add the sugar and cherries. Mix together the treacle and egg and lastly the buttermilk. Add these to the dry ingredients, stir well and turn the mixture into the prepared tin. Cook in the oven at 350°F (mark 4) for 1–1¼ hr. until risen and firm in the centre. Turn it out to cool on a wire rack.

TO PACK AND FREEZE: See notes at beginning of chapter.

TO USE: See notes.

# Ginger and apple loaf

½ lb. cooking apples, peeled and sliced
3 oz. Demerara sugar
¼ lb. golden syrup
3 oz. butter
6 oz. self-raising flour
1 level tsp. ground ginger
¼ level tsp. ground cloves
1 egg

TO MAKE: Grease an 8½-in. by 4½-in. loaf tin (top measurement). Put the apples in a pan with 1 tbsp. of the sugar and sufficient water to keep them from burning. Simmer gently until the apples are tender, then beat well and leave until cold. Put the golden syrup in a pan with the butter and remainder of the sugar, dissolve the sugar slowly then leave the mixture to cool. Sift the flour and spices into a basin. Whisk the egg, add the dissolved sugar and syrup mixture and whisk together, then add the flour. Mix well, stir in the apple pulp and beat together. Turn the mixture into the prepared tin and bake in the oven at 350°F (mark 4) for 1 hr. When cooked, allow the loaf to stand a little before turning out of the tin to cool on a wire rack.

TO PACK AND FREEZE: See notes at beginning of chapter.

TO USE: Unwrap and thaw at room temperature for 3–4 hr., or slice whilst frozen and allow 1 hr. to thaw.

# Spicy twist

*For bread dough*
1 level tsp. dried yeast
½ level tsp. sugar
5 tbsps. warm water
10 oz. strong plain flour
1 level tsp. salt
1 level tbsp. sugar
1 egg, beaten

*For filling*
2 oz. soft brown sugar
2 oz. sultanas
½ level tsp. ground allspice
½ level tsp. ground cinnamon
1 oz. butter, melted
1 egg, beaten

TO MAKE: Mix the yeast, ½ tsp. sugar and water and leave for 15–20 min. until frothy. Sift the flour and salt, add 1 tbsp. sugar and the egg. Stir in the yeast mixture and knead the dough on a floured board for about 10 min., until soft and pliable. Place it in a lightly oiled polythene bag and leave in a warm place until doubled in size.

For the filling, mix the sugar, sultanas, allspice and cinnamon together. When the dough is ready, roll it out to a rectangle about 15 in. by 8 in. Brush it with melted butter and sprinkle with the filling. Roll it up from the long side and seal the ends well. Place the roll on a baking tray and put it inside a lightly oiled polythene bag. Leave to rise until doubled in size again. Remove the polythene bag, brush with beaten egg and bake in the oven at 375°F (mark 5) for 30–35 min. Turn out to cool on a wire rack.

**TO PACK AND FREEZE:** See notes at beginning of chapter.

**TO USE:** Remove the polythene bag and re-fresh the loaf from frozen, still wrapped in foil, in the oven at 375°F (mark 5) for 50 min. Cool and serve sliced and buttered or plain.

# Apricot and cinnamon swirl

(*see colour picture facing page 176*)

*For bread dough*
2 level tsps. dried yeast
1 level tsp. sugar
$\frac{1}{4}$ pt. warm water
$1\frac{1}{4}$ lb. strong plain flour
2 level tsps. salt
2 level tbsps. sugar
2 eggs, beaten
grated rind and juice of 1 small orange

*For filling*
2 oz. Demerara sugar
1 level tbsp. ground cinnamon
4 level tbsps. apricot jam

*Makes two 1-lb. loaves*

**TO MAKE:** Mix the yeast, 1 tsp. sugar and warm water and leave for 15–20 min. until frothy. Sift together the flour and salt into a bowl. Add the sugar, eggs, orange rind and juice. Stir in the yeast mixture and knead the dough on a floured board for about 10 min., until soft and pliable. Place it in a lightly oiled polythene bag and leave in a warm place until doubled in size. Mix the sugar and cinnamon for the filling. When the dough is ready, roll it out to a rectangle about 15 in. by 8 in. Spread it with the apricot jam and sprinkle the cinnamon sugar over. Cut the dough in half cross-wise. Roll up both halves and seal the ends. Place each roll in a greased and floured, $8\frac{1}{2}$-in. by $4\frac{1}{2}$-in.

loaf tin, place them in lightly oiled polythene bags and leave to rise until almost to the top of the tin. Bake in the oven at 375°F (mark 5) for 45–50 min., until firm and golden. Turn out and cool on a wire rack.

**TO PACK AND FREEZE:** See notes at beginning of chapter.

**TO USE:** Remove overwrapping, leave loosely covered with foil and refresh from frozen in the oven at 375°F (mark 5) for 50 min. Serve sliced and buttered.

# Croissants

1 oz. fresh bakers' yeast
$\frac{1}{2}$ pt. water, less 4 tbsps.
1 lb. strong plain flour
2 level tsps. salt
1 oz. lard
1 egg, beaten
4–6 oz. butter or margarine
egg to glaze

**TO MAKE:** Blend the yeast with the water. Sift together the flour and salt and rub in the lard with the fingertips. Add the yeast liquid and egg and mix well together. Knead on a lightly floured surface for 10–15 min. until the dough is smooth. Roll the dough into a strip about 20 in. by 8 in and $\frac{1}{4}$ in. thick. Keep the edges straight and the corners square. Soften the butter or margarine with a knife to the consistency of the dough, and then divide into three. Use one part to dot over the top two-thirds of the dough, leaving a small border clear. Fold in three by bringing up the plain (bottom) third first, then folding the top third over. Turn the dough so that the fold is on the right-hand side. Seal the edges with a rolling pin. Reshape to a long strip by gently pressing the dough at intervals with a rolling pin. Repeat with the other two portions of fat. Place the dough in a greased polythene bag to prevent it forming a skin or cracking. Allow to rest in the refrigerator for 30 min. Roll out as before, and repeat the folding and rolling three more times. Place in the refrigerator for at least 1 hr. Roll the dough out to an oblong about 23 in. by 14 in. Cover again with lightly greased polythene and leave for 10 min. Trim with a sharp knife to

21 in. by 12 in. and divide in half lengthwise. Cut each strip into 6 triangles, 6 in. high and with a 6-in. base. Brush each triangle with egg glaze, then roll up loosely from the base, finishing with the tip underneath. Place the shaped croissants on ungreased baking sheets. Brush the tops with egg glaze. Put each baking sheet inside a large, lightly greased polythene bag, close and leave at room temperature for about 30 min. until the croissants are light and puffy. Brush again with egg glaze before baking in the centre of the oven at 425°F (mark 7) for about 20 min. Cool on a wire rack.

**TO PACK AND FREEZE:** Wrap in a single layer in a polythene bag or kitchen foil, or place in a rigid foil container. Seal, label and freeze.

**TO USE: Either** Leave in packaging at room temperature for 1½–2 hr. then refresh, wrapped in foil, in the oven at 425°F (mark 7) for 5 min. **Or** Place still frozen, and wrapped only in foil, in the oven at 350°F (mark 4) for 15 min.

*Maximum storage time:* 4 weeks.

# Cardamom coffee bread

1 recipe quantity bread dough
  (see Spicy twist)
2 tsps. coffee essence
2 whole cardamom seeds, crushed
beaten egg, to glaze

**TO MAKE:** Before kneading the bread dough, divide the mixture in half; knead the coffee essence and crushed cardamom seeds into 1 portion and knead the second half plain. Place each portion in a lightly oiled polythene bag and leave until doubled in size. Roll out each piece of dough to a rectangle about 15 in. by 8 in. Place the pieces on top of each other and roll up from the short end. Place the roll in a greased and floured, 8½-in. by 4½-in. loaf tin, put into a lightly oiled polythene bag and leave to double in size again. Remove the polythene bag, brush with beaten egg and cook in the oven at 375°F (mark 5) for 35–40 min., until the loaf sounds hollow when tapped. Cool rapidly on a wire rack.

**TO PACK AND FREEZE:** See notes at beginning of chapter.

**TO USE:** Remove overwrappings and refresh the loaf from frozen, still wrapped in foil, in the oven at 375°F (mark 5) for 50 min. Cool and serve sliced and buttered.

# Orange raisin ring

(*see picture opposite*)

*For dough*
2 level tsps. dried yeast
½ level tsp. sugar
5 tbsps. milk
8 oz. strong plain flour
½ level tsp. salt
½ oz. butter, melted
1 egg, beaten

*For filling*
½ oz. butter, melted
2 oz. Demerara sugar
2 oz. raisins
grated rind of 1 orange

*To complete ring from freezer*
glacé icing
shredded almonds

**TO MAKE:** Mix together the yeast, sugar, milk and 2 oz. flour, Leave in a warm place for 15–20 min. until frothy. Sift together the remaining flour and salt into a bowl. Add the melted butter and beaten egg. Mix in the yeast liquid and then turn the dough out on to a floured board and knead until it is firm, elastic and no longer sticky. Place it in a lightly oiled polythene bag in a warm place to rise until doubled in size. Knead the dough again and roll it out to a rectangle about 15 in by 8 in. Brush with the melted butter and sprinkle the sugar, raisins and orange rind on top. Roll up from the longest side and turn into a horseshoe shape. Place the ring on a greased baking sheet, put in a lightly oiled polythene bag and leave to rise again until the dough feels springy. Remove the polythene bag and bake in the oven at 375°F (mark 5) for about 25–30 min. Cool on a wire rack.

**TO PACK AND FREEZE:** See notes at beginning of chapter.

**TO USE:** Remove overwrapping, leave

*Orange raisin
ring*

loosely covered with foil and refresh from frozen in the oven at 375°F (mark 5) for 25–30 min. Cool on a wire rack. When cold, drizzle glacé icing over the horseshoe, scatter almonds over the top and leave to set. Serve sliced.

## Welsh currant bread

1 lb. 2 oz. strong plain flour
2 level tsps. salt
1½ level tbsps. dried yeast
6 fl. oz. water
1 level tsp. ground mixed spice
3 oz. margarine
3 oz. Demerara sugar
1½ lb. mixed dried fruit
1 egg, beaten
honey for glaze

**TO MAKE:** Mix together 4 oz. flour, 1 tsp. salt, the yeast and water. Leave in a warm place for 15–20 min. until the batter froths. Sift together the remaining flour, salt and the mixed spice into a large bowl. Rub in the margarine with the fingertips and add the remaining ingredients. Mix in the yeast liquid to give a fairly soft dough. Turn the dough on to a lightly floured surface and knead for 5–10 min. until smooth and no longer sticky. Put the dough in a lightly oiled polythene bag and leave to rise at room temperature for about 1 hr. Halve the dough,

knead again and place each half in an 8½-in. by 6½-in. (top measurement) greased loaf tin. Cover loosely with lightly oiled polythene bags and leave to rise until doubled in size.

Remove the tins from the polythene bags, brush with honey and bake in the oven at 375°F (mark 5) for about 50 min. until the loaves sound hollow when tapped. Turn out and cool on a wire rack.

**TO PACK AND FREEZE:** See notes at beginning of chapter.

**TO USE:** Remove overwrapping, leave covered loosely with foil and refresh from frozen in the oven at 375°F (mark 5) for about 50 min. Unwrap and cool on a wire rack, slice and serve buttered.

## Lardy cake

*(see colour picture facing page 176)*

1 level tsp. caster sugar
½ pt. warm water
2 level tsps. dried yeast
1 lb. strong plain flour
2 level tsps. salt
cooking oil
2 oz. butter
4 oz. caster sugar
1 level tsp. ground mixed spice
3 oz. sultanas or currants
2 oz. lard

**TO MAKE:** Dissolve the 1 tsp. sugar in the water, add the yeast and leave in a warm place for about 10 min. until frothing. Sift the flour and salt into a basin and stir in the yeast mixture, with 1 tbsp. oil and more water if needed to give a soft dough. Beat until smooth. Put the bowl in a polythene bag, close it and leave in a warm place for the dough to rise until doubled in size. Turn the dough on to a lightly floured surface and knead for 5–10 min. Roll it out to a strip $\frac{1}{4}$ in. thick. Cover two-thirds of the dough with small flakes of butter; combine 3 oz. caster sugar with the spice and sprinkle half of it over the same area; do the same with half the dried fruit. Fold up the dough and roll it out as for flaky pastry. Repeat the process with the lard, remaining spiced sugar and fruit. Fold and roll once more. Place the dough in a greased tin measuring 10 in. by 8 in., pressing it down so that it fills the corners. Cover loosely with an oiled poly-thene bag and leave to rise in a warm place until doubled in size. Brush with oil, sprinkle with the remaining caster sugar and mark criss-cross fashion with a knife. Bake in the oven at 425°F (mark 7) for about 30 min. Turn out and cool the cake on a wire rack.

**TO PACK AND FREEZE:** See notes at beginning of chapter.

**TO USE:** Unwrap and place the cake on a baking sheet. Refresh from frozen in the oven at 350°F (mark 4) for 40 min. Serve sliced and buttered, whilst still just warm.

# Coconut curls

$3\frac{1}{2}$ oz. butter
4 oz. caster sugar
1 oz. long shredded coconut, roughly
   chopped
1 oz. angelica, roughly chopped
1 level tsp. flour
1 tbsp. top of the milk

*Makes about 20*

**TO MAKE:** Line baking sheets with non-stick paper. Melt the butter in a small pan, add the sugar and boil for 1 min., stirring. Remove from the heat and stir in the coconut, angelica, flour and milk. Cool a little, then drop in small heaps well apart on the baking sheets. Bake in the oven at 350°F (mark 4) for about 10 min. until golden. When they are firming a little, lift with a palette knife. Cool over a rolling pin.

**TO PACK AND FREEZE:** See notes at beginning of chapter.

**TO USE:** Allow to 'come to' at room temperature for $\frac{1}{2}$ hr. before serving.

# Hot cross buns

*(see colour picture facing page 176)*

1 lb. strong plain flour
1 level tbsp. dried yeast
1 level tsp. caster sugar
$\frac{1}{4}$ pt. milk
$\frac{1}{4}$ pt. water, less 4 tbsps.
1 level tsp. salt
$\frac{1}{2}$ level tsp. ground mixed spice
$\frac{1}{2}$ level tsp. ground cinnamon
$\frac{1}{2}$ level tsp. grated nutmeg
2 oz. caster sugar
2 oz. butter, melted and cooled
1 egg, beaten
4 oz. currants
1–2 oz. chopped mixed peel

*To complete buns from freezer*
2 tbsps. milk
2 tbsps. water
$1\frac{1}{2}$ oz. caster sugar

*Makes 12 buns*

**TO MAKE:** Place 4 oz. of the flour in a large mixing bowl. Add the yeast and 1 level tsp. sugar. Warm the milk and water to about 110°F, add to the flour and mix well. Set the mixture aside in a warm place for about 30 min., until frothy. Sift together the remaining 12 oz. flour, the salt, spices and 2 oz. sugar. Stir the butter and egg into the frothy yeast mixture, add the spiced flour and the currants and peel and mix together. The dough should be fairly soft. Turn it on to a lightly floured surface and knead until smooth. Place the dough in a large, lightly oiled polythene bag, and leave it at room temperature until doubled in size—about 1–$1\frac{1}{2}$ hr. Turn the risen dough on to a floured surface and knock out the air bubbles; knead again and divide the dough into 12 pieces. Shape them into buns, using the palm of one hand. Press down hard at first on the table surface, then ease up as you turn and shape the buns. Arrange the buns well apart on floured baking sheets, place these inside lightly oiled polythene bags and allow the buns to rise at room temperature for about 30 min. Make quick cuts with a sharp knife, just cutting the surface of the dough, to make a cross on each bun. Bake in the oven at 375°F (mark 5) for 15–20 min. Cool on a wire rack.

**TO PACK AND FREEZE:** See notes at beginning of chapter.

**TO USE:** Unwrap the buns and place on baking sheets, covered loosely with foil. Refresh from frozen in the oven at 350°F (mark 4) for about 20 min. Heat the glaze ingredients together in a small pan and brush the hot buns twice with glaze, then leave to cool.

# Cherry sponge tartlets

8 oz. shortcrust pastry (i.e. made with
   8 oz. flour, etc.)
4 oz. ready-made almond paste
3 oz. soft tub margarine
3 oz. caster sugar
1 large egg
3 oz. self-raising flour
$\frac{1}{4}$ tsp. almond essence
1–2 tbsps. milk
15-oz. can red cherries, drained, stoned
   and halved

*Makes 24*

**TO MAKE:** Roll out the pastry and use it to line 24 $2\frac{1}{2}$-in. diameter bun tins. Grate the almond paste and spoon a little into the base of each pastry case. In a bowl, beat together the margarine, sugar, egg, flour, almond essence and milk. Spoon a teaspoonful of this sponge mixture over the almond paste in each tin. Position 2–3 cherry halves on the surface of each and bake in the oven at 375°F (mark 5) for about 25 min.

**TO PACK AND FREEZE:** Open freeze uncovered until firm, then pack in a rigid container, seal, label and return to the freezer.

**TO USE:** Unwrap and thaw on a wire rack for about 1 hr. at room temperature.

# Mincemeat cinnamon meltaways

(*see picture page 239*)

8 oz. plain flour
2 level tsps. cream of tartar
1 level tsp. bicarbonate of soda
$\frac{1}{2}$ level tsp. ground cinnamon
pinch of salt
4 oz. butter or margarine
4 oz. caster sugar
1 egg, beaten
$\frac{1}{2}$ lb. mincemeat
flaked almonds

*Makes about 16*

**TO MAKE:** Sift together the flour, cream of tartar, bicarbonate of soda, cinnamon and salt. Rub in the fat with the fingertips, add the sugar and mix to a soft dough with the egg. Knead the dough lightly on a floured surface and roll it out to about $\frac{1}{8}$-in. thickness on the same surface or between sheets of waxed or non-stick paper; handle the dough carefully. Stamp out bases and lids with a plain cutter, to line about 16 $2\frac{1}{2}$-in. diameter (top measurement) bun tins. Place the bases in the tins and add about 1 tsp. mincemeat to each. Top with the lids, which will seal themselves during cooking; scatter the nuts over. Bake in the oven at 400°F (mark 6) for about 15 min. Leave to cool for a short time then slip the meltaways out on to a wire rack to cool completely.

**TO PACK AND FREEZE:** Open freeze uncovered until firm. Pack in a rigid container, seal, label and return to the freezer.

**TO USE:** Unwrap and thaw at room temperature on a wire rack for about 1 hr.

# Palmiers

$7\frac{1}{2}$-oz. pkt. frozen puff pastry, thawed
1 egg white
caster sugar
$\frac{1}{4}$ pt. double cream, sweetened and
   whipped
4 tbsps. strawberry jam

*To finish palmiers from freezer*
icing sugar

*Makes 16*

**TO MAKE:** Roll out the pastry to a 10-in. by

8-in. rectangle. Beat the egg white with a fork until frothy and brush it over the pastry. Sprinkle generously with caster sugar and fold the long sides half way towards the centre; brush with egg white and dredge with caster sugar again. Fold the pastry in half lengthwise, hiding the first folds. Press it lightly and evenly and cut into 16 equal slices. Place the palmiers on a damp baking sheet. Open the tip of each and flatten the whole palmier slightly with a palette knife. Brush again with egg white and dredge with caster sugar. Bake them in the oven at 425°F (mark 7) for 10 min. turn the palmiers over with a palette knife and cook the other side for 10 min., until golden brown. Cool on a wire rack. Sandwich the palmiers in pairs with cream and jam.

**TO PACK AND FREEZE:** Open freeze the palmiers until firm, then pack in a rigid container in a single layer, seal, label and return to the freezer.

**TO USE:** Unwrap and thaw on a wire rack for 1–2 hr. at room temperature. Dredge with icing sugar before serving. Once thawed, serve as soon as possible.

*Note:* Alternatively, palmiers could be frozen unfilled, to be refreshed in the oven and filled when required.

# Congress tarts

4 oz. shortcrust pastry (i.e. made with
   4 oz. flour etc.)
raspberry jam
2 oz. butter
2 oz. caster sugar
2 oz. ground almonds
grated rind and juice of $\frac{1}{2}$ lemon
1 egg, separated

*Makes 12*

**TO MAKE:** Roll out the pastry and line 12 $2\frac{1}{2}$-in. bun tins, using a 3-in. fluted cutter. Put a little jam in the base of each pastry case. Cream the butter and sugar together until light and fluffy. Add the ground almonds, lemon rind and juice, then the egg yolk. Mix well. Stiffly beat the egg white and fold it in. Divide the mixture between the pastry cases. Roll out the pastry trimmings, cut into strips and use to make a pastry cross

on top of each tart. Bake in the oven at 425°F (mark 7) for 10 min., reduce the oven temperature to 350°F (mark 4) and cook for about a further 25 min. Cool on a wire rack.

**TO PACK AND FREEZE:** See notes at beginning of chapter.

**TO USE:** Unwrap and thaw for about 1 hr. at room temperature.

# Coffee éclairs

(*see colour picture facing page 177*)

2 oz. butter
$\frac{1}{4}$ pt. water
$2\frac{1}{2}$ oz. plain flour, sifted
2 eggs, lightly beaten

*To complete éclairs from freezer*
$\frac{1}{4}$ pt. double cream
1 tbsp. milk
3 oz. icing sugar
2 tsps. coffee essence

*Makes about 8*

**TO MAKE:** Melt the butter in the water and bring to the boil. Remove from the heat and add all the flour at once. Beat this mixture to a smooth paste, forming a ball in the centre of the pan. Allow it to cool slightly, then beat in the eggs gradually, adding just enough to give a smooth, glossy mixture of soft piping consistency. Put the mixture in a forcing bag fitted with a $\frac{1}{2}$-in. diameter plain round nozzle. Chill the bag in the refrigerator for 1 hr. Lightly grease and flour a large baking sheet. Pipe fingers of the mixture $4\frac{1}{2}$ in. long. Bake in the oven at 400°F (mark 6) for 20–25 min., until well risen, crisp and golden brown. Place on a wire rack and make a small hole in the end of each éclair to allow the steam to escape. Leave to cool. When cold, split them, leaving a hinge.

**TO PACK AND FREEZE:** Open freeze the éclair cases on a foil-lined baking sheet until firm. Wrap them in the foil and overwrap in polythene bag. Seal, label, return to freezer.

**TO USE:** Leave the éclairs in their wrapping at room temperature for 1 hr. Unwrap and refresh them on a baking sheet in the oven at 350°F (mark 4) for about 5 min. Cool on a wire rack. Whip the cream with the milk until it just holds its shape and pipe some into each éclair. Make up an icing with the sugar and coffee essence and put it in a shallow dish. Hold the base of each éclair and dip the tops through the icing, to coat. Leave to set in a cool place.

# Lemon and almond tarts

4 oz. shortcrust pastry (i.e. made with
    4 oz. flour etc.)
4 tbsps. lemon curd
2 oz. butter
2 oz. caster sugar
1 egg
2 oz. ground almonds
grated rind of $\frac{1}{2}$ lemon

*To complete tarts from freezer*
2 oz. glacé cherries, chopped
1–1$\frac{1}{2}$ tbsps. lemon juice
thinly pared rind of 1 lemon, cut into strips
4 oz. icing sugar, sifted

*Makes 12*

TO MAKE: Roll out the pastry and use it to line 12 2$\frac{1}{2}$-in. diameter (top measurement) bun tins, using a 3-in. fluted cutter. Put a little lemon curd into the base of each. Cream the butter and sugar together, beat in the egg and fold in the ground almonds and lemon rind. Divide the mixture evenly between the pastry cases. Bake tarts in the oven at 375°F (mark 5) for about 25 min. Cool on wire rack.

TO PACK AND FREEZE: Open freeze the tarts until firm, then pack in a rigid container, seal, label and return to the freezer.

TO USE: Unwrap and thaw on a wire rack for about 1 hr. Top the tarts with the chopped glacé cherries. Add the lemon juice and rind to the icing sugar and spoon it over the cherries. Leave to set.

# Melting moments

(*see colour picture facing page 176*)

4 oz. butter or margarine
3 oz. sugar
a little vanilla essence or grated lemon rind
1 egg yolk
5 oz. self-raising flour
crushed cornflakes
about 6 glacé cherries, cut in quarters

*Makes about 24*

TO MAKE: Cream the fat and sugar together until light and fluffy and beat in the flavouring and egg yolk. Work in the flour and mix to a smooth dough. Wet the hands and divide the mixture into small balls. Roll these in cornflakes, put them on greased baking sheets, press a piece of cherry in the centre of each and bake in the oven at 375°F (mark 5) for 15–20 min. Cool on a wire rack.

TO PACK AND FREEZE: See notes at beginning of chapter.

TO USE: Unwrap the cookies and thaw at room temperature for about 10 min.

# Brandy cornets

(*see colour picture facing page 176*)

2 oz. butter
2 oz. caster sugar
2 level tbsps. golden syrup
2 oz. plain flour
$\frac{1}{2}$ level tsp. ground ginger
1 tsp. brandy
grated rind of $\frac{1}{2}$ lemon

*Makes about 14*

TO MAKE: Line 2 baking sheets with non-stick paper and grease some metal cream horn tins. Melt the butter with the sugar and syrup in a pan over low heat. Stir until smooth, then remove from the heat. Sift together the flour and ginger on to a plate or paper. Stir these into the butter mixture, add the brandy and lemon rind and mix thoroughly. Leave to cool for 1–2 min. then drop 1 tsp. at a time on to the prepared baking sheets, about 4 in. apart. Bake in rotation in the oven at 350°F (mark 4) for 7–10 min. until bubbly and golden. Remove from the oven. As the biscuits begin to firm, quickly remove from the sheet, using a small palette knife, and roll round the cream horn tins. Leave in shape until set then twist gently to remove.

*Note:* If the cooked mixture becomes too brittle to handle, return the tray to the oven for a few sec. to soften.

TO PACK AND FREEZE: See notes at beginning of chapter.

TO USE: Unwrap the cornets and thaw at room temperature for about 10 min.

# Date rockies

½ lb. plain flour
1 level tsp. baking powder
½ level tsp. bicarbonate of soda
pinch salt
4 oz. margarine
4 oz. light soft brown sugar
3 oz. stoned dates, chopped
2 oz. shelled walnuts, chopped
1 large egg, beaten
a little milk

*Makes about 12*

TO MAKE: Sift together the flour, baking powder, bicarbonate of soda and salt. Rub in the margarine using the fingertips, until the mixture resembles breadcrumbs. Stir in the sugar, dates and walnuts. Add the egg and just enough milk to knit the mixture together, mixing with a fork. Using 2 forks, arrange 12 rough piles of the mixture in bun shapes, well apart on 2 greased baking sheets. Bake in the oven at 425°F (mark 7) for about 15 min. Lift off the sheets carefully and cool on a wire rack.

TO PACK AND FREEZE: See notes at beginning of chapter.

TO USE: Unwrap and thaw at room temperature for about 1 hr., or refresh from frozen in the oven at 375°F (mark 5) for 15–20 min. and serve buttered.

# Walnut cookies

(*see colour picture facing page 176*)

4 oz. butter or margarine
4 oz. light soft brown sugar
6 oz. plain flour
½ level tsp. bicarbonate of soda
½ level tsp. cream of tartar
7 shelled walnuts, halved
1½ oz. chocolate, melted

*Makes about 14*

TO MAKE: Cream together the fat and sugar. Sift the flour, bicarbonate of soda and cream of tartar together, and stir into the creamed mixture. Form into balls the size of a walnut. Place these well apart on greased baking sheets, press half a walnut on every other cookie and bake in the oven at 350°F (mark 4) for about 20 min., until pale brown.

Cool on a wire rack. Sandwich the cookies in pairs with the melted chocolate, walnut topped cookies uppermost. Leave to set.

TO PACK AND FREEZE: See notes at beginning of chapter.

TO USE: Unwrap the cookies and thaw at room temperature for about 1 hr.

# Chocolate raisin squares

(*see picture opposite*)

3 oz. butter or margarine
4 oz. caster sugar
2 eggs
3 oz. golden syrup
2 oz. plain chocolate, melted
8 oz. plain flour
1 level tsp. bicarbonate of soda
½ level tsp. cream of tartar
2 level tbsps. cocoa powder
2–3 tbsps. milk
4 oz. sultanas
2 oz. shelled walnuts, roughly chopped

*For frosting*
4 oz. butter or margarine
9 oz. icing sugar, sifted
2 oz. plain chocolate, melted
1 level tbsp. golden syrup, warmed

*Makes 16*

TO MAKE: Line a tin measuring 11½ in. by 7½ in. by 1½ in. with greased greaseproof paper. Cream together the butter and sugar until light and fluffy. Beat in the eggs and then the syrup and melted chocolate (cool, but still flowing). Sift the flour, bicarbonate of soda and cream of tartar and fold in. Blend the cocoa to a paste with a little milk, stir it into the mixture and, when well blended, add the sultanas and walnuts. Beat well. Turn into the prepared tin, level the surface and bake in the oven at 350°F (mark 4) for about 40 min. Turn out and cool on a wire rack.

Meanwhile make the chocolate frosting. Cream the butter or margarine, gradually beat in the icing sugar, together with the chocolate and syrup. Remove the paper from the cake, top it with frosting and swirl it with a knife.

TO PACK AND FREEZE: See notes at beginning of chapter.

*Chocolate
raisin
squares*

**TO USE:** Thaw according to notes and cut into squares to serve.

# Petticoat tails

(*see colour picture facing page 176*)
4 oz. butter, softened
2 oz. caster sugar
6 oz. plain flour

*To complete dish from freezer*
icing sugar

**TO MAKE:** Cream together the butter and sugar. Gradually add the flour and work together. Knead lightly. On a floured surface, roll out the dough to about $\frac{1}{8}$ in. thick. Cut a large circle, using an inverted $9\frac{1}{2}$-in. plate or pan lid as a guide, and lift it carefully (using a large palette knife) on to a baking sheet lined with non-stick paper. Use a 3-in. plain cutter to cut a smaller circle in the centre. Leave this in position and with a knife cut 10 tails through as far as the round centre. Prick all over and bake in the oven at 350°F (mark 4) for about 15 min. until pale golden brown. Cool on a wire rack.

**TO PACK AND FREEZE:** Pack carefully in foil and overwrap in a polythene bag, or place in a rigid container. Freeze.

**TO USE:** Unwrap and thaw at room temperature for about 1 hr. Place the centre round in the middle of a flat plate and overlap the tails round it to look like a skirt. Dust with icing sugar.

# Ginger buns

3 oz. butter
3 oz. caster sugar
1 tbsp. stem ginger syrup
1 large egg, beaten
4 oz. self-raising flour
2 oz. mixed peel, finely chopped
1 oz. stem ginger, chopped

*To complete buns from freezer*
4 oz. icing sugar, sifted
4 tbsps. stem ginger syrup
12 slices of ginger, for decoration

*Makes 12*

**TO MAKE:** Grease 12 $2\frac{1}{2}$-in. (top measurement) bun tins. Cream the butter and sugar with the ginger syrup until light and fluffy. Beat in the egg, a little at a time. Fold in the flour, chopped peel and chopped ginger. Divide the mixture between the prepared bun tins and bake in the oven at 350°F (mark 4) for about 30 min. Cool on a wire rack.

**TO PACK AND FREEZE:** See notes at beginning of chapter.

**TO USE:** Unwrap the buns, place them on a wire rack and thaw at room temperature for about 1 hr. Blend the icing sugar with the stem ginger syrup and spoon it into the centre of each bun. Decorate with a thin slice of stem ginger and leave to set.

# Flapjacks

(*see colour picture facing page 176*)

3 oz. butter
3 oz. Demerara sugar
4 oz. rolled oats

*Makes 12*

**TO MAKE:** Grease a shallow $7\frac{1}{2}$-in. square tin. Cream the butter. Mix together the sugar and oats and gradually work into the creamed butter until thoroughly blended. Press the mixture evenly into the prepared tin with a palette knife. Bake in the oven at 425°F (mark 7) for about 15 min., until golden brown; turn the tin half way through cooking to ensure even baking. Cool slightly in the tin, mark into fingers with a sharp knife and loosen round the edge. When the flapjack is firm, turn it out, break into fingers and cool on a wire rack.

**TO PACK AND FREEZE:** See notes at beginning of chapter.

**TO USE:** Unwrap and thaw on a wire rack at room temperature for about 30 min.

# Parkin

8 oz. plain flour
2 level tsps. baking powder
2 level tsps. ground ginger
2 oz. margarine
2 oz. lard
8 oz. medium oatmeal
4 oz. caster sugar
6 oz. golden syrup
6 oz. black treacle
4 tbsps. milk

**TO MAKE:** Line a tin measuring 10 in. by 8 in. by $1\frac{1}{2}$ in. deep with greased greaseproof paper. Sift together the flour, baking powder and ginger. Rub in the fats with the fingertips and add the oatmeal and sugar. Heat together the syrup and treacle, make a well in the rubbed-in ingredients and stir in the syrup, treacle and milk. Mix until

smooth, then pour into the prepared tin. Bake in the oven at 350°F (mark 4) for $\frac{3}{4}$–1 hr., until the mixture springs back when lightly pressed and has shrunk away a little from the sides of the tin. Turn out and cool on a wire rack.

**TO PACK AND FREEZE:** See notes at beginning of chapter.

**TO USE:** Remove the polythene bag and allow at least 2 days at room temperature to mature, wrapped in foil. Serve sliced and buttered or coated with a little lemon flavoured glacé icing.

# Coffee cup cakes

(*see picture opposite*)

2 oz. butter
2 oz. margarine
4 oz. light soft brown sugar
2 eggs, beaten
1–2 level tsps. instant coffee granules
1 tsp. water
5 oz. self-raising flour
3–4 oz. plain chocolate cake covering
milk chocolate cake covering

*Makes about 24*

**TO MAKE:** Line 24 deep-cup bun tins with paper cases; if 2 cases are used, one inside the other, this will stop over-browning; the support from the bun tins ensures a good shape, as well as helping to brown the cakes evenly. Cream together the butter, margarine and sugar thoroughly. Beat in the eggs a little at a time. Blend the coffee and water and lightly beat them into the creamed ingredients with the sifted flour, using a wooden spoon. Then, using a teaspoon, fill the paper cases about half-full with the cake mixture. Bake in the oven at 375°F (mark 5) for about 12 min. until well risen, level and lightly coloured. Remove the second paper case if used and cool the cakes on a wire rack. Melt the plain chocolate and use it to coat the centre of each cake. When set, melt a little milk chocolate; put it in a small paper forcing bag, snip off the tip with scissors and use to pipe a decorative squiggle over the plain chocolate.

**TO PACK AND FREEZE:** See notes at beginning of chapter.

*A selection
of small cakes:
Coffee cup cakes;
Orange top hats;
Swiss tarts;
Mincemeat
cinnamon melt-
aways*

**TO USE:** Unwrap and thaw at room temperature for about 2 hr.

## Iced apricot squares

6 oz. butter or margarine
2 oz. caster sugar
8 oz. plain flour
a few drops vanilla essence
4 oz. dried apricots, stewed, drained and
    chopped

*To complete squares from freezer*
glacé icing

**TO MAKE:** Line a 7½-in. square shallow cake tin with greased greaseproof paper. Cream the butter with the sugar until soft and fluffy. Gradually mix in the sifted flour and vanilla essence. Using the back of the hand, press half the mixture into the base of the prepared tin. Cover with the chopped apricots, in an even layer, then spread the remaining cake mixture carefully over the top. Bake in the oven at 375°F (mark 5) for about 40 min., until golden on top. Turn out on to a wire rack and cool.

**TO PACK AND FREEZE:** See notes at beginning of chapter.

**TO USE:** Leave wrapped at room temperature for about 2 hr. Drizzle the cake with glacé icing and when set, cut into squares.

## Orange top hats

(*see picture above*)

5 oz. butter
3 oz. caster sugar
1 large egg
1 level tsp. grated orange rind
5 oz. self-raising flour
1 tbsp. orange juice
4 oz. icing sugar

*To complete cakes from freezer*
icing sugar

*Makes about 12*

**TO MAKE:** Grease 12 2½-in. (top measurement) bun tins. Cream together 3 oz. butter and the caster sugar and beat in the egg and orange rind. Sift the flour and beat it in lightly, alternately with the orange juice. Half-fill the bun tins with the mixture and bake in the oven at 400°F (mark 6) for 15–20 min. Turn out the cakes and cool on a wire rack. When cold, remove a 'hat' from each cake, using a small cutter. Cream together the remaining 2 oz. butter and the icing sugar and pipe a whirl of this into the centre of each cake. Return the 'hats'.

**TO PACK AND FREEZE:** See notes at beginning of chapter.

**TO USE:** Unwrap the cakes and thaw on a wire rack at room temperature for about 2 hr. Dust with icing sugar before serving.

# Profiteroles

2 oz. butter
$\frac{1}{4}$ pt. water
2$\frac{1}{2}$ oz. plain flour
2 eggs

*To complete dish from freezer*
$\frac{1}{2}$ pt. double cream, whipped
Suchard sauce (page 67)

*Makes 20*

**TO MAKE:** Melt the butter in the water and bring to the boil. Remove from the heat and quickly add the flour all at once. Beat until the paste is smooth and forms a ball in the centre of the pan. (Take care not to over-beat or the mixture will become fatty.) Allow to cool slightly, then beat in the eggs gradually, adding just enough to give a smooth mixture of piping consistency.

Using a large plain nozzle, pipe the paste on to greased baking sheets in small dots. Bake in the oven at 425°F (mark 7) for about 20–25 min. Allow to cool and make a small hole for steam to escape.

**TO PACK AND FREEZE:** Place the buns on foil-lined baking sheets and open freeze until solid. Remove carefully from the sheets and pack in polythene bags or heavy duty foil. Seal, label, overwrap and return to the freezer.

**TO USE:** Thaw wrapped at room temperature for about 1 hr. then remove the wrappings, place the buns on baking sheets and refresh in the oven at 350°F (mark 4) for about 5 min. Alternatively, unwrap and refresh from frozen at 350°F (mark 4) for about 10 min. Cool on a wire rack. Make a small hole in each and fill with whipped cream. Pile into a pyramid on a serving plate and spoon a little Suchard sauce over. Serve any remaining sauce separately.

# Swiss tarts

4 oz. butter
1 oz. caster sugar
vanilla essence
4 oz. plain flour

*To complete tarts from freezer*
icing sugar
redcurrant jelly

*Makes 6*

**TO MAKE:** Line 6 small bun tins with paper cases. Cream the butter and sugar thoroughly until light and fluffy. Beat in a few drops of vanilla essence and gradually add the sifted flour, beating well between each addition. Place the mixture in a forcing bag fitted with a large star vegetable nozzle. Pipe the mixture into the paper cases; start at the centre and pipe with a spiral motion round the sides, leaving a shallow depression in the centre. Bake in the oven at 350°F (mark 4) for 20–25 min. Cool on a wire rack.

**TO PACK AND FREEZE:** See notes at beginning of chapter.

**TO USE:** Unwrap and thaw the tarts at room temperature for 45 min. Dredge with icing sugar and spoon a little redcurrant jelly in the centre.

# Appendix

## Using the freezer:

1. Always start with good quality foods and freeze them at peak freshness. Food can only come out of the freezer as good as you put it in.

2. Keep handling to a minimum and make sure everything is scrupulously clean. Freezing doesn't kill bacteria and germs.

3. Pay special attention to packaging and sealing. Exposure to air and moisture damages frozen foods.

4. Cool food rapidly if it has been cooked or blanched; never put anything hot—or even warm—into your freezer.

5. Freeze as quickly as possible, and in small quantities.

6. Freeze in the coldest part of the freezer, and don't pack the food to be frozen too closely together—spread it out until it is frozen.

7. Transfer newly added items to the main part of the cabinet once they've been frozen—if a freezer with separate compartments is being used.

8. For large amounts, switch to fast-freeze well in advance; remember to return the switch to normal later.

9. Maintain a steady storage temperature of $-18^\circ$C ($0^\circ$F) and don't do anything that will cause temperatures within the freezer to keep fluctuating.

10. Label and date food so that you can ensure a good rotation of stock. Ideally, keep a record and tick off items as you use them, then you can tell at a glance which supplies are getting low.

11. Defrost the freezer at a time when stocks are low and if possible on a cold day.

12. Be prepared for emergencies. Make sure you know what steps to take in case of a breakdown or power cut.

**Partly-thawed food—provided the packet hasn't been opened—may be refrozen but the quality of that food deteriorates. Never re-freeze anything once it has completely thawed —unless cooked in between. Remember that thawed food tends to deteriorate rapidly, so must be used at once.**

**Use of the fast-freeze switch:** manufacturers are not always clear in their instructions for the use of this, and we are often asked when it should be used.

*For small items* e.g. for one loaf plus one small casserole. It is probably not necessary to touch the switch at all, simply pop the food in. This should be safe for up to, say, 4 items (perhaps a sponge cake and some biscuits could also be included, but nothing very dense such as a leg of lamb).

*For a fairly small amount* e.g. a bulk bake of, say, 4 casseroles and a bulk bake of pies. Place the fast freeze switch on for about 2 hr. before you put the food in. Leave the switch on for about 4 hr. more, totalling 6 hr. in all, until the food is really solid.

*For a large amount* e.g. a half-carcass of meat purchased in bulk, fresh. Place the fast-freeze switch on for about 3 hr. beforehand to ensure the freezer is really cold. Place in the meat and leave for a further 12–18 hr., until solid, depending on the load. This timing is for a full freezer load when you are freezing the maximum your freezer will take. For half your possible maximum load you could obviously use the switch for less time.

These figures are intended only as a guide for those who do not have adequate instructions with their freezer. Only freeze one-tenth of your freezer's capacity in any 24 hr. (e.g. if you have a 10 cu. ft. freezer, you can freeze only 1 cu. ft. of space. 1 cu. ft. of freezer space holds about 20 lb. food, so if you have this size of freezer you can freeze 20 lb. food in 24 hours).

**Cooking for the freezer:**
1. When preparing a dish for the freezer, keep a light hand with the seasoning.
2. Use shallow rather than deep dishes.
3. As already directed above, cool everything as rapidly as possible, and freeze at once.
4. To ensure best results, don't keep cooked dishes in the freezer for more than 2 months, unless otherwise stated. (Dishes with bacon up to 6 weeks.)
5. Hard-boiled eggs are best added after removing dish from the freezer. Frozen hard boiled eggs tend to result in a rubbery texture when thawed.
6. Rice salads (to be served cold) become 'grainy' after thawing. These are best cooked fresh as required.
7. Jam and golden syrup do not freeze well by themselves and therefore dishes like hot steamed puddings should be packed bearing this in mind.
8. Commercially frozen cream is more reliable than buying fresh for home-freezing, unless it is absolutely fresh, contains at least 40% butter fat and is whipped before freezing.

# Packaging
1. Always expel as much air as possible. Use crumpled non-stick, waxed or freezer paper to fill vacant space in the top of a rigid container. Always keep a selection of different sized containers.
2. Always leave $\frac{1}{2}$–1 in. headspace when packing liquids. This allows for expansion on freezing.
3. Anything with solids and liquids e.g. stews, casseroles, fruit in syrup should have the solids covered by liquid. Again $\frac{1}{2}$ in. headspace is necessary.
4. When using polythene bags for liquids, fit the bag into a regular shaped rigid container (a pre-former) before filling. When the liquid is solid the pre-former can then be slipped off and the package is easy to store in the freezer.
5. Anything wrapped in foil should be overwrapped with polythene in case of puncturing the foil.
6. To enable casserole dishes to be free for use in the kitchen, line them with foil before pouring in the mixture. The foil wrapped mixture can then be removed from the dish when solid,

overwrapped with polythene, sealed, labelled and stored in the freezer—this is known as a preformed pack.

7. Freeze food in meal sized portions whenever possible. Individual foods like chops or steaks should be interleaved with waxed or non-stick paper. Do not freeze more than 2 pt. in one pack.

8. Small, shallow packs of soups, stews, casseroles etc will re-heat quicker than large, deep packs.

## Packaging materials : (see Page 245 for Suppliers)
There are many different types and makes of packaging material available and basically it is a matter of choosing those which best suit your own particular needs. Generally speaking you will need wrapping material, a good selection of different sized containers, sealers and labels. Here's what you can choose from :

### Wrappers
Foil
Waxed paper
Heavy gauge polythene bags
Polythene film
Cellophane—for interleaving only
*Note:* If thin polythene bags are used they should be placed one inside the other and used double or just used for overwrapping foil.

### Rigid containers
Foil containers—available in various shapes and sizes to include tartlet cases, pudding basins, freezer trays, plates and pie dishes. Eg. Alcan. Very useful for freezer to oven to table use.
Self sealing plastic containers—e.g. Tupperware. These have a very long life, have airtight lids and do not need additional sealing. Useful as pre-formers.
Other plastic containers; not self-sealing even though some do come with lids. Freezer tape is needed to seal the lids. May be used as pre-formers.
Waxed containers—if treated with care these can be used a second time. Range is limited and so therefore is use.
Specially toughened glass dishes—handy for mousses and desserts which are to be served in the dish.

### Sealers
Plastic covered wire twist-ties—for polythene bags.
Paper covered wire twist-ties—for polythene bags.
Adhesive freezer tape—(ordinary household adhesive tape will not stick in freezer temperatures) —for plastic or waxed containers without air tight lids; for parcels.
Heat sealer—for large numbers of packages (an iron may also be used).

### Labelling
Stick on labels—for parcels and containers.
Tie on labels—for polythene bags (some bags have a built in label area).
Waterproof felt pen—for writing.

## Cleaning and caring for the freezer
1. If it is brand new, wipe round inside with a cloth or sponge soaked in a solution of bicarbonate

of soda (1 tsp. soda to 2 pt. warm water). Plug in when freezer is dry and wait for at least 2 hr. for the temperature to drop before using.

2. Defrosting should be done when the ice is about $\frac{1}{4}$ in thick. Chest freezers need defrosting 1–2 times a year, upright freezers may need defrosting 3 times a year.

3. Newspaper or clean old sheets placed in the freezer a few hours before defrosting will provide chilled wrappings for the frozen food.

4. Help the defrosting by placing bowls of hot water inside the cabinet.

5. Ice chippings are easier to deal with than pools of water so gently scrape the ice away as soon as it is loosened.

6. Wipe freezer with bicarbonate of soda solution (see note 1); dry thoroughly before switching on the current. See note (1).

*Note:* Tape the address and telephone number of the freezer manufacturer's local service agent to the side of the freezer and the name and telephone number of a supplier of dry ice.

## Power cuts and breakdowns

1. Do NOT open the freezer door.

2. A fairly full load will stay frozen for at least 18 hr.; a lesser load for at least 8 hr.

3. In the case of breakdowns always check wiring, plugs, fuses and switches before calling a service engineer.

4. If the repair is going to take longer than your frozen food will last, check with the service depot for a replacement/loan service or food storage service.

5. Dry ice can be used if all else fails. It comes in large 25 lb. blocks and needs to be chopped into small pieces. Handle with care, use gloves and keep it well away from your skin. Place frozen food in box or tea chest type container, place cardboard on top, then tightly packed dry ice. Continue like this in layers to store all the food. This should help for at least 24 hours.

6. As long as food which has begun to thaw out still has ice crystals present it may be re-frozen. food which has thawed but is still cold to the touch should be used immediately or cooked, cooled and then frozen.

7. Pre-cooked meat, fish and poultry which has thawed should be used immediately.

8. Food which has completely thawed and ceases to feel cold should be discarded; fruit can be used for purées, sauces and jams as long as it looks and tastes alright.

## Moving house

1. Always check with the removal firm that they will handle the freezer and state whether it is empty or full.

2. Letting stocks run low before moving day may be a good idea but on the other hand a good selection of prepared dishes and frozen food is just the thing for a busy moving family.

3. If the move is to be completed within the day, the food should not suffer provided the door is kept shut and the freezer is the last item on the van and the first item off.

4. Make sure you have the new site for the freezer worked out and that there is a plug.

5. If you have to empty the freezer, pack the frozen food in a tea chest with dry ice (order in advance).

6. It is wise to consult your individual freezer manufacturer on this subject.

## Insurance

1. If you run a large well stocked freezer, you may feel safer with the stock insured.

2. Ask your own insurance agent for details. One freezer insurance specialist is:
   Ernest Linsdell Ltd, 419 Oxford Street, London, W.1.

3. Check the conditions of claim clauses.

## Suppliers of freezer packaging material by mail order

Alcan Polyfoil Ltd
90 Asheridge Road
Chesham
Buckinghamshire
HP5 2DE

(Will supply a foil dispenser and will also help in case of difficulty in obtaining supplies of containers.)

Melitta Bentz & Sons
Greatham Road
Bushey
Watford
Herts WD2 23Y

Lakeland Plastics (Windermere) Ltd
102 Alexandra Road
Windermere
Westmoreland

Lawson's (Bury St Edmunds) Ltd
1a St Andrews Street South
Bury St Edmunds
Suffolk
(Orders dispatched same day—no minimum order)

Frigicold Ltd
166 Dukes Road
Western Avenue
London W3 0TJ
(Available through stores etc but will supply direct when stockists unavailable)

## National suppliers of frozen food

Alveston Kitchens Ltd
Timothy Bridge Road
Stratford upon Avon
Warwickshire
Stratford on Avon 66102/3

Birds Eye Home Freezer Service
Birds Eye Foods Ltd
Station Avenue
Walton on Thames
Surrey
Walton on Thames 28888

Findus Ltd
St Georges House
Croydon
CR9 1NR
01-686 3031

Green Isle Ltd
PO Box 88 Barrow Road
Cabra West
Dublin 7
Dublin 302711

Edwards & Walkden Ltd
204 Central Markets
Smithfield
London EC1
01-248 7061

# BASIC KNOW-HOW

| FOOD AND STORAGE TIME | PREPARATION | FREEZING | THAWING AND SERVING |
|---|---|---|---|
| **MEAT, RAW** (leave unstuffed)<br>Beef: 8 months<br>Lamb: 6 months<br>Veal 6 months<br>Pork: 6 months<br>Freshly minced meat: 3 months<br>Offal: 3 months<br>Cured and smoked meats: 1–2 months<br>Sausages: 3 months | Use good quality, well-hung fresh meat. Removing bones will save space. Butcher in suitable quantities. Place polythene sheets between individual chops or steaks | Package carefully in heavy-quality polythene bags or foil, overwrap, seal and label. | All meats may be cooked from frozen (see page 92), but with large joints avoid over-cooking meat on outside and leaving it raw at centre. When thawing, use the refrigerator; keeping wrappings on, allow about 6 hr. per lb. Small items like chops, steaks, can be cooked frozen, but use gentle heat. Partial thawing may be necessary before egg-and-crumb coating, etc. |
| **MEAT, COOKED DISHES**<br>Casseroles, stews, curries, etc.: 2 months | Prepare as desired, see that the meat is cooked but not over-cooked, to allow for re-heating. Do not season too heavily—check this at point of serving. Have enough liquid or sauce to immerse solid meat completely. Potato, rice or spaghetti, unless otherwise stated, are best added at point of serving; same applies to garlic and celery | When mixture is quite cold, transfer to rigid cartons; for dishes with a strong smell or colour, inner-line cartons with polythene bags, use foil dishes or freeze in foil lined cook-ware | Re-heat food from cartons or polythene bags in a saucepan or casserole dish. Pre-shaped foil-wrapped mixtures can be re-heated in the original dish. When re-heating in a casserole from frozen allow at least 1 hr. for heating through in the oven at 400°F (mark 6) then if necessary reduce heat to 350°F (mark 4) for 40 min. and leave until really hot. Alternatively, heat gently in a pan, simmering until thoroughly heated |
| **MEAT, ROAST**<br>2–4 weeks | Joints can be roasted and frozen for serving cold—don't over-cook. Re-heated whole joints are not very satisfactory. Sliced and frozen cooked meat tends to be dry when re-heated | Best results are achieved by freezing whole joint, thawing, then slicing prior to serving. But small pieces can be sliced and packed in polythene if required to serve cold, or put in foil containers and covered with gravy, if to be served hot | Allow plenty of time for thawing out—about 4 hr. per lb. at room temperature, or double that time in the refrigerator, in the wrapping. Sliced meat requires less time. |

246

| Food | Preparation | Packaging | Thawing / Cooking |
|---|---|---|---|
| **MEAT LOAVES, PÂTÉS** <br> 1 month | Follow regular recipe. Package in the usual way, after cooling rapidly. Keep for minimum time | When quite cold, remove from tin, wrap and freeze | Thaw preferably overnight in the refrigerator, or for at least 6–8 hr. at room temperature |
| **POULTRY AND GAME** <br> Chicken: 12 months <br> Duck: 4–6 months <br> Goose: 4–6 months <br> Turkey: 6 months <br> Giblets: 3 months <br> Game birds: 6–8 months <br> Venison: 12 months | Use fresh birds only: prepare and draw in the usual way. Do not stuff before freezing. Cover protruding bones with grease-proof paper or foil. Hang game desired time before freezing | Pack trussed bird inside polythene bag and exclude as much air as possible before sealing. Freeze giblets separately. If wished, freeze in joints, wrap individually, and then overwrap | Thaw in wrapping preferably in refrigerator chickens up to 4 lb. approx. 16 hr. chickens over 4 lb. can take up to 24 hr. |
| **FISH, UNCOOKED** <br> Whole (Salmon, fresh-water fish) | Must be really fresh—within 12 hr. of the catch if possible—maximum 24 hr. Wash and remove scales by scraping tail-to-head with back of knife. Gut. Wash thoroughly under running water. Drain and dry on a clean cloth | For best results, place whole fish unwrapped in freezer until solid. Remove, dip in cold water. This forms thin ice over fish. Return to freezer; repeat process until ice glaze is $\frac{1}{4}$ in. thick. Wrap in heavy-duty polythene; support with a thin board | Allow to thaw for 24 hr. in a cool place before cooking. Once thawed, use promptly |
| Fish steaks <br><br> Salmon: 4 months <br> White fish: 6 months | Prepare in the usual way | Separate steaks with double layer of cellophane or waxed paper; wrap in heavy polythene | May be cooked from the frozen state <br> Salmon steaks steam from frozen for 45 min. |
| **SHELL FISH** | Advisable only if you can freeze the fish within 12 hr. of being caught | | |
| **FISH, COOKED** <br> Pies, fish cakes, croquettes, kedgeree, mousse, paellas: 2 months | Prepare according to recipe, but be sure fish is absolutely fresh. Hard-boiled eggs should be added to kedgeree before re-heating | Freeze in foil-lined containers, remove when hard, then pack in sealed bags | Either slow-thaw in refrigerator or put straight into the oven at 350°F (mark 4) to heat, depending on the type of recipe |
| **SAUCES, SOUPS, STOCKS** <br> 2–3 months | All are very useful as stand-bys in the freezer | When cold, pour into rigid containers, seal well and freeze | Either thaw for 1–2 hr. at room temperature or heat immediately until boiling point is reached |
| **PIZZA, UNBAKED** <br> Up to 3 months | Prepare traditional yeast mixture to baking stage. Wrap in foil or polythene | Package in foil or polythene and freeze as for unbaked pizza | Remove packaging and place frozen in cold oven set at 450°F (mark 8) and bake for 30–35 min. |

| FOOD AND STORAGE TIME | PREPARATION | FREEZING | THAWING AND SERVING |
|---|---|---|---|
| **PIZZA, BAKED**<br>Up to 2 months | Bake traditional yeast mixture in usual way | Package in foil or polythene and freeze as for unbaked pizza | Remove packaging and place frozen in a pre-heated oven at 400°F (mark 6) for about 20 min. or leave in packaging at room temperature for 2 hr. before re-heating as above for 10–15 min. |
| **PASTRY, UNCOOKED**<br>Shortcrust: 3 months<br>Flaky and puff: 3–4 months | Roll out to size required (or shape into vol-au-vent cases). Freeze pie shells unwrapped until hard, to avoid damage. Use foil plates or take frozen shell out of dish after freezing but before wrapping. Discs of pastry can be stacked with waxed paper between for pie bases or tops<br><br>*Note:* there is little advantage in bulk-freezing unshaped short-crust pastry, as it takes about 3 hr. to thaw before it can be rolled out. Bulk flaky and puff —prepare up to the last rolling; pack in polythene bags or heavy-duty foil and overwrap. To use, leave for 3–4 hr. at room temperature, or overnight in refrigerator | Stack pastry shapes with 2 pieces of cellophane or waxed paper between layers, so that if needed, one piece can be removed without thawing the whole batch. Place the stack on a piece of cardboard, wrap and seal | Thaw flat discs at room temperature, fit into pie plate and proceed with recipe. Unbaked pie shells or flat cases should be returned to their original container before cooking: they can go into oven from the freezer (ovenproof glass should first stand for 10 min. at room temperature); add about 5 min. to normal baking time |
| **PASTRY, COOKED**<br>Pastry cases: 6 months<br>Meat pies: 3–4 months<br>Fruit pies: 6 months | Prepare as usual. Empty cases freeze satisfactorily, but with some change in texture. Prepare pies as directed (using an aluminium foil dish). Brush pastry cases with egg white before filling. Cool completely before freezing | Wrap carefully—very fragile. Protect the tops of pies with an inverted paper or aluminium pie plate, then wrap and seal | Leave pies at room temperature for 2–4 hr., depending on size. If required hot, re-heat in the oven. Flan cases should be thawed at room temperature for about 1 hr.; refresh if wished |
| **PASTRY PIES, UNCOOKED**<br>Double crust: 3 months | Prepare pastry and filling as required. Make large pies in a foil dish or plate, or line an | Freeze uncovered. When frozen, remove small or pre-formed pies from containers and pack all | Unwrap unbaked fruit pies and place still frozen in the oven at 425°F (mark 7) for 40–60 min., |

| Item | Preparation | Packing | Thawing / Serving |
|---|---|---|---|
| | ordinary dish or plate with foil and use as a pre-former. Make small pies in patty tins or foil cases. Do not slit top crust of fruit pies before freezing | pies in foil or polythene bags | according to type and size. Slit tops of double crusts when beginning to thaw. (Ovenproof glass should first stand for 10 min. at room temperature.) Add a little to usual cooking time |
| Top crust: 3 months | Prepare pie in usual way; cut fruit into fairly small pieces and blanch if necessary; toss with sugar; or use cold cooked savoury filling. Cover with pastry. Do not slit crust | Use ovenproof glass or foil dishes. Wrap in foil or plastic film, protecting as for cooked pies | Unwrap, place in a pre-heated oven and bake, allowing extra time. Cut a vent in the pastry when it begins to thaw |
| Biscuit Pie Crust: 2 months | Not easy to handle unfilled unless the crust is pre-baked. Shape in a sandwich tin or pie plate, lined with foil or waxed paper. Add filling if suitable | Freeze until firm, then remove from tin in the foil wrapping and pack in a rigid container | Filled: serve cold; thaw at room temperature for 6 hr. |
| **PANCAKES, UNFILLED** 2 months | Add 1 tbsp. corn oil to a basic 4 oz. flour recipe. Make pancakes, and cool quickly on a wire rack. Interleave them with lightly oiled greaseproof paper or polythene film. Seal in polythene bags or foil | Freeze quickly | To thaw: leave in packaging at room temperature for 2–3 hr., or overnight in the refrigerator. For quick thawing, unwrap, spread out separately and leave at room temperature for about 20 min. To re-heat, place stack of pancakes wrapped in foil in the oven at 375°F (mark 5) for 20–30 min. Alternatively, separate pancakes and place in a lightly greased heated frying pan, allowing $\frac{1}{2}$ min. for each side |
| **PANCAKES, FILLED** 1–2 months | Only choose fillings suitable for freezing. Don't over-season | Place filled pancakes in a foil dish, seal and overwrap | Place frozen in packaging in oven at 400°F (mark 6) for about 30 min. |
| **SPONGE PUDDINGS, UNCOOKED** 2 months | Make in the usual way. Use foil or polythene basins, or line ordinary basins with greased foil | Seal basins tightly with foil, overwrap and freeze at once. To freeze pudding mixture in pre-formed foil, remove from basins when frozen, then overwrap Note: allow room at this stage for later rising | Remove packaging, cover top with greased foil and place, frozen, to steam—1½-pt. size takes about 2½ hr. Don't forget to return a pre-formed pudding mixture to its original basin |
| **SPONGE PUDDINGS, COOKED** 3 months | Prepare and cook in the usual way. Cool thoroughly, cover with foil and overwrap | Freeze quickly | As above—a 1½-pt. pudding takes about 45 min. to thaw and re-heat |

| FOOD AND STORAGE TIME | PREPARATION | FREEZING | THAWING AND SERVING |
|---|---|---|---|
| **SWEETS**<br>Mousses, creams, etc.<br>2–3 months | Make as usual: these can be frozen in new toughened table-ware glasses e.g. by Duralex | Freeze, unwrapped, in foil-lined container until firm, then remove container, place sweet in polythene bag, seal and return to freezer | Unwrap and thaw in refrigerator for about 6 hr., or at room temperature for about 2 hr. |
| **ICE CREAM**<br>3 months<br>Commercially made: 1 month | Either home-made or bought ice creams and sorbets can be stored in the freezer | Bought ice creams should be over-wrapped in moisture-proof bags before storing. Home-made ones should be frozen in moulds or waxed containers and over-wrapped | Put in freezing compartment of the refrigerator for 6–8 hr. to soften a little. Some 'soft' bought ice cream can be used from freezer, provided it is not kept in the coldest part |
| **CREAM**<br>Up to 12 months: ideally, about 4 months<br>Whipped: 3 months<br>Commercially frozen up to 1 yr. | Use only pasteurised, with a 40% butter-fat content, or more (i.e. double cream). Whipped cream may be piped into rosettes on waxed paper | Transfer cream to suitable container, e.g. waxed carton, leaving space for expansion. Freeze rosettes unwrapped; when firm, pack in a single layer in foil | Thaw in refrigerator, allowing 8 hr., or 1–2 hr. at room temperature. Put rosettes in position as decoration before thawing, or they cannot be handled |
| **CAKES, COOKED**<br>Including sponge flans, Swiss rolls and layer cakes: 6 months (Frosted cakes lose quality after 2 months; since aging improves fruit cakes, they may be kept longer) | Bake in usual way. Leave until cold on a wire rack. Swiss rolls are best rolled up in non-stick paper, if to be frozen without a filling. Do not spread or layer with jam before freezing. Keep essences to a minimum and go lightly with spices | Wrap plain cake layers separately, or together with cellophane or waxed paper between layers. Freeze frosted cakes (whole or cut) unwrapped until frosting has set, then wrap, seal and pack in boxes to protect icing | Iced cakes: unwrap before thawing, then the wrapping will not stick to the frosting when thawing. Cream cakes: may be sliced while frozen, for a better shape and quick thawing. Plain cakes: leave in package and thaw at room temperature. Un-iced layer cakes 3–4 hr. and small cakes thaw in about 1–2 hr. at room temperature: frosted layer cakes take up to 4 hr. |
| **CAKE MIXTURES, UNCOOKED**<br>2 months | Whisked sponge mixtures do not freeze well uncooked. Put rich creamed mixtures into containers, or line the tin to be used later with greased foil, add cake mixture and freeze uncovered. When frozen, remove from tin, package in foil and overwrap | Return to freezer | To thaw, leave at room temperature for 2–3 hr., then fill tins to bake. Pre-formed cake mixtures can be returned to the original tin, unwrap. Place frozen in pre-heated oven and bake in usual way, but allow longer cooking time |

| | Preparation | Packing | To thaw and serve |
|---|---|---|---|
| **SCONES AND TEABREADS**<br>6 months | Bake in usual way | Freeze in polythene bag in convenient units for serving | Thaw teabreads in wrapping at room temperature for 2–3 hr. Tea scones: cook from frozen, wrapped in foil, at 400°F (mark 6) for 10 min. Girdle scones: thaw 1 hr. Drop scones: thaw 30 min. or cover and bake for 10 min. |
| **CROISSANTS AND DANISH PASTRIES**<br>Unbaked, in bulk: 6 weeks<br>Baked 4 weeks | Unbaked: Prepare to the stage when all the fat has been absorbed, but don't give the final rolling. *Baked, see page 230* | Unbaked: Wrap in airtight polythene bags and freeze at once | Leave in polythene bag, but unseal and re-tie loosely, allowing space for dough to rise. Preferably thaw overnight in a refrigerator, or leave for 5 hr. at room temperature. Complete the final rolling and shaping, and bake |
| **BISCUITS BAKED AND UNBAKED**<br>6 months | Prepare in the usual way. Rich mixtures—i.e. with more than ¼ lb. fat to 1 lb. flour—are the most satisfactory | Either baked or unbaked, pack carefully. Wrap rolls of uncooked dough, or pipe soft mixtures into shapes, freeze and pack when firm. Allow cooked biscuits to cool before packing | Thaw uncooked rolls of dough slightly; slice off required number of biscuits and bake. Shaped biscuits can be cooked direct from frozen state: allow 7–10 min. extra cooking time. Cooked biscuits may require crisping in warm oven |
| **BREAD**<br>4 weeks | Freshly-baked bread, both bought and home-made, can be frozen. Crisp, crusty bread stores well up to 1 week, then the crust begins to 'shell off' | Bought bread may be frozen in original wrapper for up to 1 week; for longer periods, seal in foil or polythene. Home-made bread: freeze in foil or polythene bags | Leave to thaw in the sealed polythene bag or wrapper at room temperature 3–6 hr., or overnight in the refrigerator, or leave foil-wrapped and crisp in oven at 400°F (mark 6) for about 45 min. Sliced bought bread can be toasted from frozen state |
| **BOUGHT PART-BAKED BREAD AND ROLLS**<br>4 months | Freeze immediately after purchase | Leave loaf in the bag. Pack rolls in heavy-duty polythene bags and seal | To use, place frozen unwrapped loaf in oven at 425°F (mark 7) for about 40 min. Cool for 1–2 hr. before cutting. Rolls: place frozen unwrapped in oven at 400°F (mark 6) for 15 min. |

| FOOD AND STORAGE TIME | PREPARATION | FREEZING | THAWING AND SERVING |
|---|---|---|---|
| **SANDWICHES** 1–2 months | Most types may be frozen, but those filled with hard-boiled eggs, tomatoes, cucumber or bananas tend to go tasteless and soggy | Wrap in foil, then in polythene bag | Thaw unwrapped at room temperature or in the refrigerator. Times vary according to size of pack. Cut pinwheels, sandwich loaves, etc., in portions when half-thawed |
| **TOASTED SANDWICHES** Up to 2 months | Use white or brown bread—cheese, ham, fish are all suitable, but avoid salad foods; season lightly | Interleave for easy separation. Wrap in foil or polythene bags | Place frozen unwrapped sandwiches under a hot grill; thawing will take place during toasting |
| **MARMALADE** 6 months; useful if it's not convenient to make marmalade when Seville oranges are in season | Wash, dry and freeze Seville oranges whole, or prepare marmalade to cooked pulp stage—i.e. before addition of sugar | Pack whole oranges in polythene bags; pulp in suitable containers | Thaw, still wrapped, in fridge, allowing 6–8 hr. per lb. for whole fruit and 9–12 hr. per lb. for pulped, then finish cooking |
| **BUTTER** Salted: 3 months Unsalted: 6 months | Always buy fresh stock (farmhouse butter must be made from pasteurised cream) | Overwrap in foil in $\frac{1}{2}$–1 lb. quantities | Allow to thaw in refrigerator |
| **COMMERCIALLY FROZEN FOODS** Up to 3 months as a rule *Note*: The times quoted by the manufacturers are often less than those given for home-frozen foods, because of the handling in distribution, before the foods can reach your own freezer | No further preparation, etc., needed, except for ice cream, which should be overwrapped if it is to be kept for longer than 3 weeks | Follow directions on packet | |
| **HERBS** Up to 6 months | Wash and trim if necessary. Dry thoroughly | Freeze in small bunches in a rigid foil container or polythene bag. Alternatively, herbs, especially parsley, can be chopped before freezing | Can be used immediately. Crumble whilst still frozen |
| **EGGS** separated 8–10 months | Freeze only fresh eggs—yolks and whites separately | Pack in waxed or rigid containers. Yolks—mix with $\frac{1}{2}$ level tsp. salt or sugar; whites need no addition. Label carefully | Thaw in refrigerator or rapidly thaw at room temperature for about $1\frac{1}{2}$ hr. |

# Index

*Recipes using frozen food are in italics*